I ımatic and Hydraulic Systems

Pneumatic and Hydraulic Systems

W. Bolton

Butterworth-Heinemann
Linacre House, Jordan Hill, Oxford OX2 8DP
A division of Reed Educational and Professional Publishing Ltd

A member of the Reed Elsevier plc group

OXFORD BOSTON JOHANNESBURG
MELBOURNE NEW DELHI SINGAPORE

First published 1997

British Library Cataloguing in Publication Data
A catalogue record for this book is available from the British Library

ISBN 0 7506 3836 2

Printed and bound in Great Britain

Contents

Preface

Aims

- To provide an accessible, readable introduction to pneumatic and hydraulic systems for control engineering which is suitable for college and university students.

- To show how pneumatic and hydraulic systems can be used in developing control systems and provide the reader with the skills to design, assemble and check systems.

Audience

This book more than covers the GNVQ in Engineering Unit Hydraulics and Pneumatics, and also covers much of the GNVQ in Engineering Unit Programmable Logic Controllers. In addition it is relevant to students taking City and Guilds courses, training courses used for introducing or updating practising engineers and first-year degree courses.

Format

The book is copiously illustrated. Worked examples are used to illustrate the application of principles at many places within chapters. At the end of each chapter there are further problems. Answers to all the problems are given at the end of the book.

Content

The range of material covered in this book is seen as appropriate for an introductory course in pneumatics and hydraulics applied to the development of control systems.

Chapter 1 deals with basic principles of engineering science of relevance to pneumatics and hydraulics.

Chapter 2 is devoted to the equipment used for the generating of fluid power and its distribution, covering compressors, air treatment, air receivers and compressor control, pneumatic plant layout, hydraulic pumps and circuits.

Chapter 3 provides a consideration of the form of directional control valves, pressure control valves and flow control valves.

Chapter 4 is a discussion of fluid power actuators, linear and rotary actuators being considered.

Chapter 5 introduces common mechanical, pneumatic and electrical examples of position sensors.

Chapter 6 is a consideration of electro-pneumatics and hydraulics, including solenoid valves and electrical relays.

Chapter 7 brings together the principles of control valves from Chapter 3, actuators from Chapter 4 and sensors from Chapter 5 in a consideration of the control of cylinders to implement control strategies. This includes directional control, speed control, hydro-pneumatic systems, pilot operation, automatic operation, shutdown systems, sequential control circuits and cascade control.

Chapter 8 is an introduction to logic systems and Boolean algebra. AND, OR, NOT, NAND, NOR and INHIBITION gates and combinations of such gates are considered and their implementation with valve circuits. The De Morgan laws and Karnaugh maps are used as aids to simplification.

Chapter 9 shows how control systems can be developed with solenoid valves, sensors and programmable logic controllers, the development of ladder programs being discussed in some detail. PLC internal relays, timers, counters, set and reset, shift registers and master control relays are included.

Chapter 10 is a review of maintenance and fault-finding procedures that are used with pneumatic and hydraulic systems and with PLC-controlled systems.

W. Bolton

1 Pneumatic and hydraulic principles

1.1 Introduction

This chapter defines basic terms and reviews the basic principles of fluid power systems. The term *fluid* is used for either gases or liquids, this being because both can flow freely. Gases and liquids under pressure can be used to transmit energy over long distances, such systems being referred to as *fluid power systems*. *Pneumatics* is the term used when compressed air is the fluid and *hydraulics* when the medium is a liquid, typically oil, under pressure.

Hydraulic systems tend to be used at much higher pressures than pneumatic systems and consequently can produce much larger forces and torques. Both, however, tend to be more readily used for large forces and torques than electromagnetic systems such as electric motors. Often pneumatics and hydraulics are combined with electrical/electronic systems, these being used to provide the control signals. This and programmable logic controllers (PLCs) are discussed in Chapter 9.

1.2 Pressure

Pressure is defined as the force per unit area:

$$\text{pressure} = \frac{\text{force}}{\text{area}} \qquad [1]$$

The SI unit is the pascal (Pa), one pascal being the pressure exerted by a force of 1 N acting on an area of 1 m^2. The pascal is a small unit and the prefixes kilo and mega are generally required for the types of pressures encountered in pneumatics and hydraulic systems.

The *atmospheric pressure* varies with both location and time, but for most pneumatic calculations it can be regarded as being constant at 10^5 Pa. Pressures are often expressed in terms of multiples of this pressure. For convenience 10^5 Pa is termed a pressure of 1 bar. Thus, for example, 10 bar is a pressure ten times that of the atmospheric pressure:

$$10 \text{ bar} = 10 \times 10^5 = 10^6 \text{ Pa} = 10^3 \text{ kPa} = 1 \text{ MPa}$$

When pressure is measured relative to the atmospheric pressure it is termed the *gauge pressure*, a pressure measured from absolute zero pressure being termed the *absolute pressure*:

$$\text{absolute pressure} = \text{gauge pressure} + \text{atmospheric pressure} \qquad [2]$$

Hence a gauge pressure of 2×10^5 Pa is an absolute pressure of $2 \times 10^5 + 1 \times 10^5 = 3 \times 10^5$ Pa.

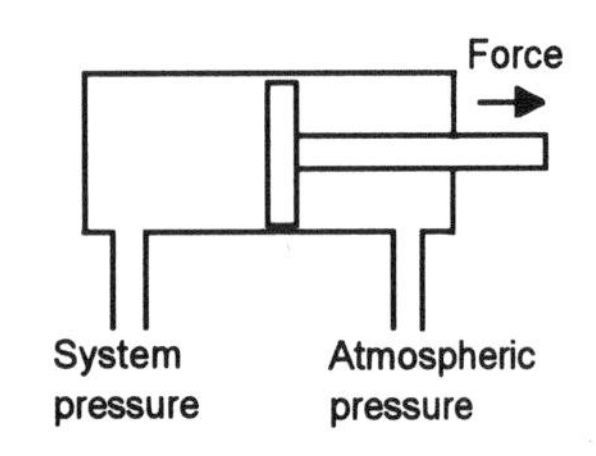

Figure 1.1 *Example*

Example

A pneumatic cylinder (Figure 1.1) has an internal cross-sectional area of 5×10^{-3} m^2 and a piston which is required to exert a force of 3 kN. What gauge pressure is required in the system?

The system pressure acts on one side of the piston and the atmospheric pressure on the other side. The net force acting on the piston is thus that due to the difference in the two pressures. Since the pressure is that due to the system on one side of the piston and the atmospheric pressure on the other side, the difference in the two pressures is the gauge pressure of the system. Using equation [1]:

$$\text{system gauge pressure} = \frac{3 \times 10^3}{5 \times 10^{-3}} = 6 \times 10^5 \text{ Pa}$$

This is 600 kPa = 0.6 MPa = 6 bar.

1.2.1 Pressure head

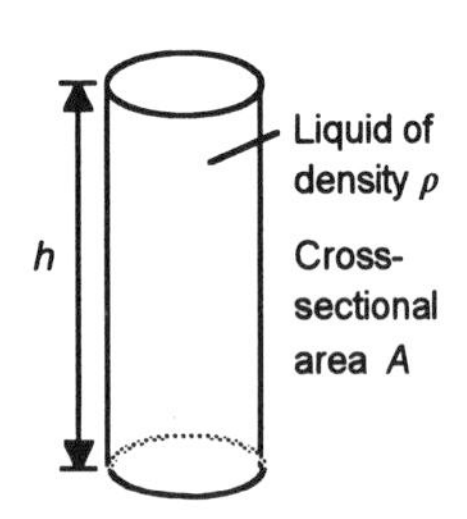

Figure 1.2 *Pressure due to a liquid column*

A column of fluid (Figure 1.2) of height h will exert a pressure at its base due to the weight of the fluid above. For a column of cross-sectional area A, the volume of this fluid will be Ah. If the fluid has a density ρ, the mass will be $Ah\rho$ and its weight $Ah\rho g$. Thus the pressure due to this column of fluid is:

$$\text{pressure} = h\rho g \qquad [3]$$

1.2.2 Transmission of force by fluids

Blaise Pascal, in about the year 1650, determined laws governing how fluids transmit power. These are:

1 Provided the effect of the weight of a fluid can be neglected, the pressure is the same throughout an enclosed volume of fluid at rest.

2 The static pressure acts equally in all directions at the same time.

3 The static pressure always acts at right angles to any surface in contact with the fluid.

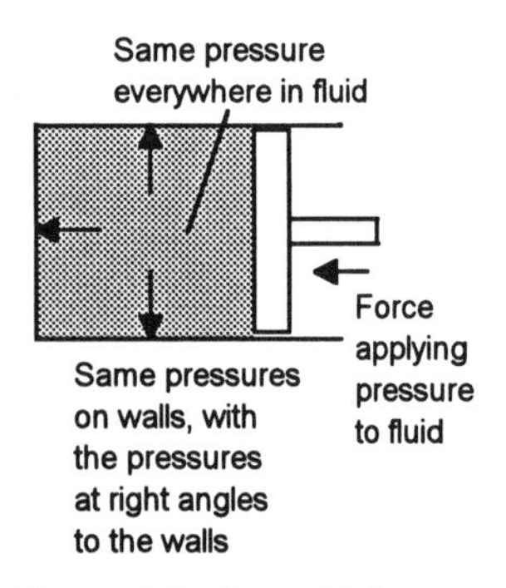

Figure 1.3 *Pascal's laws*

As a consequence of these laws, when a pressure is applied to one end of an enclosed volume of fluid (Figure 1.3) the pressure is transmitted equally and undiminished to every other part of the fluid.

Example

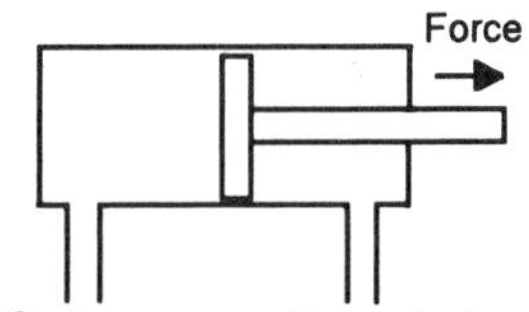

A system applies a gauge pressure of 4 kPa to the cylinder shown in Figure 1.4. If the piston has an area of 0.2 m^2, what will be the force acting on it?

Figure 1.4 *Example*

The pressure will be the same throughout the volume of fluid enclosed in the cylinder and acting equally in all directions. Thus the gauge pressure acting on the piston will be 4 kPa. The pressure is at right angles to the piston surface. Thus the force on the piston is given by equation [1] as:

$$\text{force} = \text{pressure} \times \text{area} = 4 \times 10^3 \times 0.2 = 800 \text{ N}$$

1.3 Gas laws

Air can be considered to be a reasonable approximation to an ideal gas for the range of pressures and temperatures occurring with pneumatic systems and thus obey the *ideal gas laws*. These are:

1 *Boyle's law*
In an ideal gas in which the mass and temperature remain constant, the volume V varies inversely as the absolute pressure p, i.e.:

$$pV = \text{a constant} \quad [4]$$

2 *Charles' law*
In an ideal gas in which the mass and the pressure remain constant, the volume V varies directly as the absolute temperature T, i.e.:

$$\frac{V}{T} = \text{a constant} \quad [5]$$

Absolute temperatures are measured on the kelvin scale. To convert from temperatures on the Celsius scale, a reasonable approximation is to add 273. Thus 0°C = 273 K.

3 The *pressure law*
In an ideal gas in which the mass and volume remain constant, the pressure p varies directly as the absolute temperature T, i.e.:

$$\frac{p}{T} = \text{a constant} \quad [6]$$

The combination of the three gas laws results in the *general gas equation*:

$$\frac{pV}{T} = \text{a constant} \quad [7]$$

The constant is for a particular mass of a particular gas and thus the equation can be written for a mass m of gas as:

$$pV = mRT \qquad [8]$$

with R, termed the *characteristic gas constant*, being the value of the constant for 1 kg of a particular ideal gas.

Example

A container has a volume of 0.10 m^3 and is filled with compressed air at a gauge pressure of 600 kPa and a temperature of 40°C. If the atmospheric pressure is 101 kPa, determine the air pressure in the container when the air cools to 20°C. Neglect any change in dimensions of the container as a result of the temperature change.

The volume is assumed to be constant, thus equation [6] is the relevant equation. This can be written as:

$$\frac{p_1}{T_1} = \frac{p_2}{T_2}$$

where p_1 and T_1 are the initial pressure and temperature and p_2 and T_2 the final pressure and temperature. Thus:

$$p_2 = \frac{p_1 T_2}{T_1} = \frac{(600 + 101) \times 10^3 \times (273 + 40)}{273 + 20} = 749 \times 10^3 \text{ Pa}$$

This is a gauge pressure of 648 kPa.

1.3.1 Expansion and compression of gases

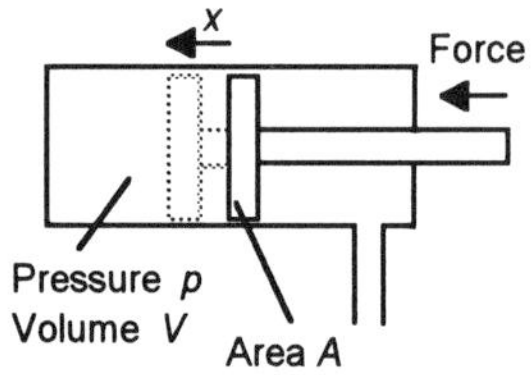

Figure 1.5 *Compressing a gas*

Consider a piston being used to compress a gas in a cylinder (Figure 1.5). If a constant force F is applied and moves the piston in a distance x then the work done is Fx. The change in volume of the gas will be Ax. Thus:

$$\text{work done} = Fx = pAx = p \times \text{change in volume} \qquad [9]$$

The expansion or compression of a gas is said to be *isothermal* if it takes place at a constant temperature. In such a case Boyle's law is obeyed. But with a piston in a cylinder when work is done there is a transfer of energy into or out of the gas and so its temperature changes. The change can only be considered to approximate to an isothermal change if the piston is moved slow enough for the compression or expansion to be so slow that energy from the surroundings has time to flow into or out of the gas and so the temperature remains constant.

However, when there is no flow of energy either into or out of a gas during expansion or compression there will be a temperature change. Such a change is said to be *adiabatic*. With an adiabatic change:

$$pV^{\gamma} = \text{a constant} \qquad [10]$$

where γ is the ratio of the specific heats at constant pressure and constant volume for the gas. For dry air this ratio is about 1.4. Adiabatic conditions occur when expansion or compression take place quickly and there is no time for energy to enter or leave the gas.

We can combine equation [10] with the general gas equation [7] to give:

$$\frac{pV^\gamma}{(pV/T)} = V^{\gamma-1}T = \text{a constant} \qquad [11]$$

or:

$$\frac{pV^\gamma}{(pV/T)^\gamma} = p^{1-\gamma}T^\gamma = \text{a constant} \qquad [12]$$

Example

An air cylinder is used to cushion the opening of a door, the movement forcing a piston to compress air in a cylinder. If the cylinder initially contains air at atmospheric pressure and the length of the cushioning stroke is restricted to 80 mm with the full stroke being 150 mm, what will be the maximum pressure in the cylinder? Take γ to be 1.4 and the atmospheric pressure to be 101 kPa.

Assuming that the opening of the door is quick enough for the compression to be adiabatic, equation [10] gives:

$$p_1V_1^\gamma = p_2V_2^\gamma$$

where p_1 and V_1 are the initial pressure and volume and p_2 and V_2 the final pressure and volume. Thus:

$$101 \times 10^3 \times (0.150A)^{1.4} = p_2(0.070A)^{1.4}$$

where A is the cross-sectional area of the cylinder. Thus p_2 = 294 kPa or a gauge pressure of 193 kPa.

1.4 Flow through pipes

Consider the flow of a fluid through a pipe (Figure 1.6) and the rate at which fluid flows through the section AA. If the fluid has an average velocity v then in a time t the fluid will have moved a distance vt. Thus in a time t the volume of fluid passing through section AA is vtA, where A is the cross-sectional area of the pipe. Thus the volume rate of flow Q is:

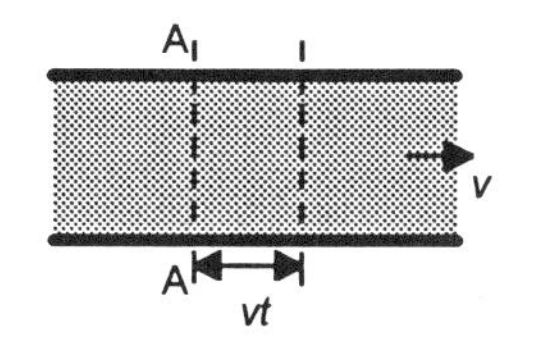

Figure 1.6 *Flow through a pipe*

$$\text{volume rate of flow } Q = vA \qquad [13]$$

For a fluid of density ρ, the mass rate of flow is:

$$\text{mass rate of flow} = Q\rho = \rho vA \qquad [14]$$

Example

Air of density 1.24 kg/m^3 flows through a pipe of diameter 300 mm at a mass rate of flow of 3 kg/s. Determine the mean velocity of the flow.

Using equation [14]:

$$\text{mean velocity} = \frac{3}{1.24 \times \frac{1}{4}\pi 0.300^2} = 34.2 \text{ m/s}$$

Example

Determine the pipe size needed to carry air at a volume rate of flow of 12 dm^3/s and velocity 9 m/s.

Using equation [13]:

$$A = \tfrac{1}{4}\pi d^2 = \frac{Q}{v} = \frac{12 \times 10^{-3}}{9}$$

Hence d = 42 mm.

1.4.1 Pressure drop

When a fluid flows through a pipe, frictional forces at the walls and turbulence will mean that energy is expended. As a consequence the work done 'pushing' the fluid into the pipe is greater than the work that can be done by the fluid exiting from the pipe and thus there is a pressure drop. For air flow through pipes, the pressure drop p per unit length of pipe is found to be approximately proportional to the square of the mean fluid velocity or, by equation [13], (volume rate of flow Q/pipe cross-sectional area)^2. Thus doubling the volume rate of flow quadruples the pressure drop per unit length of pipe.

Problems

1 A pneumatic cylinder (as in Figure 1.1) has an internal cross-sectional area of 5×10^{-3} m^2 and has a piston which is required to exert a force of 2 kN. What gauge pressure is required in the system?

2 A hydraulic press (Figure 1.7) has one cylinder with a piston of diameter 150 mm and another cylinder with a piston of diameter 125 mm. What force has to be applied to the smaller diameter piston to allow the larger diameter piston to exert a force of 15 kN?

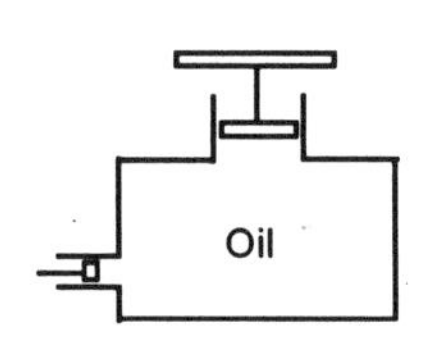

Figure 1.7 *Problem 2*

3 A container is filled with compressed air at a gauge pressure of 800 kPa and a temperature of 45°C. If the atmospheric pressure is 101 kPa, determine the air gauge pressure in the container when the air cools to 20°C. Neglect any change in dimensions of the container as a result of the temperature change.

4 A compressed air tank has a volume of 0.5 m^3 and contains air at a gauge pressure of 1.96 MPa and a temperature of 50°C. If the atmospheric pressure is 103 kPa, determine the mass of air in the tank. The gas constant for air is 0.287 kJ/kg K.

5 Air at an absolute pressure of 300 kPa and a temperature of 25°C is compressed adiabatically. If the temperature rises to 180°C, what is the new pressure? Take γ as 1.4.

6 Air at an absolute pressure of 700 kPa and having a volume of 0.015 m^3 is expanded adiabatically to a pressure of 140 kPa. Determine the final volume of the air. Take γ as 1.4.

7 Air at a temperature of 30°C and an absolute pressure 200 kPa passes through a pipe with a mean velocity of 20 m/s. Determine the mass rate of flow. (Hint: use $pV = mRT$ to obtain the density, taking R to be 287 J/kg K.

8 Determine the pipe size needed to carry air at a volume rate of flow of 16 dm^3/s and velocity 8 m/s.

9 Determine the pipe size needed to carry air at a volume rate of flow of 20 dm^3/s and velocity 9 m/s.

10 Air of density 1.24 kg/m^3 flows through a pipe of diameter 200 mm at a mass rate of flow of 1 kg/s. Determine the mean velocity of the flow.

2 Production and distribution of fluid power

2.1 Introduction

This chapter is concerned with the equipment used for the generation of fluid power and its distribution. For pneumatics this involves a consideration of common forms of compressors, conditioning of the air to remove contaminants, the use of air receivers to enable constant air pressures to be maintained, the layout of pneumatic power systems and the sizing of pipes used. With hydraulics, the types of pumps commonly used, conditioning, the role of accumulators and plant layout are discussed.

2.1.1 Free air

In discussing compressors and air distribution, the term *free air* is often used. Free air is defined as air at normal atmospheric pressure, i.e. at a pressure of 101.3 kPa. Compressors, for example, are specified in terms of the *free air delivered* (f.a.d.), this being the volume a given quantity of compressed air would occupy at atmospheric pressure and the same temperature. This enables comparisons between compressors to be more easily made. The gas laws (see Section 1.3) can be used to convert 'free air' to other pressures and/or temperatures or vice versa.

Example

A compressor has a rated output of 3 m^3/min free air delivery. What will be the output at an absolute pressure of 700 kPa and the same temperature?

Since the temperature is unchanged from the free air condition and only the pressure changes, we can use Boyle's law (Chapter 1, equation [4]):

$$101.3 \times 3 = 700V$$

Hence the output is 0.43 m^3/min at 700 kPa.

Example

A compressor is required to deliver 0.2 m^3/min of compressed air at an absolute pressure of 500 kPa. What compressor output is required in terms of free air?

Using Boyle's law (Chapter 1, equation [4]):

$$500 \times 0.2 = 101.3V$$

Hence the free air output is 0.99 m³/min.

2.2 Compressors

Compressors can be broadly classified as being in two groups:

1 *Positive displacement compressors*
Successive volumes of air are isolated and then compressed to a higher pressure. There are essentially two forms of positive displacement compressor, reciprocating and rotary (see Sections 2.2.1, 2.2.2 and 2.2.3 for the most common forms).

2 *Dynamic compressors*
These are rotary continuous machines in which a high-speed rotating element accelerates the air and converts the resulting velocity head into pressure.

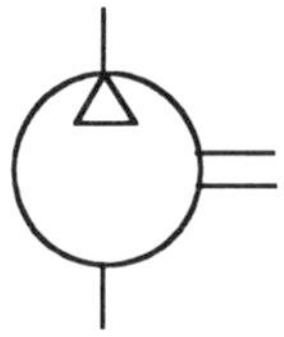

In this book the discussion is restricted to the more common forms of compressor and hence to just positive displacement types. Figure 2.1 shows the symbol used for a compressor with fixed capacity.

Figure 2.1 *Symbol for a compressor*

2.2.1 Reciprocating compressors

Reciprocating compressors, or *piston compressors* as they are often termed, are positive displacement compressors involving pistons moving in cylinders. Figure 2.2 shows the basic form of a *single-acting, single-stage, vertical, reciprocating compressor*.

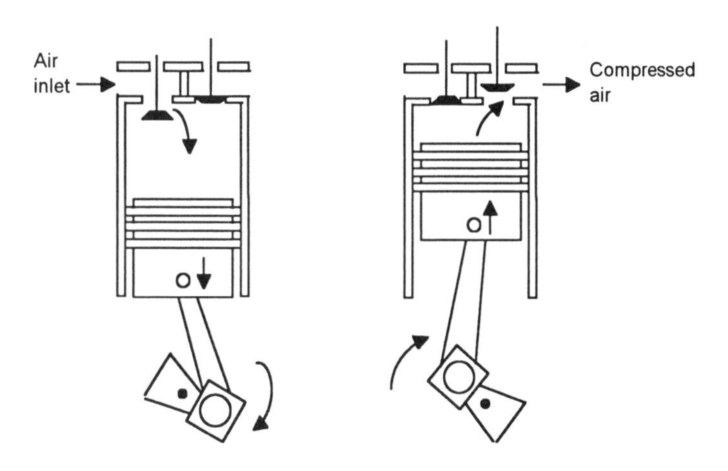

Figure 2.2 *Single-acting, single-stage, vertical, reciprocating compressor*

On the air intake stroke, the descending piston causes air to be sucked into the chamber through the inlet valve. When the piston starts to rise again, the trapped air forces the inlet valve to close and so becomes compressed. When the air pressure has risen sufficiently, the outlet valve opens and the trapped air flows into the compressed air system. After the piston has reached the top dead centre it then begins to descend and the cycle repeats itself. Both the valves are spring loaded.

The compressor in Figure 2.2 is termed *single-acting* because one pulse of air is produced per piston stroke; *double-acting* compressors are designed to produce pulses of air on both the up and down strokes of the piston. It is termed *single-stage* because the compressor goes directly from atmospheric pressure to the required pressure in a single operation.

For the production of compressed air at more than a few bar, two or more stages are generally used. Normally two stages are used for pressures up to about 10 to 15 bar and more stages for higher pressures. Thus with a two stage compressor we might have the first stage taking air at atmospheric pressure and compressing it to, say, 2 bar and then the second stage compressing this air to, say, 7 bar. Because the temperature of the air is increased quite significantly when it is compressed, an intercooler is used between the two stages. Figure 2.3 shows the basic form of a *single-acting, two-stage, reciprocating compressor*.

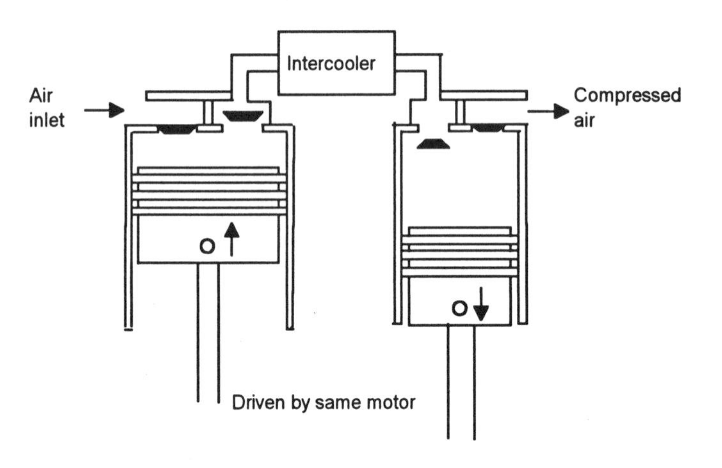

Figure 2.3 *Single-acting, two-stage, reciprocating compressor*

Considerable savings in power consumption can be obtained by the use of multistage compressors with intercooling rather than just a single-stage compressor. The work done in compressing a gas is given by equation [9] in Chapter 1 as the pressure multiplied by the change in volume. Thus if we have, for the trapped air in a compressor, the pressure/volume graph shown

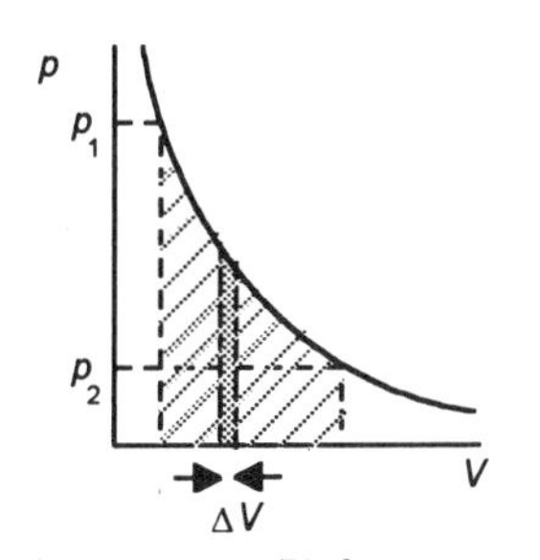

Figure 2.4 *p/V diagram*

in Figure 2.4, the work done in producing a volume change ΔV at a pressure p is $p\Delta V$ and thus the area of the strip under the graph. Thus when the pressure is changed from p_1 to p_2, the work done is the area under the pressure/volume graph between those pressures. With a two-stage compressor, cooling takes place at an intermediate pressure and thus the pressure/volume diagram (Figure 2.5) shows a drop in volume at the intermediate pressure as a result of the air being cooled. As a consequence there is a power saving, this being indicated by the shaded part of the diagram. Power savings of the order of 15% are achievable.

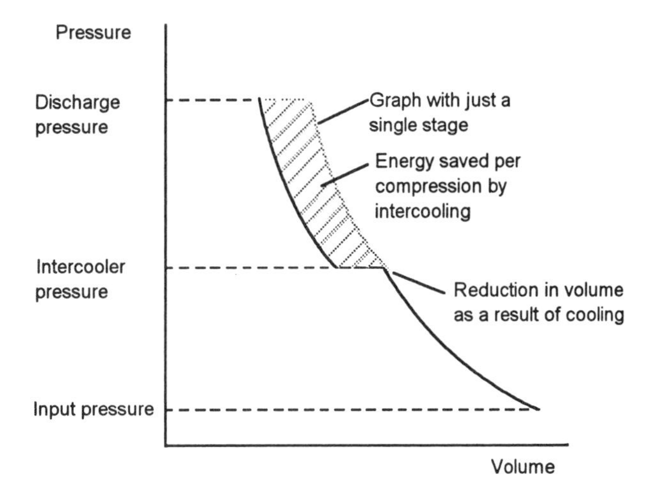

Figure 2.5 *p/V diagram for a two-stage compressor*

Figure 2.6 shows the basic form of a *double-acting, two-stage, reciprocating compressor*. With a double-acting compressor, both sides of a piston are used to compress the air. This arrangement results in larger air delivery and greater efficiency.

Reciprocating piston compressors can be used as a single-stage compressor to produce air pressures up to about 12 bar and as a multistage compressor up to about 140 bar. Typically, air flow deliveries tend to range from about 0.02 m^3/min free air delivery to about 600 m^3/min free air delivery. A characteristic of this form of compressor is that the air is compressed in pulses and some device, usually a receiver (see Section 2.4), is needed to smooth pulsations. The reciprocating piston compressor is supplied in both lubricated and non-lubricated form. With the lubricated form, the crankcase is oil filled and oil is fed to valves, bearings and other sliding points. A consequence of this is that oil inevitably contaminates the compressed air. This can largely be removed by air-line filters. Oil-free piston compressors have dry crankcases and piston rings with bearings and other sliding components made of graphite or PTFE or some other similar substance which has a low coefficient of friction.

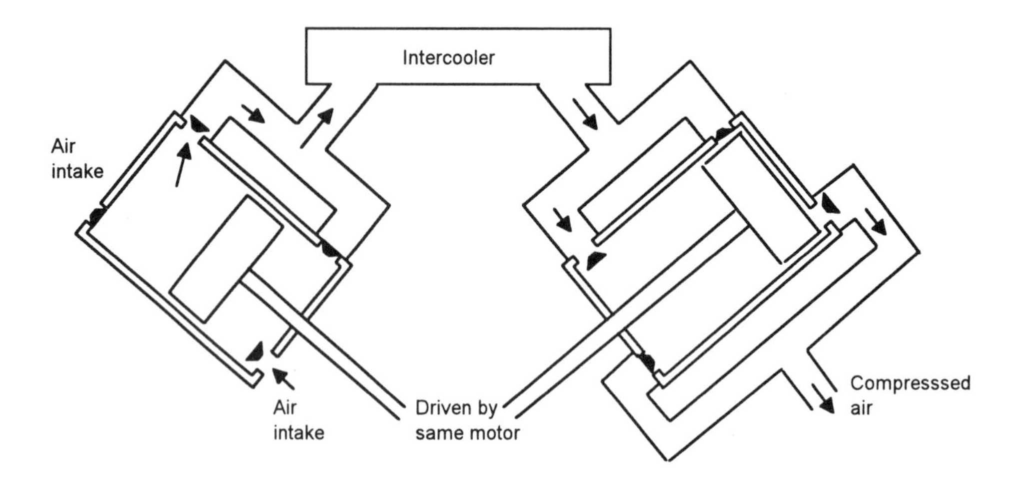

Figure 2.6 *Double-acting, two-stage, reciprocating compressor*

2.2.2 Rotary compressors

The *rotary vane compressor* has a rotor mounted eccentrically in a cylindrical chamber (Figure 2.7). The rotor has blades, the vanes, which are free to slide in radial slots. The rotation causes the vanes to be driven outwards against the walls of the cylinder. As the rotor rotates, air is trapped in pockets formed by the vanes and as the rotor rotates so the pockets become smaller and the air is compressed. Compressed packets of air are thus discharged from the discharge port. Oil is injected into the chamber to assist in sealing the clearance gaps between the vanes and walls of the cylinder and between the vanes and the rotor. It also helps to dissipate the heat resulting from the compression of the air.

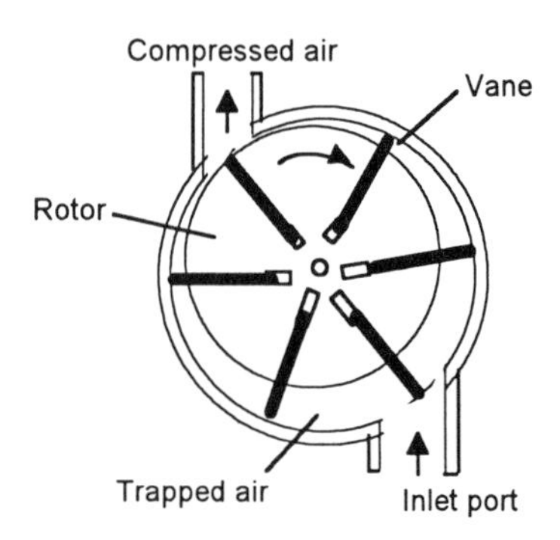

Figure 2.7 *Rotary vane compressor*

Single-stage, rotary vane compressors typically can be used for pressures up to about 800 kPa with flow rates of the order of 0.3 m^3/min to 30 m^3/min free air delivery.

2.2.3 Screw compressors

The *rotary screw compressor* (Figure 2.8) has two intermeshing rotary screws which rotate in opposite directions. As the screws rotate, air is drawn into the casing through the inlet port and into the space between the screws. Air is then trapped in a cavity between adjacent threads and the casing with the ends of the cavity sealed by the screws intermeshing. As the screws rotate, the air is moved along the length of the screws and compressed as the space becomes progressively smaller. Compressed air then emerges from the discharge port. Such compressors can be supplied as oil flooded or oil free. The oil-flooded type has oil injected into the chamber and assists in sealing the clearance gaps between the screws and walls of the cylinder and between the screws. It also helps to dissipate the heat resulting from the compression of the air. With the oil-free type, no internal air cooling takes place and the air has to be cooled after it exits the compressor. Typically, single-stage, rotary screw compressors can be used for pressures up to about 1000 kPa with flow rates of between 1.4 m^3/min and 60 m^3/min free air delivery.

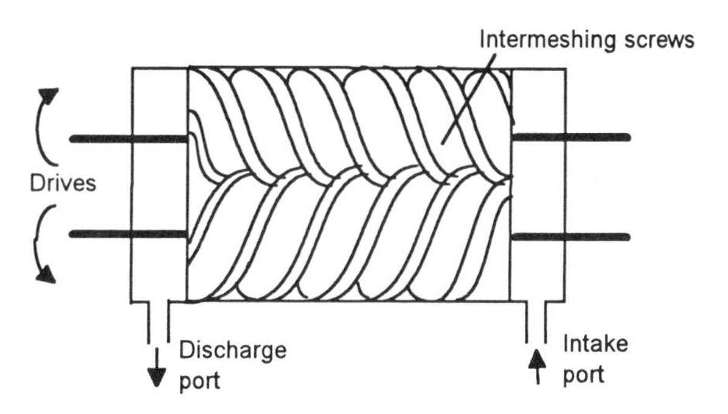

Figure 2.8 *Screw compressor*

2.2.4 Compressor volumetric efficiency

With a reciprocating compressor there has to be some clearance at the end of the compression stroke between the piston head and the cylinder. It is also impossible to fully charge the cylinder with air at the end of the suction stroke. In addition there may also be some leakage of air through the inlet valve at the changeover from suction to compression and also leakage across the piston from the high-pressure to the low-pressure side. Because of this, the volume of air delivered by the compressor is less than the swept volume of the cylinder. This is also true of other forms of compressor. Thus we specify a volume efficiency as:

$$\text{volumetric efficiency} = \frac{\text{volume of free air delivered per minute}}{\text{swept volume of cylinder per minute}} \quad [1]$$

2.3 Air treatment

The air leaving the compressor can be hot and contain contaminants such as oil from the compressor, moisture and dirt particles. Thus *after-coolers*, *dryers* and *filters* are used to give contaminant-free air at the ambient temperature, with *lubricators* to add controlled lubricants to the air in order to lubricate pneumatic devices.

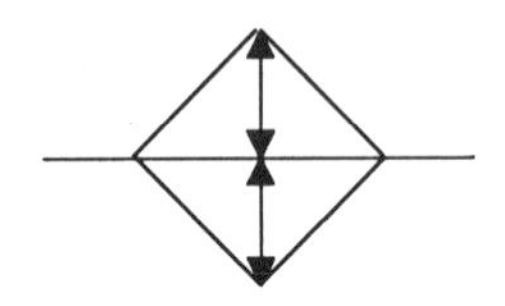

Figure 2.9 *Symbol for a cooler*

2.3.1 After-coolers

After-coolers are heat exchangers similar to intercoolers, Figure 2.9 showing the general symbol for a cooler. They can be either air or water cooled. A water cooler consists essentially of a series of pipes through which the cooling water flows, the air moving through the space round the pipes and so becoming cooled. The aim of using a cooler is to get the air to no higher than about 15°C above the temperature of the atmospheric air at the compressor intake. The air leaving the compressor will contain water vapour and the reduction in temperature produced by the cooler condenses some of this vapour. This condensate has to be drained. For this purpose automatic drain valves are generally used. These allow the liquid to escape when it reaches a particular level in the valve. Figure 2.10 shows the symbols for drains and a *float-operated drain trap*, which when the water rises above a certain level causes the float to open the valve.

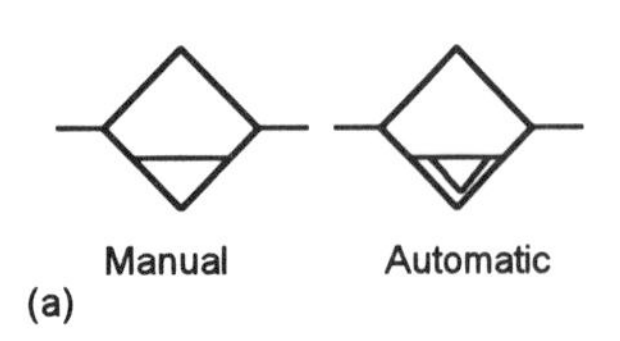

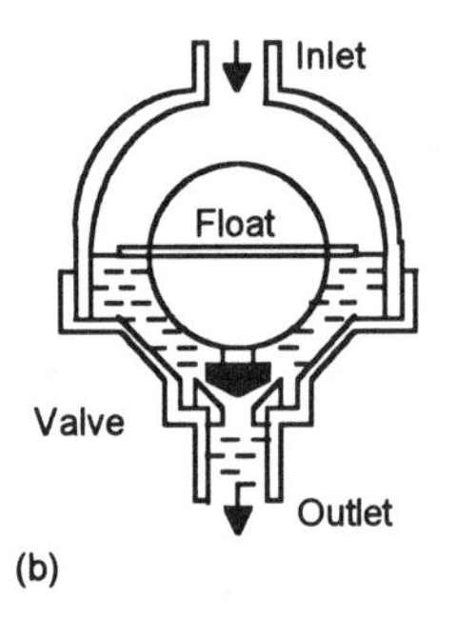

Figure 2.10 *(a) Symbols, (b) automatic drain*

2.3.2 Dryers

Water in compressed air can cause many problems, e.g. corrosion and malfunction of pneumatic components and problems if the compressed air is used in manufacturing processes. In discussing the quantity of water vapour in air, the following terms are frequently used:

1 *Saturation*
Air is said to be saturated when it contains the maximum amount of water vapour which it can hold at a particular temperature. Figure 2.11 shows the water content needed to saturate air at different temperatures. For example, at –20°C air is saturated by 1 g/m^3 but at 40°C saturation is not achieved until there is 50 g/m^3.

2 *Relative humidity*
The relative humidity is a measure of the amount of water vapour present in a sample of air at a particular temperature, being defined as:

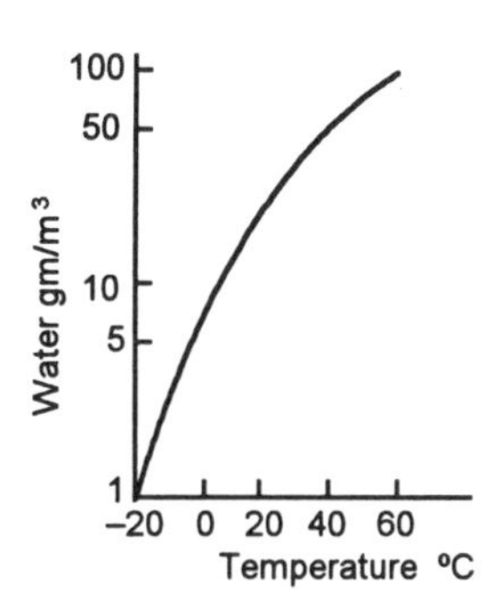

Figure 2.11 *Saturation*

$$\text{relative humidity} = \frac{\text{mass of water vapour present in sample}}{\text{mass of water vapour needed to saturate it}} \times 100\% \quad [2]$$

If we have air at 20°C containing 5 g/m^3 of water then, since saturation requires about 15 g/m^3 at that temperature, the relative humidity is $(5/15) \times 100\% = 33\%$.

3 *Dew point*
The dew point is the temperature at which a sample of air becomes saturated. Thus, using Figure 2.11, if we take air containing 1 g/m^3 of water it will have to be cooled to –20°C before it becomes saturated and liquid starts to condense out. Thus –20°C is the dew point of that sample. Figure 2.11 can thus be considered to be a graph of dew point for different amounts of water vapour. Table 2.1 gives some dew point values.

Table 2.1 *Dew point values*

Water content g/m^3	Dew point °C
4.9	0
6.8	5
9.4	10
12.8	15
17.3	20
23.1	25
30.4	30
39.6	35

The amount of water vapour needed to saturate a sample of air depends on the temperature. It also depends on the pressure because increasing the pressure on a sample of air will reduce its volume. To illustrate the effect of pressure, consider a compressor which takes in air at an absolute pressure of 101 kPa with a relative humidity of 70% and a temperature of 25°C. The air at the discharge port of the compressor is delivered at 4 m^3/min with a temperature of 35°C and an absolute pressure of 700 kPa. We can use the gas laws (equation [7], Chapter 1) to determine the change in the volume of air.

$$\frac{p_1 V_1}{T_1} = \frac{p_2 V_2}{T_2}$$

where p_1, V_1 and T_1 are the inlet pressure, volume and temperature of the amount of air passing through the compressor in one minute and p_2, V_2 and T_2 are the outlet pressure, volume and temperature of that air. Thus:

$$V_2 = \frac{p_1 V_1 T_2}{p_2 T_1} = \frac{101 \times 4 \times 308}{700 \times 298} = 0.60 \text{ m}^3$$

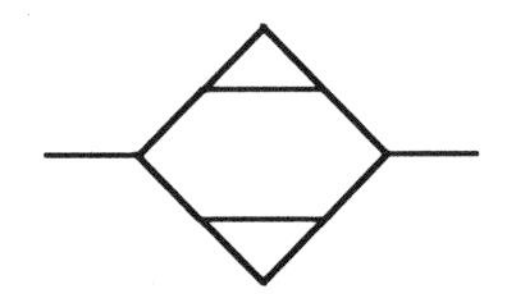

Figure 2.12 *Symbol for an air dryer*

At 25°C the amount of water needed to saturate 1 m^3 is 23.1 g. Thus a relative humidity of 70% means that 1 m^3 of the inlet air holds 16.2 g. Each minute, since 4 m^3 enters the compressor, 64.8 g of water vapour enter. At the discharge this volume of air has been reduced to a volume of 0.60 m^3 but at a higher temperature of 35°C. At 35°C the amount of water needed to saturate 1 m^3 is 39.6 g/m^3. Thus 0.60 m^3 can hold 23.8 m^3 and so each minute we must have 64.8 – 23.8 = 41.0 g of water condensing out.

The aim of a dryer is to reduce the amount of water vapour in the compressed air to a level which will prevent water condensing out in the system supplied with the compressed air, Figure 2.12 showing the general symbol. There are three main types of air dryer:

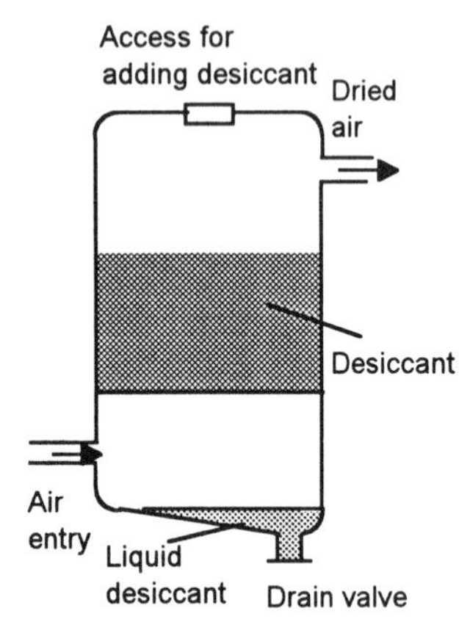

Figure 2.13 *Chemical absorption dryer*

1 *The chemical absorption dryer*
This involves passing the air through a chemical, the desiccant, which absorbs some of the water vapour in the air (Figure 2.13). During absorption, the chemical slowly liquefies and seeps to the bottom of the vessel where it is drained off. The dryer requires periodically to be recharged with the chemical. The dew point of air leaving this type of dryer is typically about 5°C.

2 *The adsorption dryer*
This consists of two pressure chambers containing a desiccant chemical (Figure 2.14). The air to be dried is passed through one chamber and the moisture removed while the other chamber containing spent chemicals is being regenerated by heating and the passage of a purge air stream. After a time the air flow is switched from one chamber to the other.

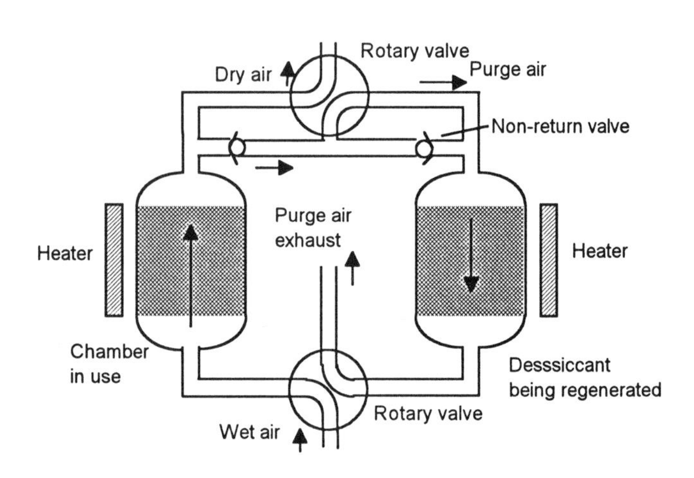

Figure 2.14 *Adsorption dryer*

The changeover from drying compressed air in one chamber to regeneration in that chamber and the passage of the air through the other regenerated chamber takes place automatically on a time-cycle basis. Adsorption dryers can achieve dew points of –20°C or lower.

3 *The refrigeration dryer*
This involves cooling the wet air down to a temperature of about +2°C and so condensing out much of the water vapour to give a dew point of that temperature. Figure 2.15 shows the basic principle. The incoming wet air is first cooled by passing through a heat exchanger where it is cooled by the outgoing dry air, warming that air in the process. It then passes through the refrigerant heat exchanger where it is further cooled. Some typical temperatures at the various points in the sequence are shown on the figure.

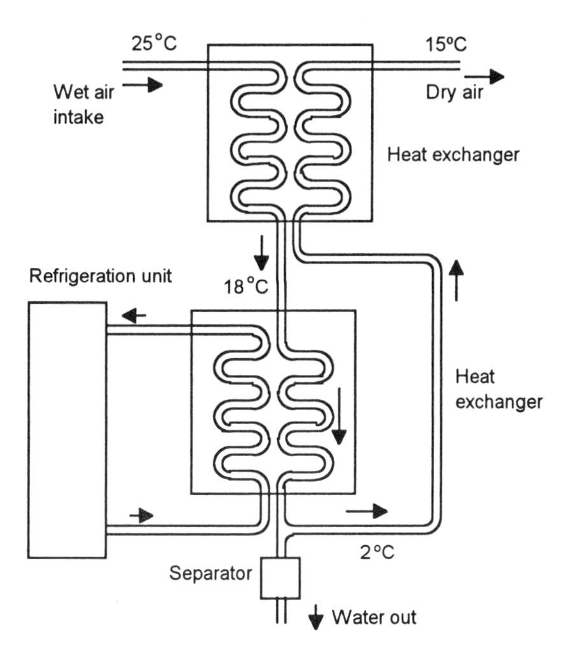

Figure 2.15 *Refrigeration dryer*

2.3.3 Filters

Filters are used to remove dirt before the air enters a compressor. Such filters can be *dry filters* with replaceable cartridges or *wet filters* where the air is bubbled through an oil bath and then passed through a wire mesh filter. The dirt particles become attached to the oil drops and are removed by the wire mesh. Both types of filter require regular servicing.

Filters are also used with the air emerging from a compressor. Here they are used to remove free droplets of water and oil, together with particles of dirt. Figure 2.16(a) shows the basic symbols for a filter and Figure 2.16(b) shows a typical air-line filter which also incorporates a water trap, the symbol used being a composite of those for a filter and a water trap. The entering compressed air is deflected by a cone which forces the air into a swirling motion. This causes the heavier water particles and dirt to be thrown outwards to the wall of the filter. They then fall past the baffle and into the trap bottom where they can be drained. Smaller particles are removed by the air passing through a filter element.

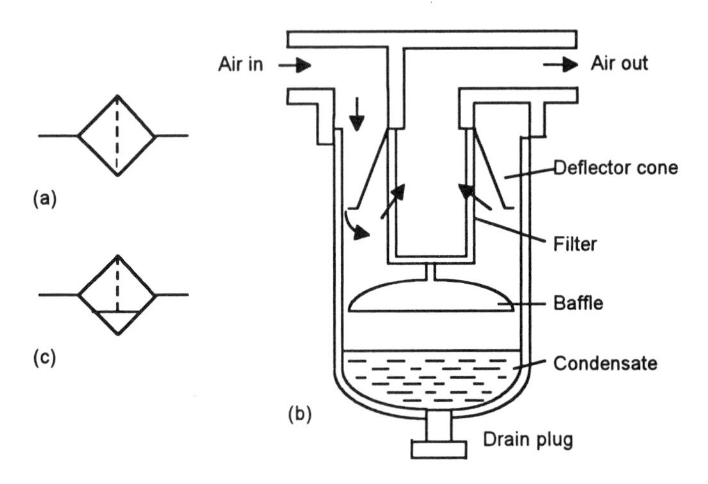

Figure 2.16 *(a) Symbol for a filter, (b) air-line filter with manual water trap, (c) symbol for filter with manual water trap*

Air filters are rated in terms of the largest size of spherical particles that will pass through the filter. Particle sizes are measured in terms of micrometres (generally termed microns). Dust particles are generally larger than 10 μm, while smoke and oil particles are about 1 μm.

2.3.4 Lubricator

To ensure that sliding parts in pneumatic devices such as cylinders and valves are lubricated, a carefully controlled amount of a specially selected lubricant is often added to compressed air by means of a *lubricator*. This results in an oil fog or mist (micro fog) being carried by the compressed air. The term oil fog and oil mist/micro fog lubricators are often used. Oil mist tends to result in smaller oil drops suspended in the air. Figure 2.17 shows the basic principles of a typical air mist lubricator and its symbol. It relies on the principle that when air flows through a constriction there is a pressure drop and this causes oil to be sucked up its tube and into the air stream. The resulting oil-laden air is sprayed onto a baffle plate which causes further oil-drop fragmentation and then flows up to the discharge port, larger drops of oil dropping out of the air stream.

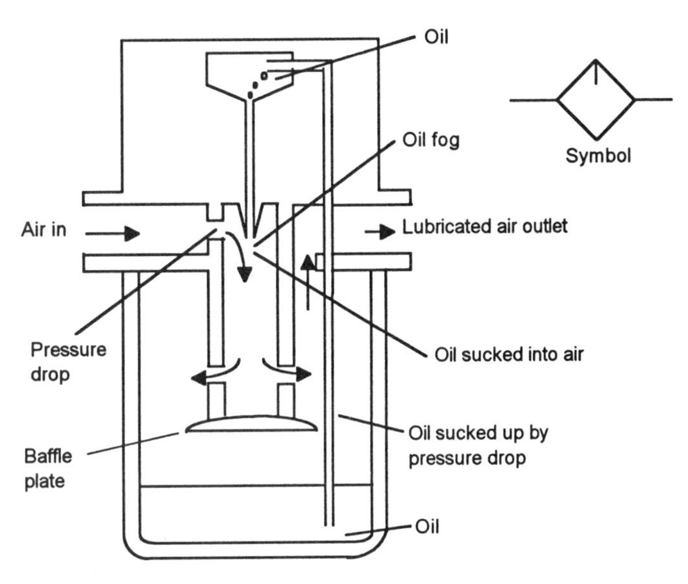

Figure 2.17 *Oil mist lubricator*

2.3.5 Pressure regulators

Generally the air pressure from the supply is set to be greater than that required for a pneumatic system and pressure regulation is used to keep the pressure constant regardless of the flow. A *pressure-reducing valve*, termed a *pressure regulator*, is used for this purpose. Pressure-reducing valves occur in two basic forms:

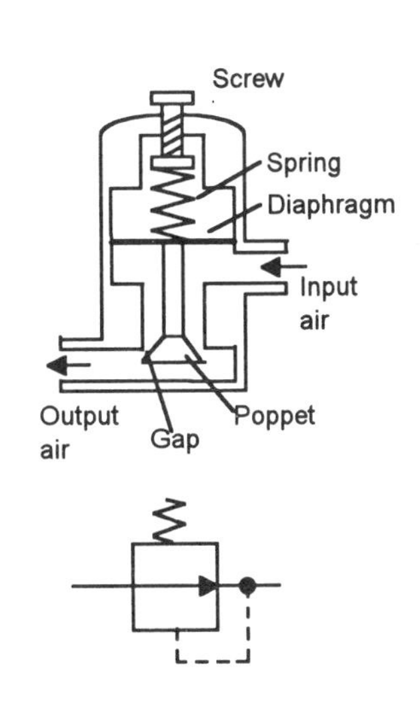

Figure 2.18 *Non-relieving pressure regulator*

1 *Non-relieving pressure regulator*
Figure 2.18 shows the basic form of such a valve and its symbol. The air enters at pressure p_1 and flows through a constriction, the gap between the poppet and the valve case, and so a pressure drop occurs with the result that the outlet pressure p_2 is less than the input pressure p_1. By controlling the size of this gap the outlet pressure can be controlled. This is achieved by a diaphragm, on one side of which a force is exerted by a spring and on the other side by the air pressure. The required pressure is set by adjusting the spring so that these forces are in equilibrium. If the input pressure falls, the spring force exceeds that due to the air pressure and the diaphragm is forced down. This moves the poppet down and increases the gap between the poppet and the valve case. As a result the pressure drop through the gap between the poppet and the case decreases and the discharge pressure from the valve thus increases. If the input pressure rises, the spring force becomes less than

that due to the air pressure and the diaphragm is forced upwards. This reduces the gap and increases the pressure drop.

2 *Relieving pressure regulator*
Most regulator valves have an in-built pressure relief function which allows air to be vented to the atmosphere if the pressure in the system on the output side of the regulator rises above some predetermined value. With this type of valve, the diaphragm has a relief orifice in its centre which is closed when the outlet pressure is acceptable but opens and allows air to vent to the atmosphere if the outlet pressure is too high. Figure 2.19 shows the basic form and symbol of such a regulator.

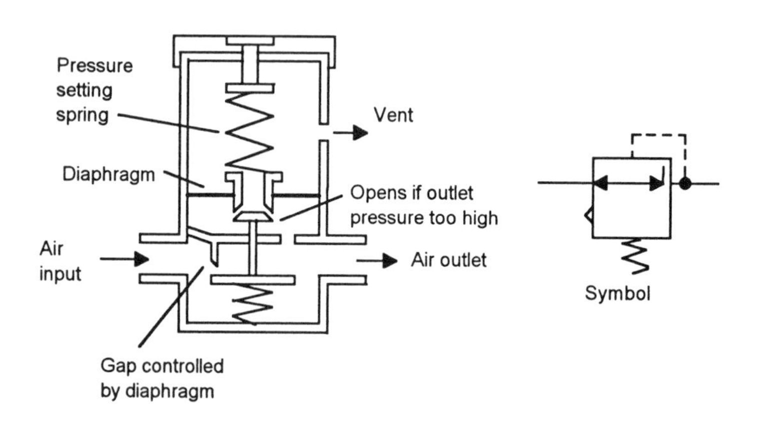

Figure 2.19 *Relieving pressure regulator*

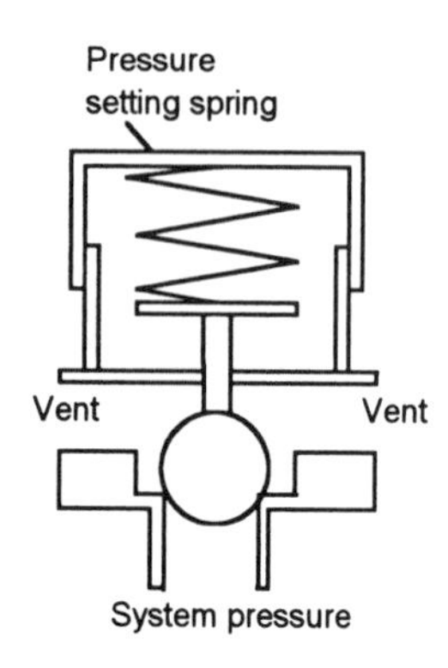

Figure 2.20 *Pressure relief valve*

The pressure regulation can be direct, the pressure being altered by an operator adjusting a screw, or *pilot operated* where the regulator is automatically controlled by the air flow.

Another valve which should be mentioned in this context is the *pressure relief valve* (Figure 2.20). This is employed as a back-up device which will open and vent air if the pressure in the system becomes too high. The ball valve is held closed by the spring, the force exerted on the ball by the spring being greater than the force exerted by the system air pressure. However, if the pressure in the system rises, the force it exerts on the ball rises and at some predetermined value it exceeds the force exerted by the spring and the ball valve opens and vents air from the system.

2.3.6 Service units

The term *service unit* is used for a combination of a filter, moisture separator, pressure regulator, pressure indicator and lubricator, such combinations frequently being required. Figure 2.21 shows the individual component symbols and the composite symbol used for the service unit.

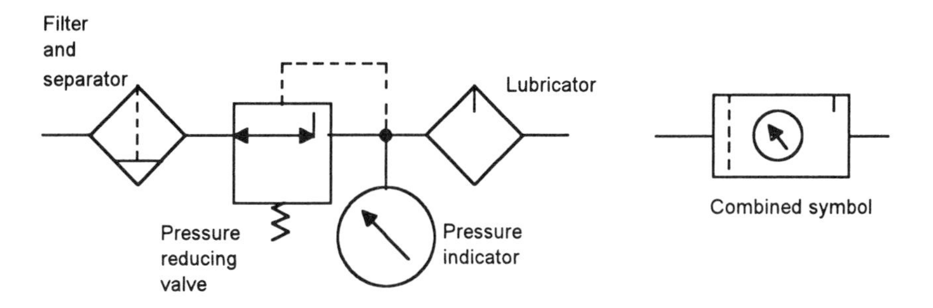

Figure 2.21 *Service unit*

2.4 Air receivers and compressor control

Compressors are normally operated with an *air receiver*. The purposes served by the air receiver are to store compressed air and so eliminate the need for the compressor to run continuously, to smooth out the pulsing flow of air from the compressor, to act as an emergency supply to the system in the event of power failure and to assist with air cooling and thus enable condensate to drop out before the air enters the distribution system. The size of the receiver and the allowable variations in the supply pressure determine the frequency with which the compressor switches on and off. Air receivers are usually cylindrical with a safety relief valve to guard against high pressures, a manually operated or automatic condensate drain, isolation valves to enable the receiver to be isolated, a pressure gauge and sometimes a temperature gauge. Figure 2.22 shows an air receiver, together with its accessories, and its symbolic representation.

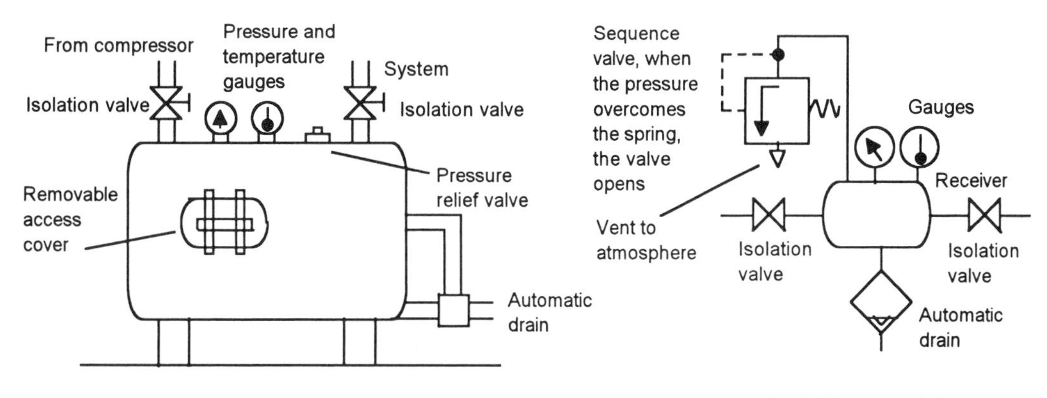

Figure 2.22 *Air receiver and its accessories*

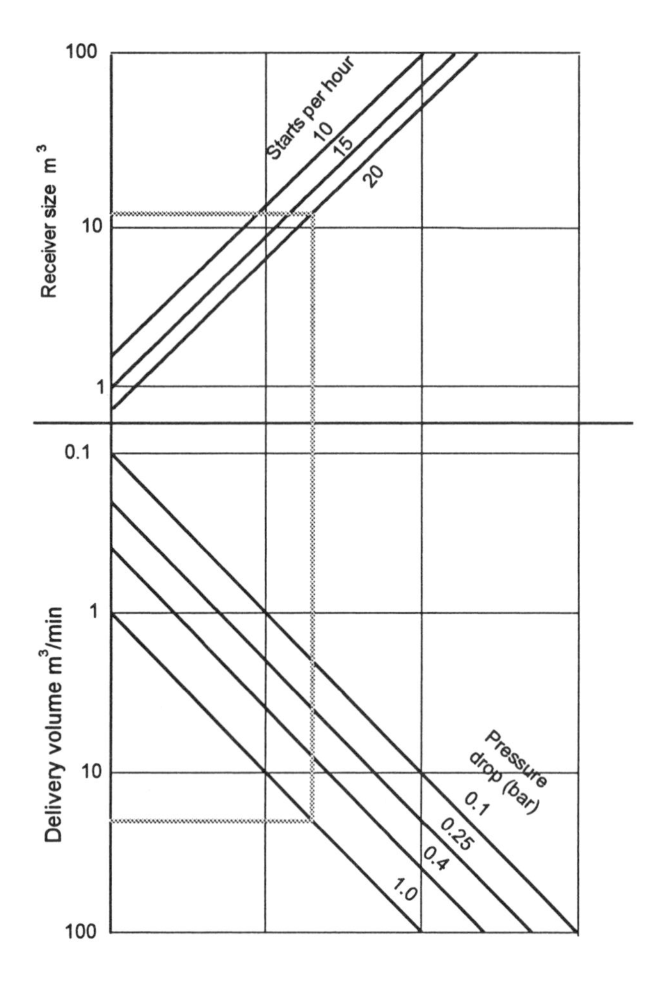

Figure 2.23 *Air receiver sizing chart*

2.4.1 Air receiver sizing

Charts and formulas are available for the determination of the size of an air receiver. Figure 2.23 shows the type of chart supplied by a manufacturer with a compressor. It is used by first selecting the delivery volume of free air required per minute; determining the point where the horizontal line through that delivery volume intersects the pressure drop required of the receiver, i.e. the difference in the pressure in the receiver between when the compressor is switched on and when it is switched off; taking the vertical line from that point to where it intersects the allowable number of starts per hour of the compressor; then taking the horizontal line from that point to where it intersects the receiver size axis and hence obtain the required value for the size. Alternatively the size can be determined from first principles by the use of the gas laws, the following example illustrating this.

Example

A pneumatic system requires an average delivery volume of 20 m^3/min free air delivery. The air compressor has a rated free air delivery of 25 m^3/min and a working gauge pressure of 7 bar, being controlled to switch off load when the receiver gauge pressure rises to 7 bar and back on load when the receiver gauge pressure has dropped to 6 bar. Determine a suitable receiver capacity if the maximum allowable number of starts per hour for the compressor is 20.

20 starts per hour means that the average time between starts is 3 min. In 3 min the system demand is 3×20 m^3 free air. To supply this the compressor must run for $(3 \times 20)/25 = 2.4$ min. Thus the compressor is off load for 0.6 min. During this time the receiver has to supply the system while the gauge pressure in the receiver falls from 7 to 6 bar. The volume of air supplied from the receiver in this time is $0.6 \times 20 =$ 12 m^3 free air.

If we assume that the temperature is constant, Boyle's law gives for the volume of free air in a receiver of volume V containing air at a gauge pressure of 7 bar:

$$\text{volume of free air} = \frac{(7+1)V}{1} = 8V\ \text{m}^3$$

The volume of free air in the receiver containing air at a gauge pressure of 6 bar is:

$$\text{volume of free air} = \frac{(6+1)V}{1} = 7V\ \text{m}^3$$

Thus the volume of free air delivered from the receiver as the pressure falls from 7 bar to 6 bar is $8V - 7V = 1V$ m^3. Thus $1V = 12$ m^3 and so the receiver volume required is 12 m^3.

2.4.2 Compressor control

Control of the compressor is necessary in order to maintain the pressure in the air receiver. The simplest method of control is *start/stop control* where the compressor is started when the receiver pressure falls to some minimum pressure and stopped when the pressure in the receiver has risen to the required value. An electrical pressure switch can be used to monitor the pressure and provide the signals to start and stop the compressor. Another method is to run the compressor continuously and use an exhaust valve on the outlet side of the compressor to vent air when the required pressure is reached. This is termed *exhaust regulation*. This is wasteful since the compressor is still using power while compressing air which is just vented to the atmosphere. Other control techniques involve controlling the input to a compressor. With one method a valve is used to close the intake to the compressor when the air receiver pressure reaches the required value. The compressor is then not compressing any air and so the only power used is the friction of the free-running compressor. Figure 2.24 illustrates these three types of control.

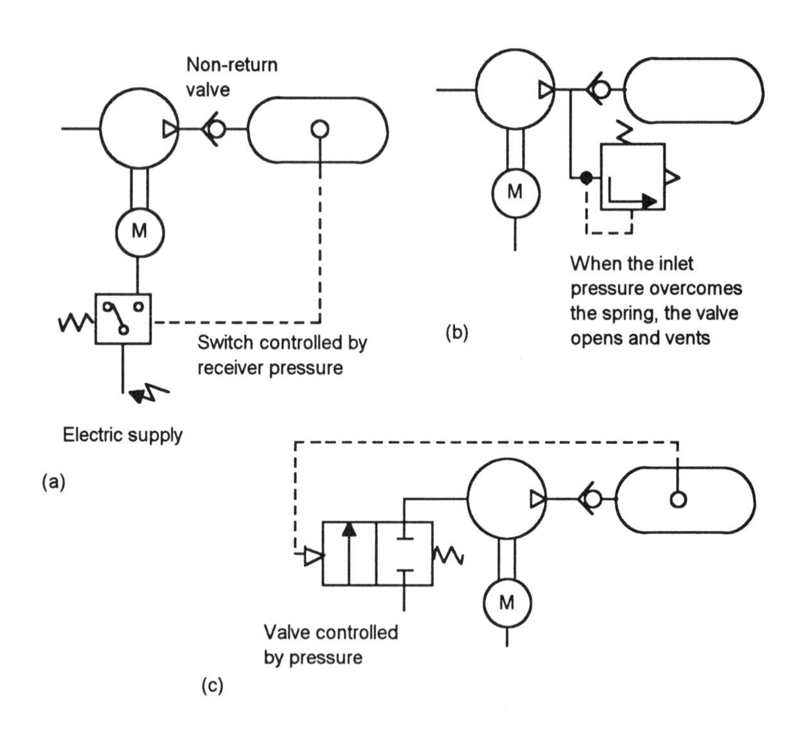

Figure 2.24 *Compressor control: (a) start–stop, (b) exhaust regulation, (c) intake regulation*

2.5 Plant layout

The production of compressed air is likely to involve the following basic elements: an inlet filter to remove dirt particles from the air entering the compressor, then the compressor followed by a cooler and a filter with water trap, then an air receiver followed a dryer (Figure 2.25). Ideally all water and oil should have separated out before the air enters the pneumatic system. However, this seldom happens and thus further separation is likely to occur in the distribution pipe work. To assist drainage, the pipe work should be given a fall of about 1 or 2 in 100 in the direction of the flow and provided with suitable drainage points. The main pipe is generally in the form of a ring, this helping to lower the velocity of the air in the main and so reduce the chance of the air picking up already deposited moisture and reducing the pressure drop (Figure 2.26). Any branch line should be taken off the top of the main pipe to prevent water in the main pipe from running into it and provided with a drain at the bottom (Figure 2.26).

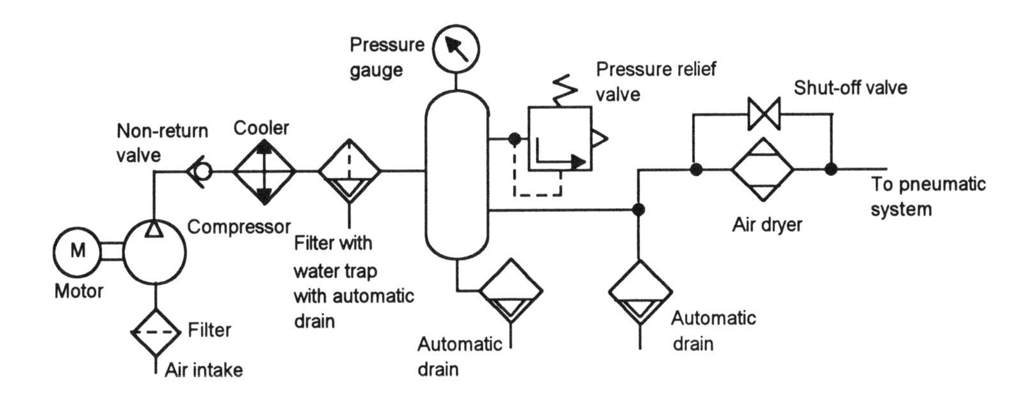

Figure 2.25 *Typical air production grouping*

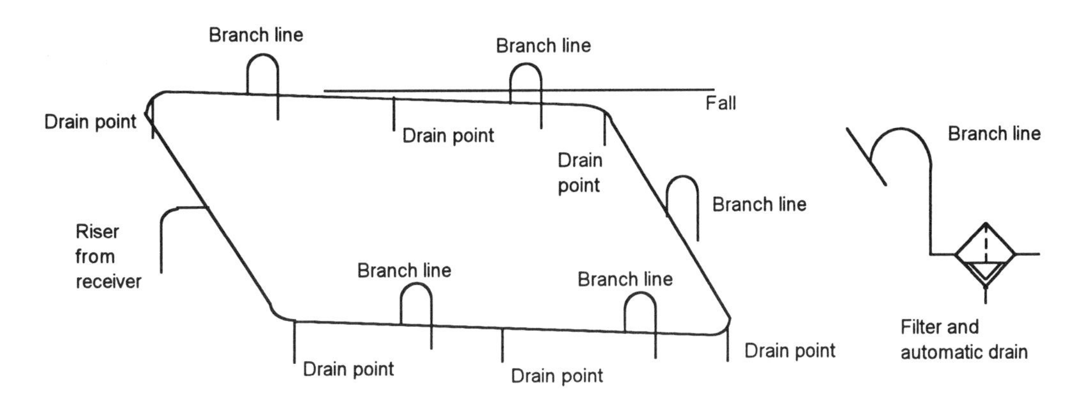

Figure 2.26 *Ring main system with detail of a branch line*

2.5.1 Pipe sizing

Charts and nomographs are available for determining the size of pipes required to meet particular situations, e.g. a particular pipe length, rates of free air flow, absolute system pressure and a required pressure loss. Figure 2.27 shows an example. To use the nomograph:

1 Draw a line from the required pipe length through the required free air flow and mark the point (X) where it intersects the reference line.

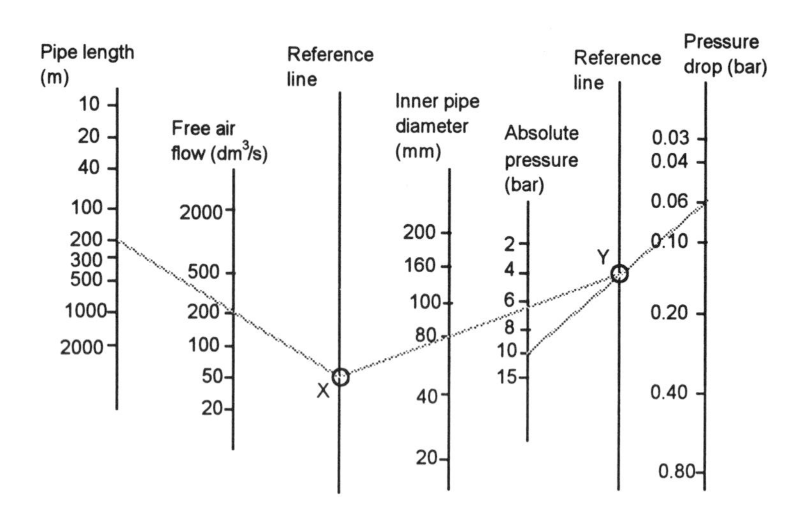

Figure 2.27 *Nomograph for steel pipe sizing*

2 Draw a line from the required pressure drop through the mean absolute pressure. Mark the point (Y) where it intersects the second reference line.

3 Draw a line through points X and Y and read the value of internal diameter of the pipe where the line intersects the scale. That is the required pipe size.

For example, if we have a pipe length of 200 m and a required free air flow of 200 dm^3/s, then reference point X is obtained. If we require the pressure drop to be 0.06 bar when the working pressure is 10 bar then reference point Y is obtained. Joining X and Y gives an inner pipe diameter of 80 mm.

Steel pipes are very commonly used and are available in a number of standard sizes. Standard sizes include nominal bores of 6.0, 8.0, 10.0, 15.0, 20.0, 25.0, 32.0, 40.0, 50.0, 65.0, 80.0, 100.0, 125.0 and 150.0 mm. A pressure drop of 1 bar per 100 m occurs with a nominal bore of 20.0 mm and a working pressure of 7 bar with a flow rate of 5.4 dm^3/s, with a working pressure of 10 bar with a flow rate of 6.5 dm^3/s. With a nominal bore of 50 mm, a pressure drop of 1 bar per 100 m occurs with a working pressure of 7 bar at a flow rate of 49.6 dm^3/s and 10 bar at 60.1 dm^3/s. With a nominal bore of 100 mm, a pressure drop of 1 bar per 100 m occurs with a working pressure of 7 bar at a flow rate of 368 dm^3/s and 10 bar at 446 dm^3/s.

Pressure drops due to pipe fittings such as a 90° bend or a tee or a valve can be determined by the use of tables which show the pressure drop in terms of equivalent pipe lengths of nominal bores. Table 2.2 gives some examples.

Table 2.2 *Typical equivalent pipe lengths in metres for fittings*

Fitting	Nominal inner pipe diameter in mm				
	25	40	50	80	100
90° elbow	0.5	0.8	1.0	1.8	2.4
90° bend	0.25	0.5	0.6	1.0	1.2
Globe valve	1.4	2.4	3.4	5.2	7.3
Gate valve	0.2	0.3	0.4	0.6	0.9
Run of standard tee	0.4	0.7	0.85	1.3	1.6
Side outlet tee	1.4	2.4	2.7	4.6	5.7

Example

Using Table 2.2, determine the total equivalent pipe length for a 50 mm internal pipe diameter system having six 90° bends, two elbow fittings, four tee-connections (90° turn) and two gate valves.

The equivalent pipe lengths are: six 90° bends $6 \times 0.6 = 3.6$ m, two elbow fittings $2 \times 1.0 = 2.0$ m, four tee-connections $4 \times 2.7 = 10.8$ m, two gate valves $2 \times 0.4 = 0.8$ m. The total is 17.2 m.

2.6 Hydraulic pumps

The hydraulic pump takes oil from a tank and raises its pressure and hence causes it to flow through the hydraulic circuit and back down in pressure again. Figure 2.28 shows the symbol used for a pump, this being similar to that used for the air compressor but a solid arrowhead being used to indicate hydraulic flow rather than the open arrowhead used to indicate air flow. Hydraulic pumps are specified by the flow rate they can deliver and the maximum pressure they can withstand.

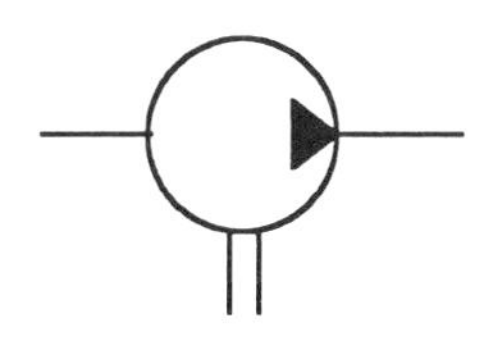

Figure 2.28 *Symbol for a pump*

If a pump forces fluid along a pipe of cross-sectional area A against a pressure p and moves it a distance x in a time t (Figure 2.29) then the work done in that time is:

$$\text{work done in time } t = \text{force} \times \text{distance} = pAx$$

Thus the power, i.e. rate of working, is:

$$\text{power} = \frac{pAx}{t}$$

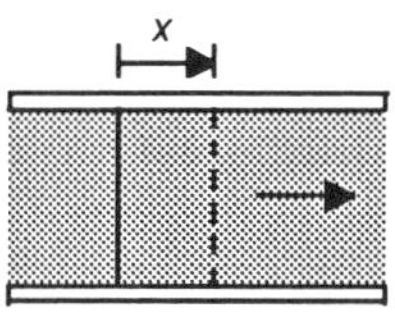

But Ax is the volume moved in time t and thus Ax/t is the volume flow rate Q. Hence:

Figure 2.29 *Moving fluid*

$$\text{power} = pQ \qquad [3]$$

This is the power required to deliver fluid at this pressure and rate of flow.

Example

Determine the power required to pump oil at a pressure of 5 MPa and rate of 0.1 m^3/min.

Using equation [3]:

$$\text{required power} = pQ = 5 \times 10^6 \times 0.1/60 = 8.3 \times 10^3 \text{ W} = 8.3 \text{ kW}$$

2.6.1 Pump efficiency

Hydrostatic pumps are positive displacement devices in that a definite volume of fluid is swept out from the pumping chamber for every revolution of the drive shaft, the volume being termed the *capacity* of the pump. If there are no losses of fluid due to leakages then the volume Q_p delivered per unit time is:

$$Q_p = C_p n_p \qquad [4]$$

where C_p is the capacity of the pump and n_p the number of revolutions per unit time of the pump shaft. The capacity can be calculated from the geometry of the pump.

No pump is perfect, there being invariably some leakage of oil. Thus the actual delivered volume per unit time Q_a is less than that indicated by equation [4]:

$$Q_a = Q_p - Q_l \qquad [5]$$

where Q_l is the leakage volume per unit time. The rate of leakage depends on the viscosity μ of the fluid, the pressure p and the clearance spaces through which the oil can leak. We can represent this as:

$$Q_l = k\frac{p}{\mu} \qquad [6]$$

where k is a constant for a particular pump. The *volumetric efficiency* of the pump is the fraction of the volume per second which is actually delivered. It is thus:

$$\text{volumetric efficiency} = \frac{Q_a}{Q_p} = \frac{Q_p - Q_l}{Q_p}$$

$$= 1 - \frac{Q_l}{Q_p} \qquad [7]$$

Using equation [6], this can be expressed as:

$$\text{volumetric efficiency} = \left(1 - \frac{kp}{\mu C_p n_p}\right) \times 100\% \qquad [8]$$

The volumetric efficiency at a constant rate of revolution thus decreases with increasing pressure. The volumetric efficiency at a constant pressure increases with increasing rate of revolution of the pump shaft.

For a rotating shaft of radius r, the distance travelled by a point on its surface in one revolution is $2\pi r$ and if there are n_p revolutions per second then the distance travelled per second is $2\pi r n_p$. If the shaft is being acted on by a torque T_p then, since we have $T_p = Fr$ with F the tangential force, the work done per second by the rotating shaft, i.e. the power, is $F \times 2\pi r n_p = 2\pi n_p T_p$. Thus the shaft power input to the compressor is:

$$\text{shaft power} = 2\pi n_p T_p \qquad [9]$$

where T_p is the pump shaft torque. The total oil flow rate Q_p through the pump per second is given by equation [4] as $C_p n_p$ and all this reaches a pressure p. Thus, using equation [3], the power delivered is $C_p n_p p$. Hence the mechanical efficiency of the pump is:

$$\text{mechanical efficiency} = \frac{C_p n_p p}{2\pi n_p T_p} = \frac{C_p p}{2\pi T_p} \qquad [10]$$

Thus at a constant speed of rotation of the shaft, the mechanical efficiency increases as the pressure increases.

The overall efficiency of the pump is the ratio of the power in delivered fluid to the shaft power. Thus, using equations [4], [5] and [9]:

$$\text{overall efficiency} = \frac{(Q_p - Q_l)p}{2\pi n_p T_p} \qquad [11]$$

Using equations [7] and [10] this can be written as:

$$\text{overall efficiency} = \text{volumetric efficiency} \times \text{mechanical efficiency} \qquad [12]$$

Example

The volumetric efficiency of a pump is 0.8 and it has a capacity of 1.6×10^{-4} m^3/revolution. Determine the actual delivery flow rate when the shaft speed is 1000 revolutions per minute and the delivery pressure is 10 MPa.

Using equation [4]:

$$Q_p = C_p n_p = 1.6 \times 10^{-4} \times 1000 = 0.16 \text{ m}^3/\text{min}$$

Since the volumetric efficiency is Q_a/Q_p then:

$$Q_a = 0.8 \times 0.16 = 0.128 \text{ m}^3/\text{min}$$

2.6.2 Types of pumps

There are three types of hydraulic pump in common use:

1 *Gear pump*
This consists of two close-meshing gear wheels which rotate in opposite directions (Figure 2.30). Fluid is forced through the pump as it becomes trapped between the rotating gear teeth and the housing and so it transferred from the inlet port to be discharged at the outlet port. Such pumps generally operate at pressures below about 15 MPa and at 2400 rotations per minute. The maximum flow capacity is about 0.5 m^3/min. Leakage occurs between the teeth and the casing and between the interlocking teeth. This limits the volumetric efficiency generally to about 0.90. Overall efficiency is about 0.6 to 0.8.

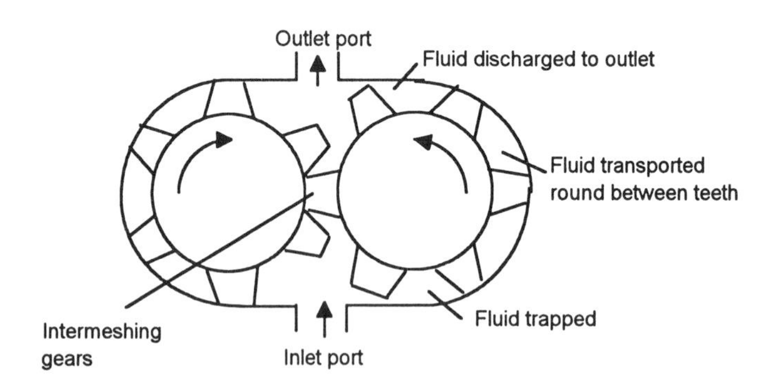

Figure 2.30 *Gear pump*

Another version of the gear pump is the *lobe pump*. With this version the gears have been replaced by lobes (Figure 2.31). Here fluid is trapped between the lobes and the casing and transported round from the inlet port to the outlet port.

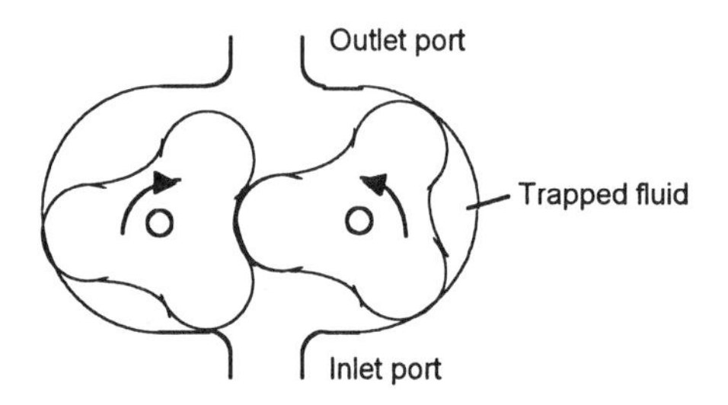

Figure 2.31 *Lobe pump*

2 *Vane pump*
The vane pump has spring-loaded sliding vanes slotted in a driven rotor (Figure 2.32). As the rotor rotates, the vanes follow the contours of the casing. This results in fluid becoming trapped between successive vanes and the casing and transported round from the inlet port to outlet port. The leakage is less than with the gear pump and a volumetric efficiency of the order of 0.95 is attainable.

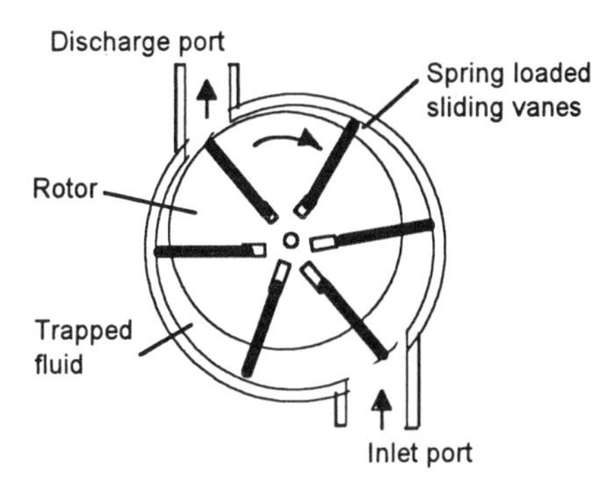

Figure 2.32 *Vane pump*

3 *Piston pump*
A very simple piston pump just involves taking a sample of air and then compressing it by moving a piston. The bicycle pump is an obvious example of such a pump. Piston pumps used in hydraulics can take a number of forms. Figure 2.33 shows a common form, the *radial piston pump*. The cylinder block rotates round the stationary cam and this causes hollow pistons, with spring return, to move in and out. The result is that fluid is drawn in from the inlet port and transported round for ejection from the discharge port.

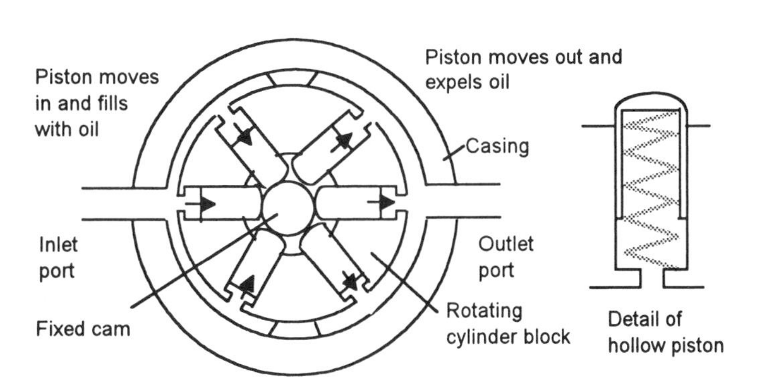

Figure 2.33 *Radial piston pump*

An alternative form of piston pump is the *axial piston pump* (Figure 2.34). With this pump the pistons move axially rather than radially. The pistons are arranged axially in a rotating cylinder block and made to move by contact with the swash plate. This plate is at an angle to the drive shaft and thus as the shaft rotates they move the pistons so that air is sucked in when a piston is opposite the inlet port and expelled when it is opposite the discharge port.

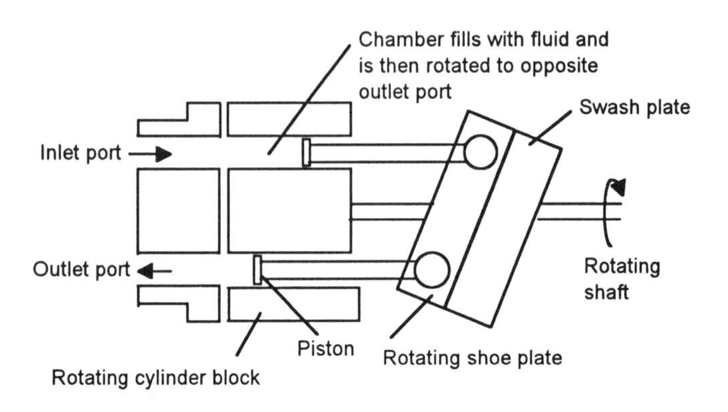

Figure 2.34 *Axial piston pump with swash plate*

Piston pumps have a high volumetric efficiency, radial up to about 95% and axial up to about 98%, and can be used at higher hydraulic pressures than the gear or vane pumps. For example, a radial piston pump can be used up to about 60 MPa.

2.7 Hydraulic circuit

Figure 2.35 shows the basic elements of a hydraulic circuit. The pump is driven by a motor which takes hydraulic fluid, via a filter, from a supply tank. The fluid is pumped into the system via a non-return valve, this being to prevent the fluid being forced back into the pump as a result of pressure pulses in the system. A pressure relief valve is used to keep the system at a safe pressure level. An accumulator is included primarily to smooth pressure pulses developed in the pump and cope with high transient flow demands by providing additional flow for short periods. A shut-off valve is included to enable the accumulator to be safely discharged, while further shut-off valves are included to enable the load part of the system, just the simple actuator shown in this figure, to be isolated from the pump part in case it has to be disconnected and likewise to isolate the pressure indicator. The circuit is completed by the hydraulic fluid being returned to the supply tank. Filters are placed in various parts of the circuit to control contamination.

The following sections give more details of the basic elements of supply tank, filters, pressure relief valve and accumulator included in the hydraulic circuit.

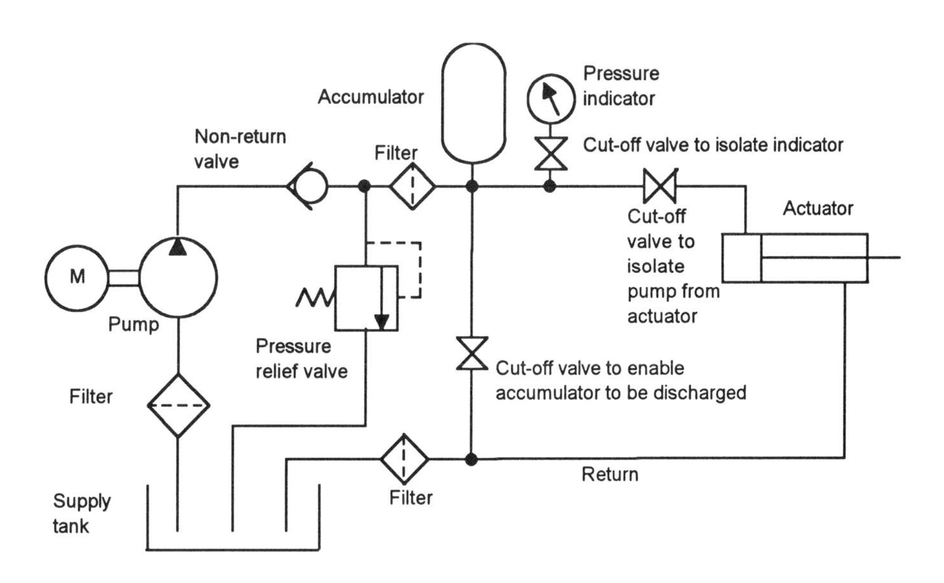

Figure 2.35 *Hydraulic circuit*

2.7.1 Supply tank

The supply tank must contain enough hydraulic fluid to meet the circuit demand and still leave some reserve in the tank. Typically this means a volume at least three times the volume of fluid delivered in one minute. The tank also serves as a heat exchanger to coil the fluid and allow time for contaminants to settle out.

2.7.2 Filters

Fluid contamination in a hydraulic circuit can cause valves to stick, seals to fail and wear of components. Sources of contamination include debris in the hydraulic fluid as a result of storage prior to use, dirt particles in the hydraulic circuit when it is assembled, particles picked up from the working environment, sludge arising from a high running temperature, etc. Filters are commonly used in the hydraulic circuit:

1 In the inlet line for the fluid from the tank to the pump to protect the pump.

2 In the pressure line after the pump to protect actuators and valves.

3 In the return line to limit the particles being returned to the tank.

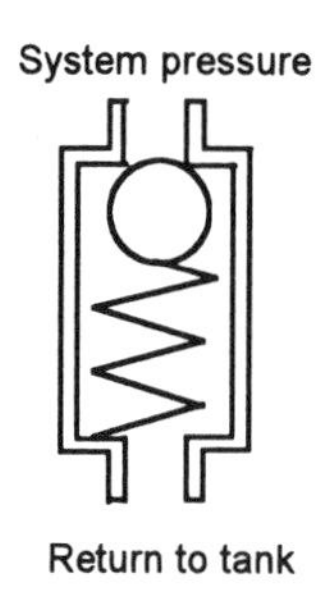

Figure 2.36 *Simple pressure relief valve*

2.7.3 Pressure relief valve

Figure 2.36 shows a simple form of *pressure relief valve* that can be used to regulate the pressure in a hydraulic circuit. The forces on the ball are those arising from the pressure and the spring. As the pressure increases so the force on ball due to the pressure increases, until at some level the force is greater than that exerted by the spring and so the ball is forced down and the pressure is relieved as fluid escapes. The pressure at which relief occurs is set by the amount of spring compression.

2.7.4 Accumulator

The accumulator in a hydraulic circuit has the functions of:

1 Smoothing pressure pulses developed in the pump.

2 Controlling shock pressure loading due to such events as rapidly closing valves.

3 Coping with high transient flow demands by providing additional flow for short periods.

4 Providing standby power for situations such as pump failure in order to allow equipment to return to a safe condition.

5 Allowing for expansion when there is an increase in temperature.

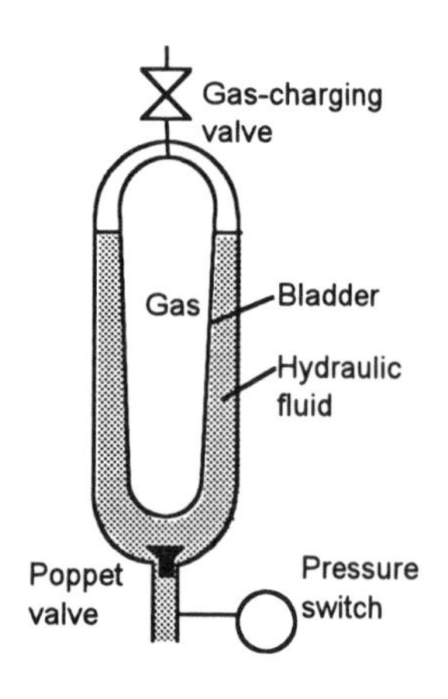

Figure 2.37 *Gas-pressurise accumulator*

Most accumulators are the gas-pressurised type (Figure 2.37). Gas within a bladder is in the chamber containing the hydraulic fluid. Smaller and older types may be a spring-loaded design. Figure 2.38 shows the symbols for both types.

Consider the gas-pressurised accumulator. If we assume that the expansion or contraction of the gas in the bladder takes place slowly enough for the temperature to remain constant then we can apply Boyle's law to the gas. Initially we charge the bladder with gas at a pressure p_1 and volume V_1, there being no hydraulic fluid in the accumulator and thus the bladder has expanded to fill the entire accumulator and so V_1 is the accumulator volume. If we now switch on the hydraulic pump and admit hydraulic fluid to the accumulator to give its maximum pressure and the gas pressure changes to p_2 and the volume to V_1, then Boyle's law gives:

$$p_1V_1 = p_2V_2$$

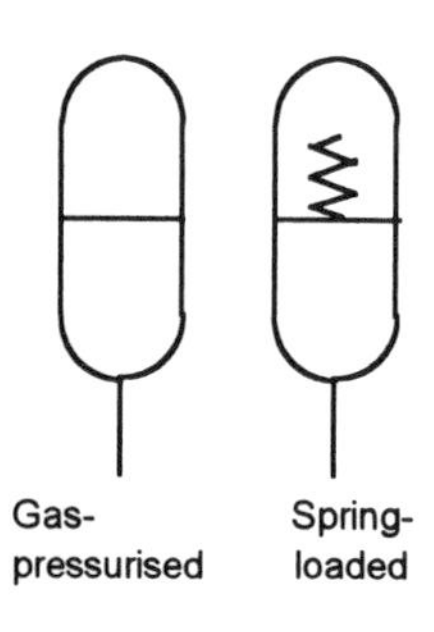

Figure 2.38 *Symbols*

If hydraulic fluid is then removed from the accumulator the gas pressure drops to p_3 and the volume increases to V_3. Applying Boyle's law gives:

$$p_2V_2 = p_3V_3$$

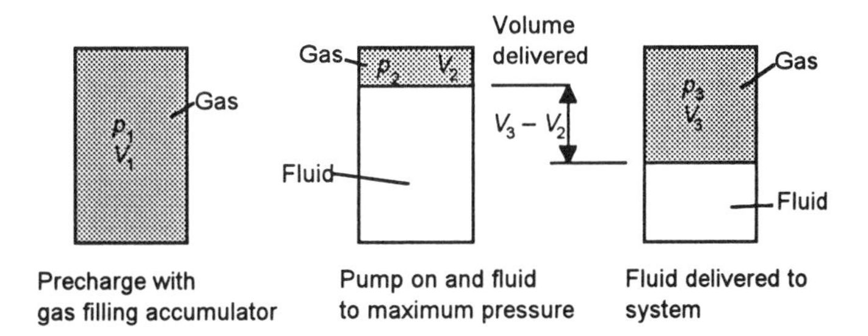

Figure 2.39 *Volumes of gas and hydraulic fluid in the accumulator*

The change in volume $V_3 - V_2$ is the volume of hydraulic fluid removed from the accumulator (Figure 2.39). Thus:

$$\text{fluid delivered } V_{out} = V_3 - V_2 = \frac{p_1 V_1}{p_3} - \frac{p_1 V_1}{p_2} \qquad [13]$$

Thus:

$$\text{accumulator size } V_1 = V_{out} \frac{\frac{p_2}{p_1}}{\frac{p_2}{p_3} - 1} \qquad [14]$$

If, however, the gas is compressed or expanded rapidly and there is not time for heat to enter or leave the gas, then the adiabatic equation pV^γ has to be used instead of Boyle's law. In this case, if the accumulator is rapidly charged and discharged we have:

$$p_1 V_1^\gamma = p_2 V_2^\gamma = p_3 V_3^\gamma$$

and so:

$$\text{fluid delivered } V_{out} = V_3 - V_2 = \frac{p_1^{1/\gamma} V_1}{p_3^{1/\gamma}} - \frac{p_1^{1/\gamma} V_1}{p_2^{1/\gamma}}$$

$$\text{accumulator size } V_1 = V_{out} \frac{\left(\frac{p_2}{p_1}\right)^{1/\gamma}}{\left(\frac{p_2}{p_3}\right)^{1/\gamma} - 1} \qquad [15]$$

Often, however, when the accumulator is used to supplement pump delivery and cope with sudden demands we have the accumulator charged slowly and discharged rapidly when, say, a piston moves in a cylinder. Then, for the discharge we have:

$$p_2 V_2^\gamma = p_3 V_3^\gamma$$

and for the charging by the pump:

$$p_1V_1 = p_2V_2$$

Then:

$$\text{fluid delivered } V_{out} = V_3 - V_2 = \frac{p_2^{1/\gamma} p_1 V_1}{p_3^{1/\gamma} p_2} - \frac{p_1 V_1}{p_2}$$

$$\text{accumulator size } V_1 = V_{out} \frac{\left(\frac{p_2}{p_1}\right)}{\left(\frac{p_2}{p_3}\right)^{1/\gamma} - 1} \quad [16]$$

Where the accumulator is used to supplement pump delivery, the practice is to use a precharge pressure which is 90% of the pressure when the accumulator is delivering fluid, i.e. $p_1 = 0.90p_3$.

Example

Determine the size of the accumulator necessary to supply 5 dm^3 of hydraulic fluid between absolute pressures of 200 bar and 100 bar if the accumulator is initially charged to a pressure of 90 bar.

Assuming isothermal conditions, equation [14] gives:

$$\text{accumulator size } V_1 = V_{out} \frac{\frac{p_2}{p_1}}{\frac{p_2}{p_3} - 1} = 5 \frac{\frac{200}{90}}{\frac{200}{100} - 1} = 11.1 \text{ dm}^3$$

If, however, we had assumed adiabatic conditions, with $\gamma = 1.4$, then equation [15] gives:

$$\text{accumulator size } V_1 = V_{out} \frac{\left(\frac{p_2}{p_1}\right)^{1/\gamma}}{\left(\frac{p_2}{p_3}\right)^{1/\gamma} - 1} = 5 \frac{\left(\frac{200}{90}\right)^{1/1.4}}{\left(\frac{200}{100}\right)^{1/1.4} - 1}$$

The accumulator size under these conditions is 13.8 dm^3.

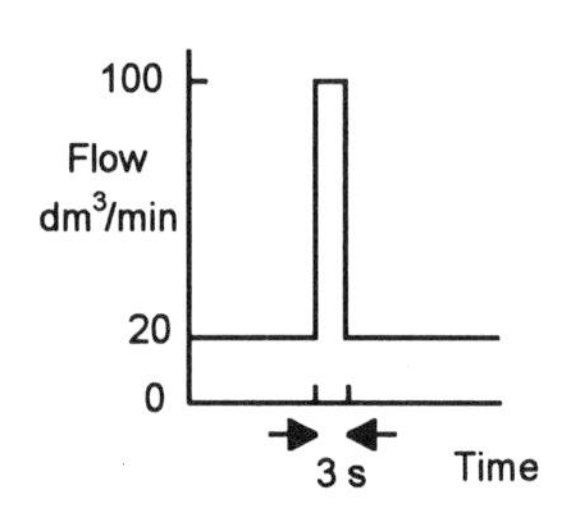

Figure 2.40 *Example*

Example

Figure 2.40 shows the requirement for hydraulic fluid flow in a circuit involving an actuator. Determine the size of accumulator that would enable the demand to be met if the pump supplies 20 dm^3/min at an absolute pressure of 20 MPa and when the actuator operates the pressure drops to 13 MPa. The precharge absolute pressure of the accumulator is 11 MPa.

The volume that needs to be supplied by the accumulator in 3 s is 80 × (3/60) = 4 dm³. The discharge time of 3 s is fairly short and so it seems likely that conditions will more adiabatic than isothermal. Thus, if we also assume that the charging is adiabatic, equation [15] gives:

$$\text{accumulator size } V_1 = V_{\text{out}} \frac{\left(\frac{p_2}{p_1}\right)^{1/\gamma}}{\left(\frac{p_2}{p_3}\right)^{1/\gamma} - 1} = 4 \frac{\left(\frac{20}{11}\right)^{1/1.4}}{\left(\frac{20}{13}\right)^{1/1.4} - 1}$$

An accumulator size of 17 dm³ is thus required.

If, however, we had assumed that the charging took place isothermally and only the discharge was adiabatic, equation [16] gives:

$$\text{accumulator size } V_1 = V_{\text{out}} \frac{\left(\frac{p_2}{p_1}\right)}{\left(\frac{p_2}{p_3}\right)^{1/\gamma} - 1} = 4 \frac{\left(\frac{20}{11}\right)}{\left(\frac{20}{13}\right)^{1/1.4} - 1}$$

An accumulator size of 20.2 dm³ is thus required.

2.8 Pneumatic and hydraulic symbols

Symbols have been indicated for representing the various components already discussed in this chapter. More symbols are introduced in later chapters, in particular Chapter 3 giving those for control valves and Chapter 4 those for actuators. However, Figure 2.41 shows some general items regarding symbolic representation.

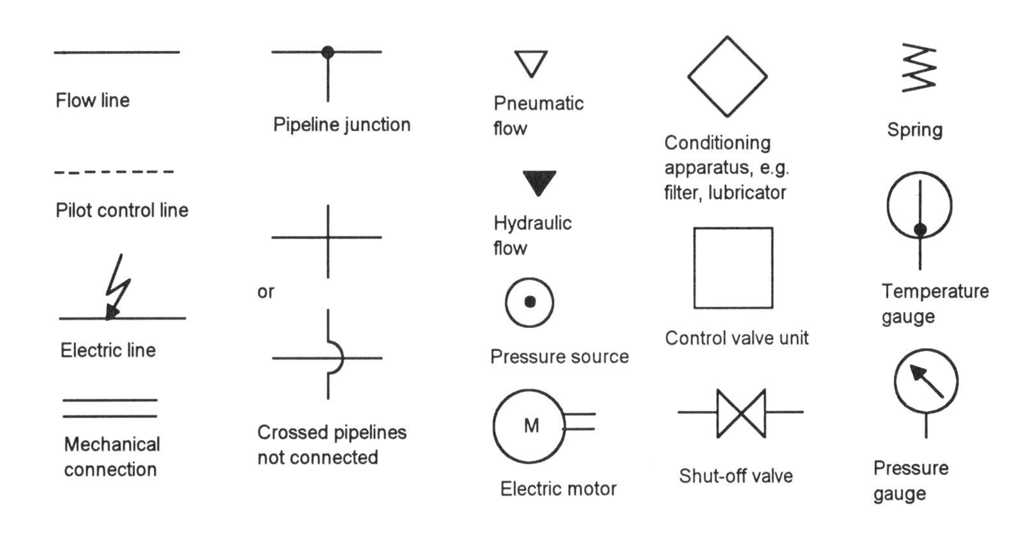

Figure 2.41 *Symbols*

Problems

1 A compressor has a rated output of 0.5 m^3/min free air delivery. What will be the output at an absolute pressure of 500 kPa and the same temperature?

2 A compressor is required to deliver 0.6 m^3/min of compressed air at an absolute pressure of 600 kPa. What compressor output is required in terms of free air?

3 What are the main advantages of using (a) a two-stage piston reciprocating compressor rather than achieving the required pressure by just using a single-stage, (b) a double-acting rather than a single-acting piston reciprocating compressor?

4 What are the reasons for rotary vane compressors being predominantly oil injected?

5 A compressor takes in air at an absolute pressure of 101 kPa with a relative humidity of 60% and a temperature of 20°C. The air at the discharge port of the compressor is delivered at 2 m^3/min with a temperature of 35°C and an absolute pressure of 800 kPa. Determine the amount of condensed water in the output per minute.

6 A dryer produces air with a dew point of 5°C. What will be the mass of water vapour per cubic metre in that air?

7 Using the chart given in Figure 2.23, determine the size of receiver required when the pneumatic system requires 10 m^3/min free air delivery, the air compressor switches the compressor on when the receiver pressure drops to 5 bar and off when it rises to 6 bar and the number of starts per hour is 20.

8 A pneumatic system requires an average delivery volume of 20 m^3/min free air delivery. The air compressor has a rated free air delivery of 35 m^3/min, a working gauge pressure of 6 bar and is controlled to switch off load when the receiver gauge pressure rises to 6 bar and back on load when the receiver gauge pressure has dropped to 5 bar. Determine a suitable receiver capacity if the maximum allowable number of starts per hour for the compressor is 20.

9 A pneumatic system requires an average delivery volume of 20 m^3/min free air delivery. The air compressor has a rated free air delivery of 35 m^3/min, a working gauge pressure of 7 bar and is controlled to switch off load when the receiver gauge pressure rises to 6 bar and back on load when the receiver gauge pressure has dropped to 5.2 bar. Determine a suitable receiver capacity if the maximum allowable number of starts per hour for the compressor is 20.

10 A pneumatic system has an average delivery volume of 7 m^3/min free air delivery with a pressure which is allowed to fluctuate between 7 bar and

6 bar. The air compressor delivers 10 m^3/min and the air receiver has a volume of 6 m^3. Determine the number of starts per hour for the compressor.

11 State three functions of an air receiver in a pneumatic system.

12 Explain, by means of a sketch, the correct method of tapping an air main.

13 Explain the reason for a 'fall' in an air line.

14 Using Figure 2.27, determine the diameter of an air distribution pipe which is to carry air at a free air flow rate of 200 dm^3/s if a pressure drop of 10 kPa is permitted. The maximum working pressure for the system is 800 kPa and the equivalent pipe length is 200 m.

15 Using Figure 2.27, determine the diameter of an air distribution pipe which is to carry air at a free air flow rate of 100 dm^3/s if a pressure drop of 6 kPa is permitted. The maximum working pressure for the system is 400 kPa and the equivalent pipe length is 200 m.

16 Using Table 2.2, determine the total equivalent pipe length for a 80 mm internal pipe diameter system having five 90° bends, two elbow fittings, three tee-connections (90° turn) and two gate valves.

17 The volumetric efficiency of a pump is 0.8 and it has a capacity of 1.2×10^{-4} m^3/revolution. Determine the actual delivery flow rate when the shaft speed is 1200 revolutions per minute and the delivery pressure is 7 MPa.

18 A pump has a mechanical efficiency of 0.9 and a volumetric efficiency of 0.8 when running at a particular speed and pressure. What is the overall efficiency?

19 A pump has a volumetric efficiency of 0.8 when operating at 1000 revolutions per minute and a pressure of 10 MPa. What will be the volumetric efficiency when it is operating at 1500 revolutions per minute and a pressure of 14 MPa?

20 State reasons for using an accumulator with a hydraulic circuit.

21 Determine, assuming isothermal conditions, the size of the accumulator necessary to supply 12 dm^3 of hydraulic fluid between absolute pressures of 180 bar and 160 bar if the accumulator is initially charged to a pressure of 120 bar.

22 An accumulator has a size of 0.5 m^3 and a precharge absolute pressure of 10 bar. How much hydraulic fluid has to be pumped into the

accumulator to raise the pressure to 28 bar? Assume the oil is pumped slowly.

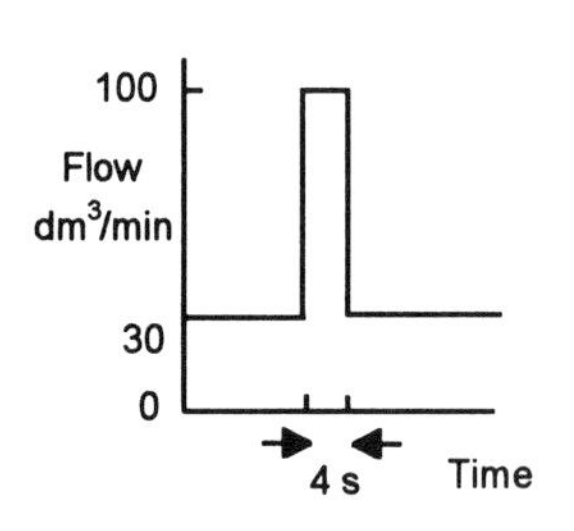

Figure 2.42 *Problem 23*

23 Figure 2.42 shows the requirement for hydraulic fluid flow in a circuit involving an actuator. Determine the size of accumulator that would enable the demand to be met if the pump supplies 30 dm^3/min at a pressure of 20 MPa and when the actuator operates the pressure drops to 12 MPa. The precharge pressure of the accumulator is 9 MPa.

24 A pump has a delivery of 0.4 dm^3/s at a gauge pressure of 70 bar. Determine the size of accumulator that would enable a demand for hydraulic fluid of 0.8 dm^3 over a period of 0.1 s if the gauge pressure is not to drop by more than 10 bar. The minimum time between demands is 40 s. Take the precharge pressure to be 90% of the working pressure.

25 A pump has a delivery of 0.2 dm^3/s at a gauge pressure of 75 bar. Determine the size of accumulator that would enable a demand for hydraulic fluid of 0.5 dm^3 over a period of 0.1 s if the gauge pressure is not to drop to less than 10 bar. The minimum time between demands is 30 s. Take the precharge pressure to be 90% of the working pressure.

3 Control valves

3.1 Introduction

The main purpose of a valve in a pneumatic or hydraulic circuit is to control outputs. Valves can be divided into a number of groups according to what they control:

1 *Directional control valves*
A directional control valve on the receipt of some external signal, which might be mechanical, electrical or a fluid pressure pilot signal, changes the direction of, or stops, or starts the flow of fluid in some part of the pneumatic/hydraulic circuit.

2 *Pressure control valves*
These are used to control the pressure in part of the pneumatic/hydraulic circuit.

3 *Flow control valves*
These are used to control the rate of flow of a fluid through the valve.

This chapter considers all three types of valve.

3.2 Directional control valves

A directional control valve on the receipt of some external signal, which might be mechanical, electrical or a fluid pilot signal, changes the direction of, or stops, or starts the flow of fluid in some part of the pneumatic/hydraulic circuit. They can be used to carry out such functions as:

1 Controlling the direction of motion of an actuator (see Chapter 4 for more details).

2 Selecting alternative flow paths for fluid.

3 Stopping and starting the flow of fluid.

4 Carrying out logic functions such as AND, OR, NAND, etc. For example, some action may be initiated if event A and event B both occur but not if neither event nor only one event occurs (see Chapter 8).

3.2.1 Symbols

The basic symbol for a control valve is a square. With a directional control valve two or more squares are used, with each square representing the switching positions provided by the valve. Thus Figure 3.1(a) represents a

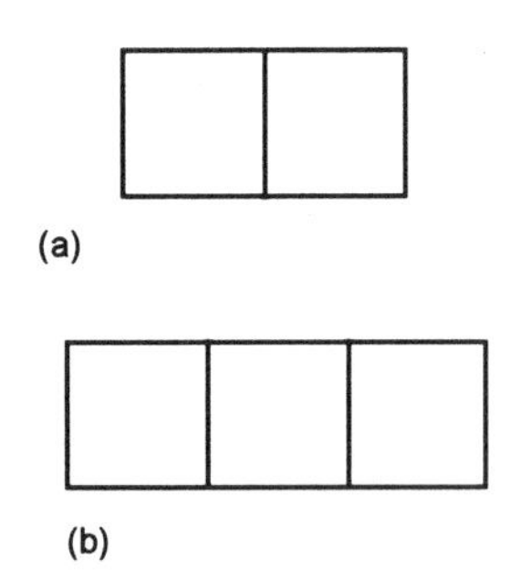

Figure 3.1 *(a) Two position, (b) three position*

valve with two switching positions, Figure 3.1(b) a valve with three switching positions.

Lines in the boxes are used to show the flow paths with arrows indicating the direction of flow (Figure 3.2(a)) and shut-off positions indicated by lines drawn at right angles (Figure 3.2(b)). The pipe connections, i.e. the inlet and outlet ports to the valve, are indicated by lines drawn on the outside of the box and are drawn for just the 'rest/initial/neutral position', i.e. when the valve is not actuated (Figure 3.2(c)). With a two-position valve, this is generally to the right-hand position, with a three-position valve to the central position. You can imagine each of the position boxes to be moved by the action of some actuator so that it connects up with the pipe positions to give the different connections between the ports.

The ports of a valve are shown on the outside of the box for the initial position and labelled by a number or letter according to its function. Table 3.1 gives the standard numbers and symbols used.

Table 3.1 *Port labels*

Port	Numbering system	Lettering system
Pressure supply port	1	P
Exhaust port(s)	3	R, one exhaust, 3/2 valve
	5, 3	R, S, two exhausts, 5/2 valve
Signal outputs	2, 4	B, A
Pilot line opens flow 1 to 2	12	Z, single pilot 3/2 way valve
	12	Y, 5/2 way valve
Pilot line opens flow 1 to 4	14	Z, 5/2 valve
Pilot line flow closed	10	Z, Y
Auxiliary pilot air	81, 91	Pz

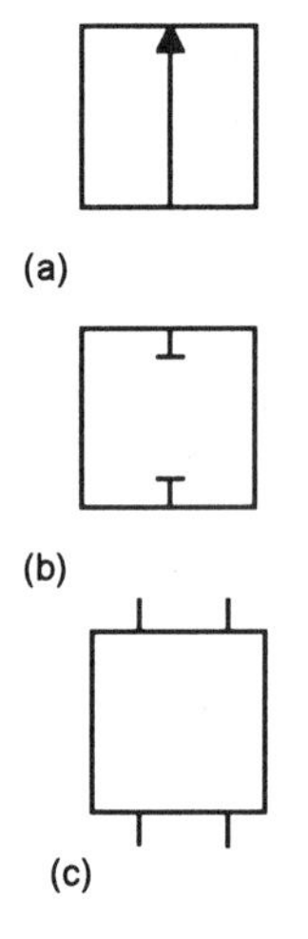

Figure 3.2 *(a) Flow path, (b) shut-off, (c) initial connections*

Directional control valves are described by the number of ports and the number of positions. Thus a 2/2 valve has 2 ports and 2 positions, a 3/2 valve 3 ports and 2 positions, a 4/2 valve 4 ports and 2 positions, a 5/3 valve 5 ports and 3 positions. Figure 3.3 shows some commonly used examples and their switching options.

Figure 3.3(a) shows a 2/2 valve that is normally closed and that when activated connects the pressure port to the output port. It is thus an off–on switch.

Figure 3.3(b) shows a 2/2 valve that is normally open with the pressure port connected to the output port. When activated it closes both the ports and so switches the pressure to the output off. It is thus an on–off switch.

Figure 3.3(c) shows a 3/2 valve that normally has the pressure to the output off and the output exhausting via the exhaust port. When activated, pressure is applied to the output port and the exhaust port closed.

Figure 3.3(d) shows a 3/2 valve that normally has the pressure to the output and the exhaust port closed. When activated, the pressure is switched off from the output and the output exhausts via the exhaust port.

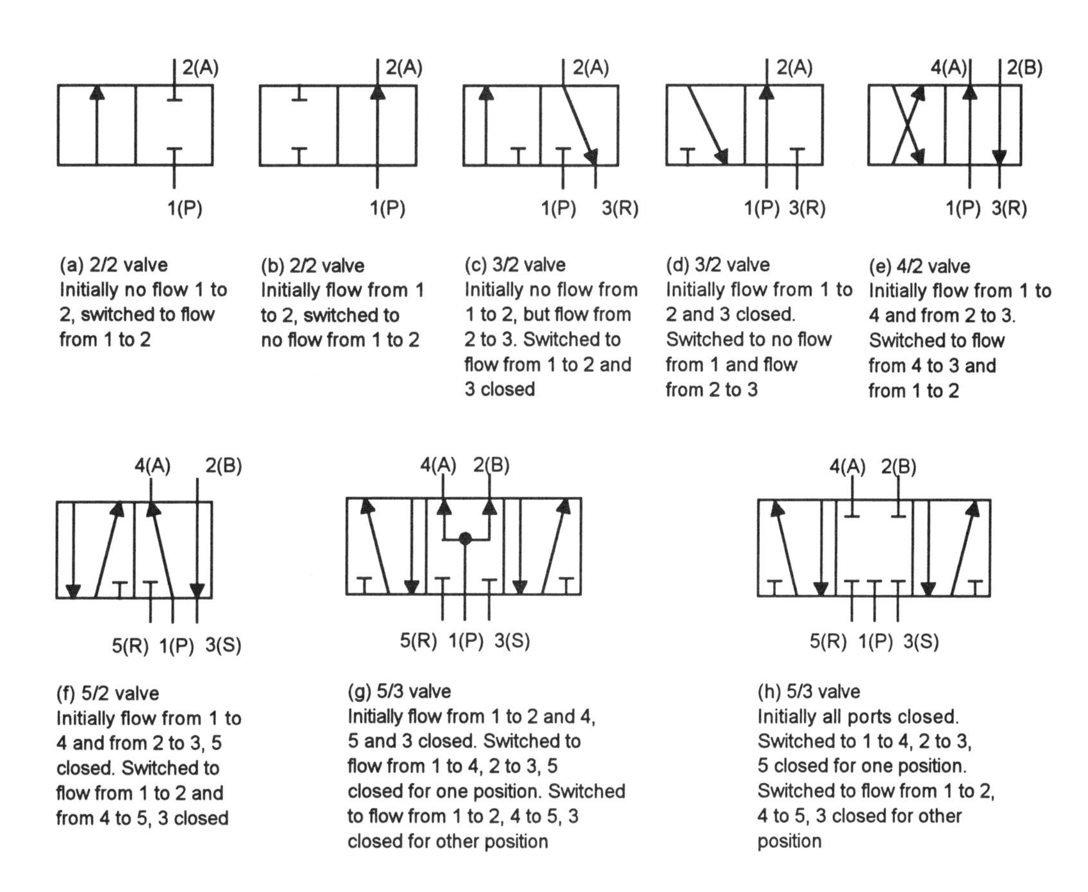

Figure 3.3 *Commonly used direction valves*

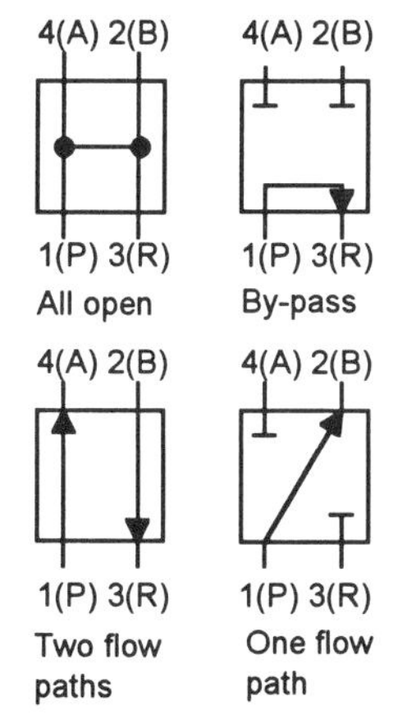

Figure 3.4 *Examples of centre positions*

Figure 3.3(e) shows a 4/2 valve that normally has the pressure applied to output 4 while output 3 exhausts through the exhaust port. When activated, the pressure switches to output 3 while output 4 exhausts through the exhaust port.

Figure 3.3(f) shows a 5/2 valve that normally has the pressure applied to output 4 while output 2 exhausts through exhaust port 3. When activated, the pressure switches to output 2 while output 4 exhausts through exhaust port 5.

Figure 3.3(g) shows a 5/3 valve that normally has the pressure applied to both outputs 2 and 4. When switched to one position, the pressure is applied to just output port 4 with output 2 exhausting through exhaust port 3. When switched to the other position, the pressure is applied to just output port 2 with output port 4 exhausting through exhaust port 5. Figure 3.3(h) shows a 5/3 valve with the same switching connections but that is normally switched off rather than on as in (g). Figure 3.4 shows some other examples of centre position connections.

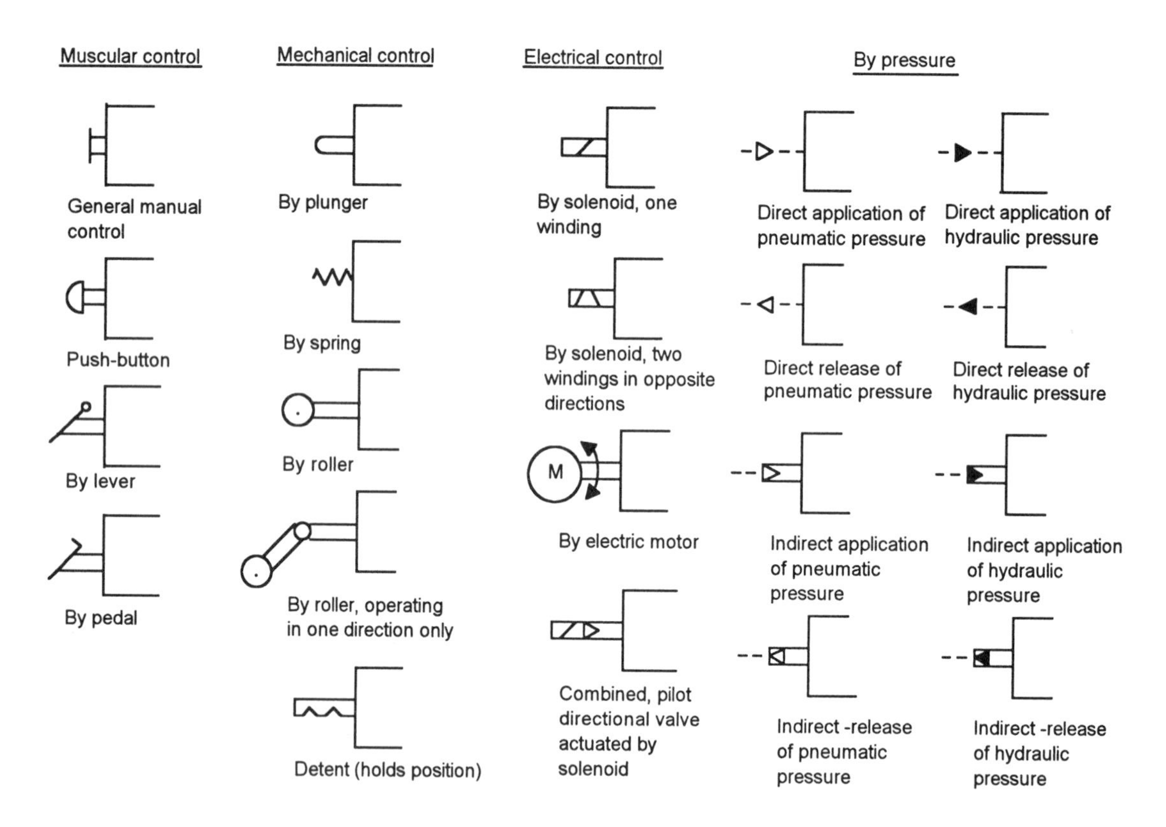

Figure 3.5 *Valve actuation methods*

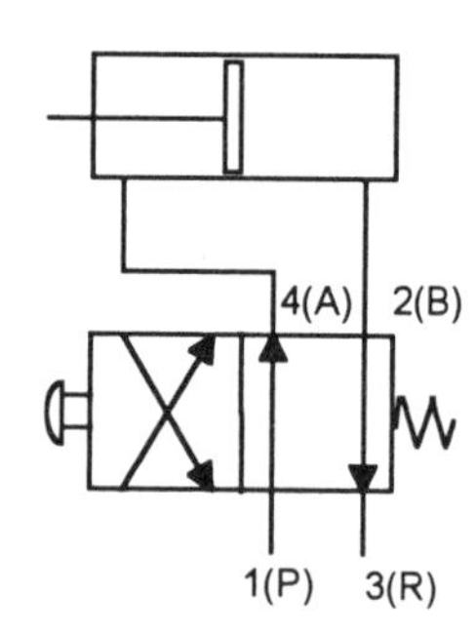

Figure 3.6 *4/2 valve*

Figure 3.5 shows the symbols for commonly used valve actuation methods. Thus a complete valve symbol might be as shown in Figure 3.6. This represents a 4/2 valve which normally has the pressure applied to output port 4 and the other output port 2 is connected to the exhaust port. When the push-button is pressed, the valve switches to its other position and the pressure is now applied to the output port 2 and the output port 4 is connected to the exhaust port. When the push-button is released, the spring causes the initial position to be re-obtained. The net result is that normally the piston is at the right-hand end of the cylinder and when the push-button is pressed it moves to the left-hand end. When the button is released, the piston returns to the right-hand end.

Figure 3.7 shows another example of a complete valve symbol. This represents a 4/3 valve which is spring centred. In that position the valve is off with no pressure being applied to either outlet port. When solenoid X is activated, the pressure is applied to output 4 and output 2 is connected to the exhaust. When solenoid Y is activated, the pressure is applied to output 2 and output 4 is connected to the exhaust. In the absence of either solenoid being activated, the springs return the valve to the closed position.

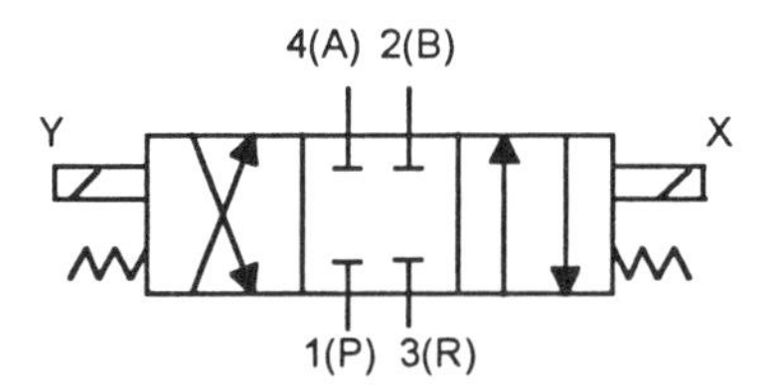

Figure 3.7 *4/3 valve*

Example

State what happens for the pneumatic circuit shown in Figure 3.8 when the push-button is pressed and then released.

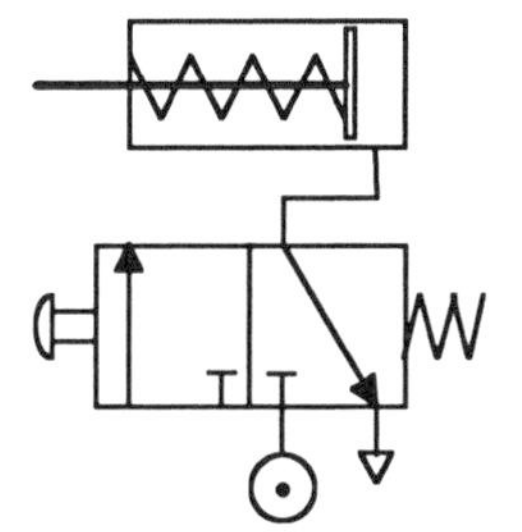

Figure 3.8 *Example*

The right-hand box gives the initial position. This indicates that the pressure source, i.e. the circle with the dot in the middle, is connected to a closed port and the output from the right-hand end of the cylinder is connected to the exhaust port, i.e. the open triangle. When the push-button is pressed the connections between the ports become those indicated in the left-hand box. The pressure source is then connected to the output port and hence to the right-hand end of the cylinder. The pressure can then be used to force the piston back against its spring and move it from left to right. When the push-button is released, the connections between the ports become those in the right-hand box and the right-hand end of the cylinder is exhausted. The piston then moves back from left to right.

3.2.2 Shuttle valves

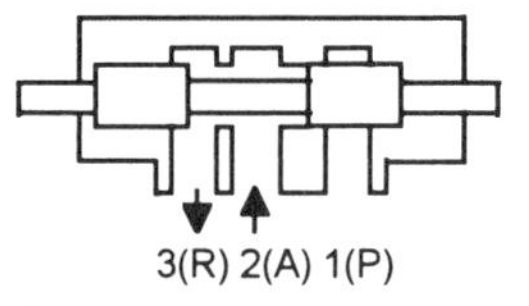

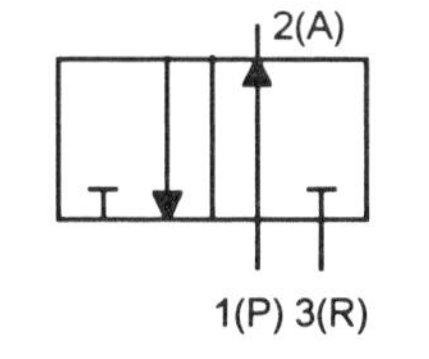

Figure 3.9 *3/2 shuttle valve*

The most common form of directional control valve is the *shuttle* or *spool valve*, though the *poppet valve* and *rotary valve* are sometimes encountered. Shuttle valves have a spool moving horizontally within the valve body. Raised areas, termed *lands*, block or open ports to give the required valve operation. The following are examples of such valves.

Figure 3.9 illustrates these features with a 3/2 valve. In the first position, the shuttle is located so that its lands block off the 3 port and leave open, and connected, the 1 and 2 ports. In the second position, the shuttle is located so that it blocks of the 1 port and leaves open, and connected, the 2 and 3 ports. The shuttle can be made to move between these two positions by manual, mechanical, electrical or pressure signals applied to the two ends of the shuttle.

Figure 3.10 shows an example of a 4/3 shuttle valve. It has the rest position with all ports closed. When the shuttle is moved from left to right, the pressure is applied to output port 2 and port 4 is connected to the exhaust port. When the shuttle moves from right to left, pressure is applied to output port 4 and port 2 is connected to the exhaust port.

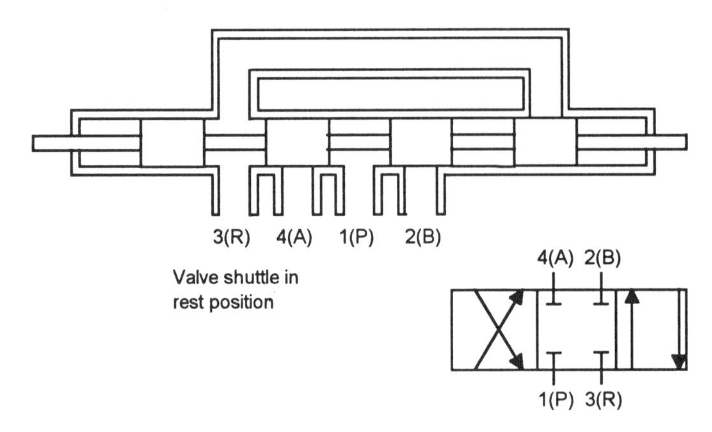

Figure 3.10 *4/3 shuttle valve*

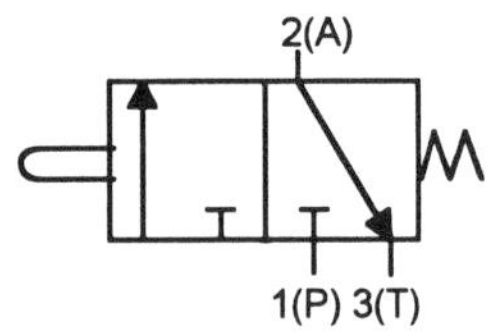

Figure 3.11 *3/2 valve*

The following two examples of shuttle valves incorporate the means of controlling the shuttle. Figure 3.11 shows a 3/2 valve which is actuated by contact being made with a plunger and returned to its rest position by a spring. Figure 3.12 shows a 5/3 valve which is spring centred and actuated by pneumatic pilot pressures.

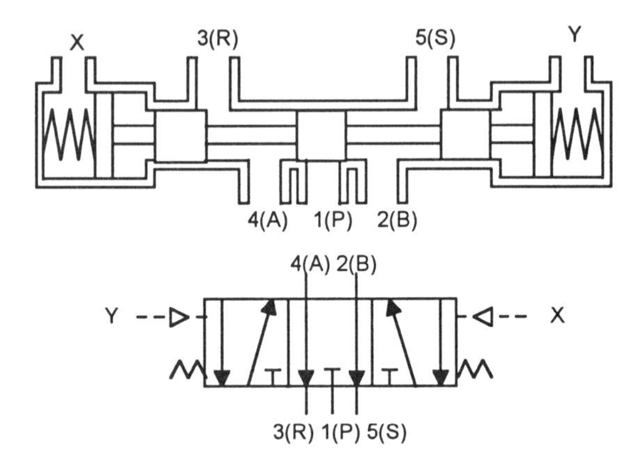

Figure 3.12 *5/3 valve*

Spool valves can be used for flow reversal, only small actuating forces are required and can be built for more than two switching positions. The small clearances between the spool and the casing mean that the slightest amount of contamination can result in spool seizure and so filtration and lubrication of air is necessary. They must also be mounted on a flat surface to avoid spool seizure as a consequence of the valve body warping to even a slight extent.

3.2.3 Poppet valves

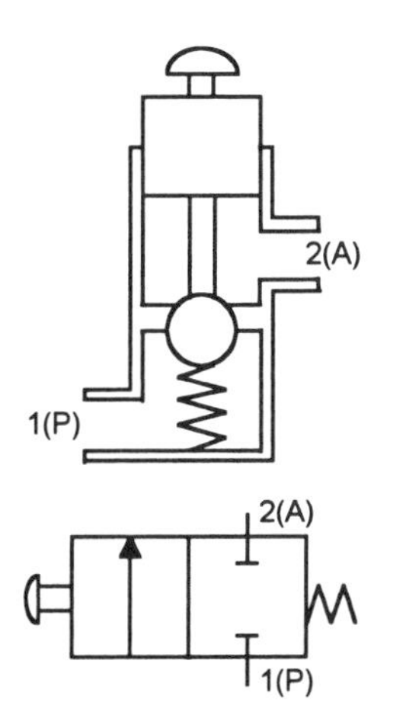

Figure 3.13 *2/2 poppet valve*

In a *poppet valve*, balls, discs or cones are used to control flow. Poppet valves offer several advantages over shuttle valves. They are simple, cheap, robust, virtually leak free, have low wear rates, rapid response and do not require lubrication. However, a major disadvantage is the force required to operate them is relatively large.

Figure 3.13 shows an example of 2/2 poppet valve and its symbol. The application of pressure forces the ball against its seat and closes the valve. When the push-button is pressed, the ball is moved downwards and so opens the valve, allowing the pressure to the output. When the push-button is released, the spring and pressure forces the ball back into its seat.

Figure 3.14 shows an example of a 3/2 poppet valve and its symbol. The spring keeps the ball against its seat and so the pressure port closed and the output is connected to the exhaust. When pilot pressure is applied to the valve, the ball is moved downwards and so allows pressure to be applied to the output and closes the exhaust. When the pilot pressure ceases, the spring and the pressure force the ball back into its seat.

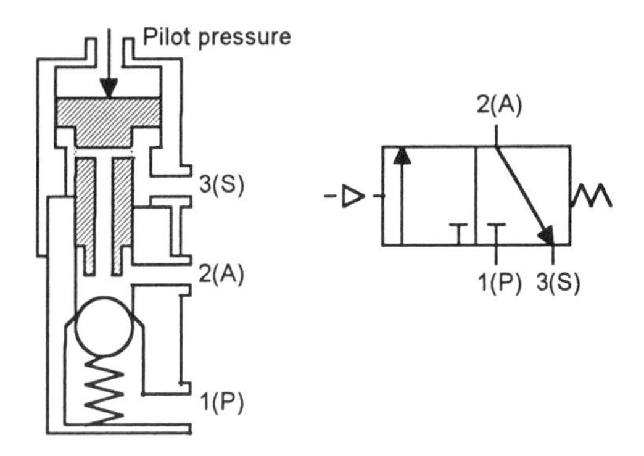

Figure 3.13 *3/2 poppet valve*

3.2.4 Rotary valve

Rotary valves have a rotating spool which aligns with holes in the valve casing to give the required operation. They are compact, simple and only require low operating forces. They operate, however, with only low pressures and thus are limited in their use, being mainly used as manually operated machine-mounted valves to provide a pilot signal to a main power valve.

Figure 3.15 shows an example of such a valve. It is a 4/3 valve. In the rest position all ports are closed. In the left-hand position, pressure is supplied to port 2 and port 4 exhausts. In the right-hand position, pressure is supplied to port 4 and port 2 exhausts.

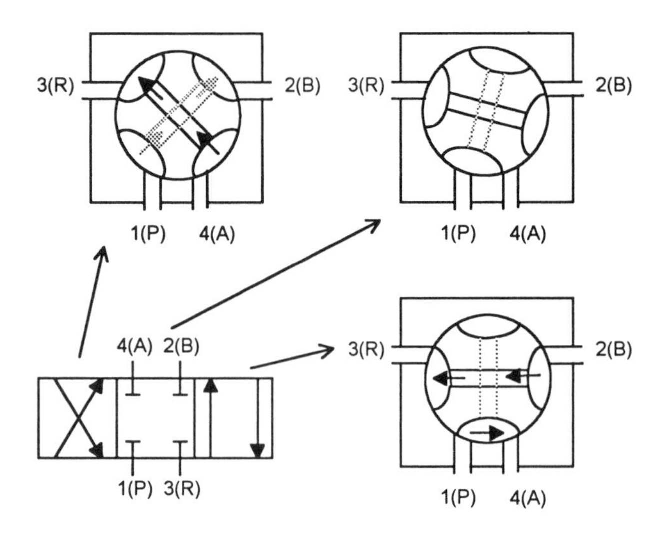

Figure 3.15 *Rotary 4/3 valve*

3.3 Pressure control valves

There are three main types of pressure control valves:

1 *Pressure regulating valves*
This controls the operating pressure in a circuit and keeps the pressure constant irrespective of any pressure fluctuations that occur in the system. These were discussed in Section 2.3.5 and examples given in Figures 2.18 and 2.19.

2 *Pressure limiting valves*
These are used as safety mechanisms to limit the pressure below some set value. They are often referred to as *pressure relief valves*. An example was given in Figure 2.20.

3 *Pressure sequence valves*
The pressure sequence valve senses the pressure of an external line, compares it with a preset value and gives a signal when the preset limit is reached.

Figure 3.16 shows the symbols used for the various forms of pressure control valves.

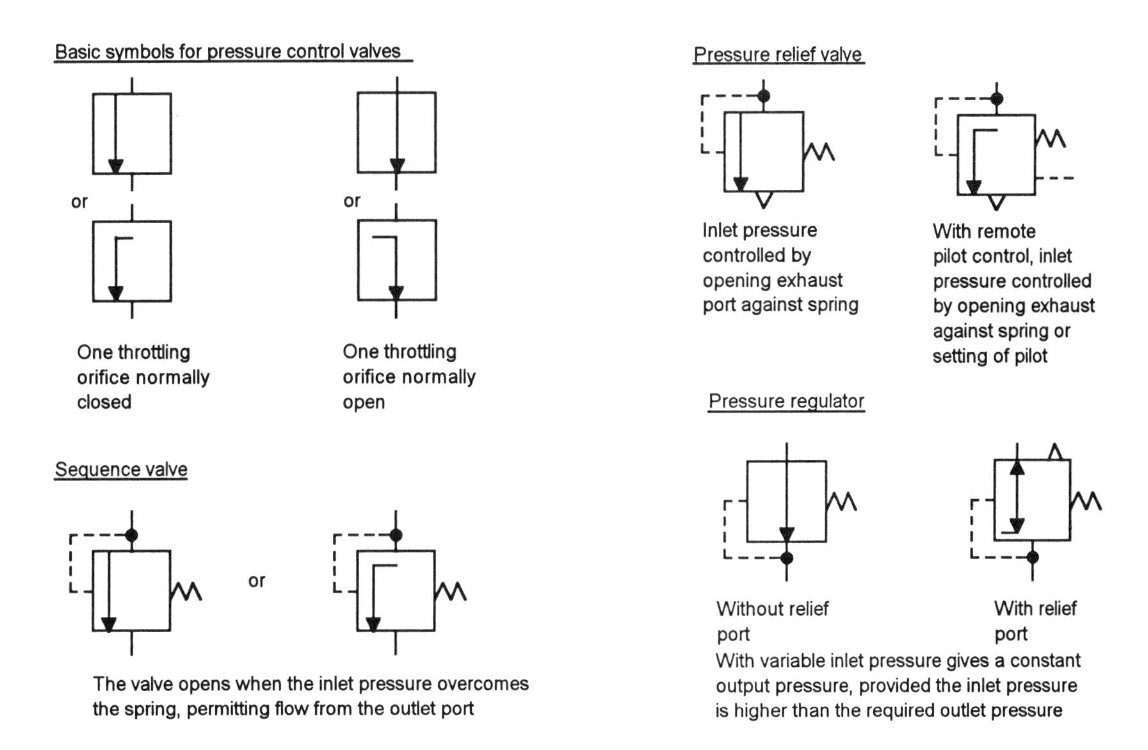

Figure 3.16 *Symbols for pressure control valves*

Figure 3.17 shows how the sequence valve can be set by the inlet pressure to the valve. We can, however, have the valve set by an external pressure. In such a form, the sequence valve can be considered as the valve combination shown in Figure 3.18. If the pressure X exceeds the value set by the spring force, the valve opens and the pilot pressure is applied to the 3/2 part of the combination valve. If this pilot pressure is great enough, the outlet port 2 which was connected to the exhaust port 3 is switched to allow the pressure from port 1 to be applied to the output.

Such a pressure sensing valve is used where a specific pressure is required for some machine function. Thus an actuator might be required to be actuated and start some other operation when the clamping pressure on some machine component reaches a particular value. Thus, with Figure 3.18, when pressure X reaches the required value the outlet port 2 is connected to the pressure port 1 and thus an operation, triggered by that pressure, starts up.

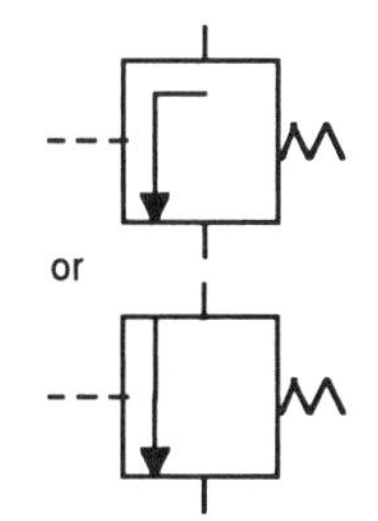

Figure 3.17 *Sequence valve set by external source*

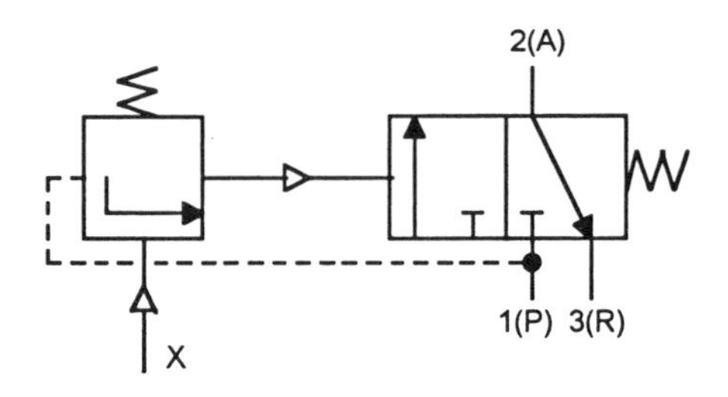

Figure 3.18 *Pressure sequence valve*

3.4 Non-return valve

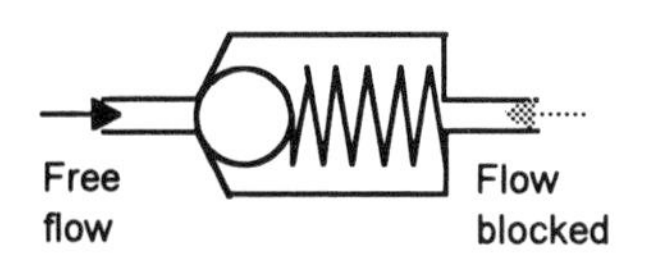

Figure 3.19 *Non-return valve*

The *non-return valve* or *check valve* is used to allow fluid to flow through the valve in one direction only, blocking the flow in the reverse direction. The non-return valve may be found as an element of other valves, e.g. flow control valves (see Section 3.5). The simplest form of non-return valve is shown in Figure 3.19. The pressure flow from left to right causes the ball to move and allow the fluid through the valve. The pressure flow from right to left forces the ball to close the valve and so prevent flow in that direction. Figure 3.20 shows the symbols used for non-return valves.

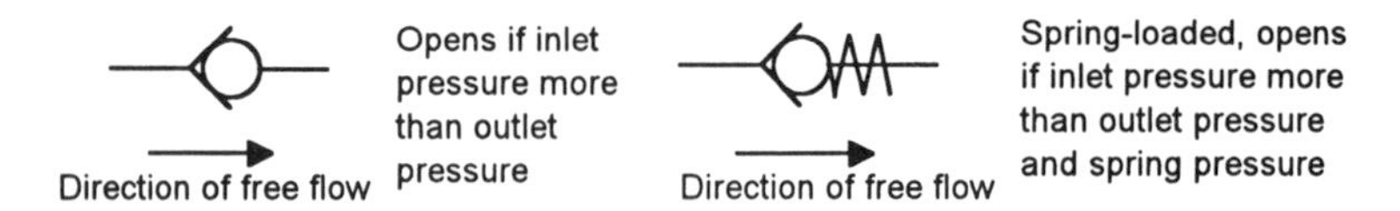

Figure 3.20 *Symbols for non-return valves*

3.4.1 Shuttle valve

The *shuttle valve* is a double non-return, or double check, valve. It allows pressure in a line to be obtained from one source or another. Figure 3.21 shows the basic principle of such a valve and its symbol. When air arrives at the left-hand side X it forces the ball to move and block the right-hand port Y. Thus the air from X now leaves through the output port. When air enters from the right-hand side Y it forces the ball to move and block the left-hand port X. Thus the air from Y now leaves through the output port.

Figure 3.22 shows an application of a shuttle valve. When either of the two push-buttons on the directional valves are pressed, air is admitted to the cylinder and causes the piston to move to the right. The shuttle valve gives the OR function (see Chapter 7 for more details). It prevents any cross flow between the two inputs. Note that a T-connector could not be used in place of the shuttle valve, since it would provide a path for the pressure input to one valve to exhaust through the other when one of the push-buttons is pressed.

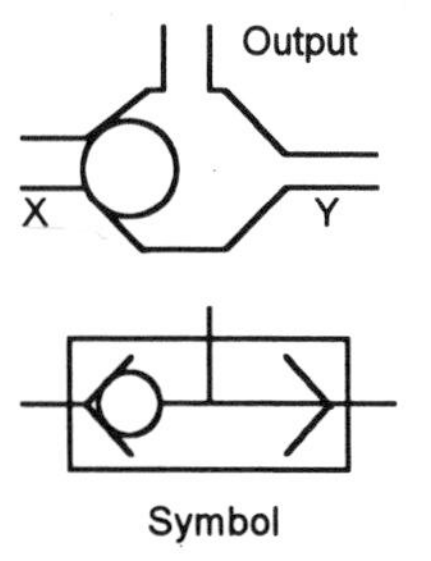

Figure 3.21 *Shuttle valve*

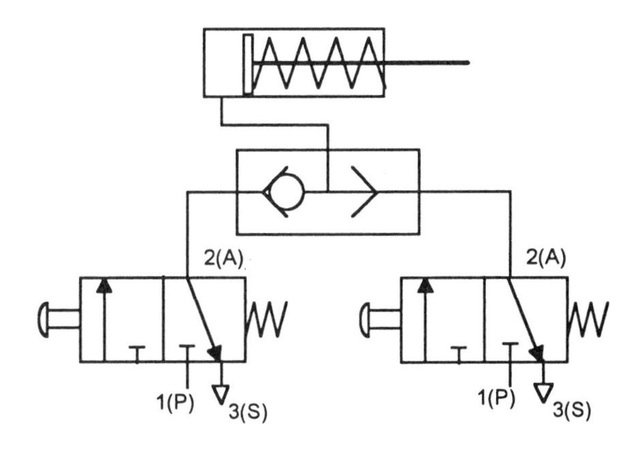

Figure 3.22 *Application of a shuttle valve*

Example

Explain the operation of the pneumatic circuit shown in Figure 3.23 when push-buttons A or B are pressed.

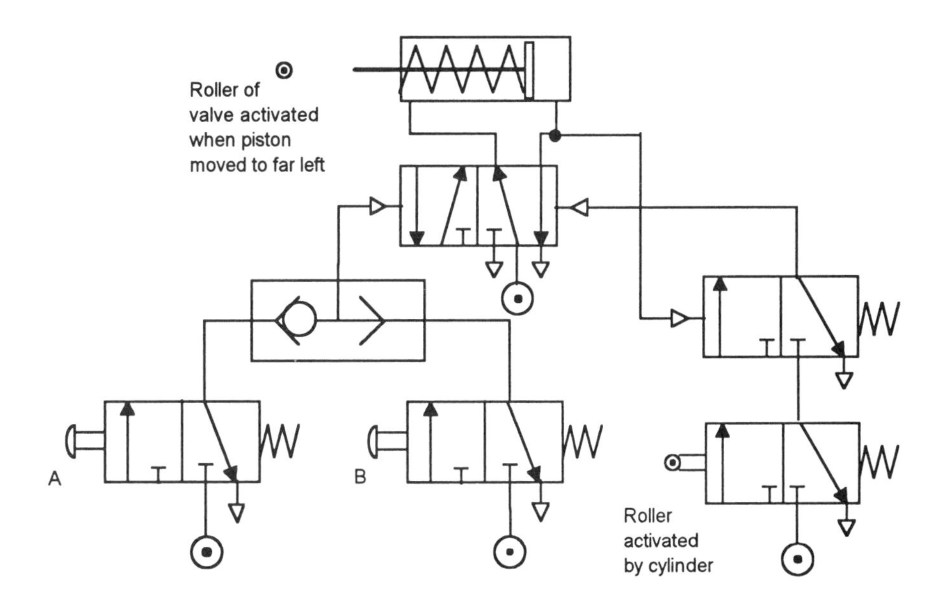

Figure 3.23 *Example*

The left-hand group of valves gives an OR situation so that when either push-button is pressed, pilot pressure is applied to the 5/2 valve so that it supplies compressed air to the right-hand end of the cylinder. This causes the piston to move to the left. It continues in that direction until it actuates the roller. This switches the bottom right 3/2 valve so that it supplies compressed air to the 3/2 valve above it. This has already received the pilot signal from the pressure in the right-hand end of the cylinder and so it supplies pilot pressure to the 5/2 valve which causes it to switch and so apply compressed air to the left-hand side of the cylinder. The overall result is an automatically retracting piston.

3.4.2 Rapid exhaust valve

The *rapid exhaust valve* is used to rapidly vent cylinders. Figure 3.24 shows the basic principle of the valve, a typical way it might be used with a cylinder and the symbol used. The valve has a movable disc which allows pressure to be applied to the output port (Figure 3.24(a)). When the output is exhausting, the disc is moved to block the pressure port (Figure 3.24(b)) and so allow the exhausting air to rapidly flow through a large exhaust port. Thus, with the arrangement shown in Figure 3.24(d), the exhaust air does not have to flow through the directional valve but escapes direct to the atmosphere. As a consequence the cylinder is rapidly vented.

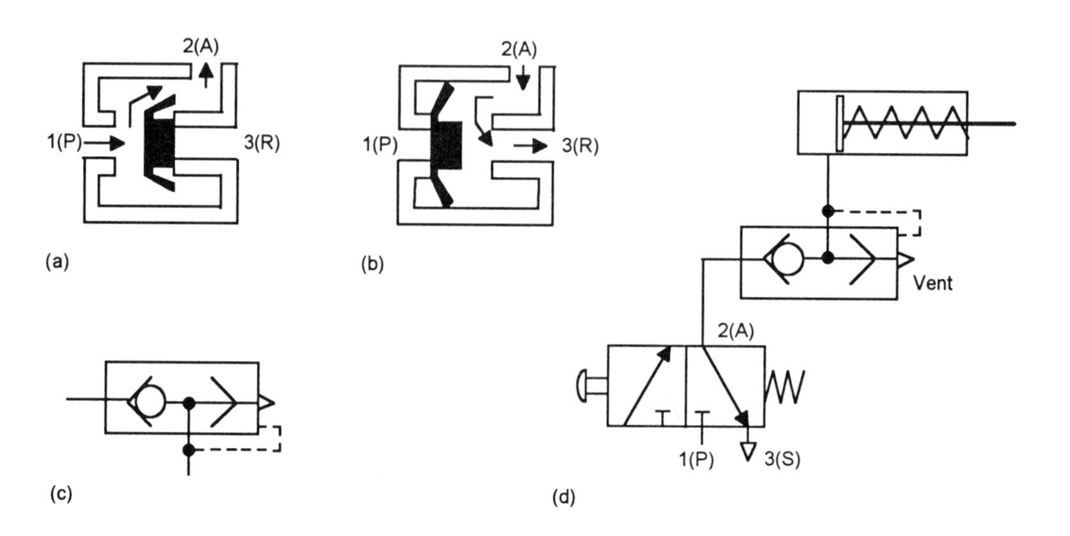

Figure 3.24 *Rapid exhaust valve: (a) and (b) basic principles of operation, (c) symbol, (d) a typical application*

3.5 Flow control valves

The *flow control valve* is used to restrict the flow of fluid in a particular direction in order to control the rate of flow. They might, for example, be used to control the speed of the piston in a cylinder so that it does not exceed some maximum speed value or cushion the end of the stroke so that there is not a sudden deceleration. They work by placing a variable restriction in the flow path. Figure 3.25 shows the basic principle and the symbols used. The valve has a throttling screw which can be used to adjust the size of the restriction to give the required flow. The one-way flow control valve permits flow adjustment in one direction only, the other valves permitting flow adjustment for the flow of air in both directions through the valve.

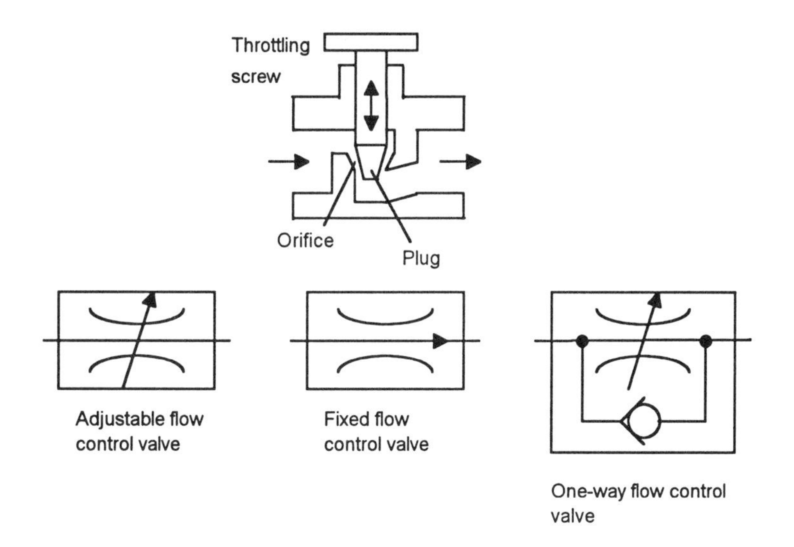

Figure 3.25 *Flow control valve*

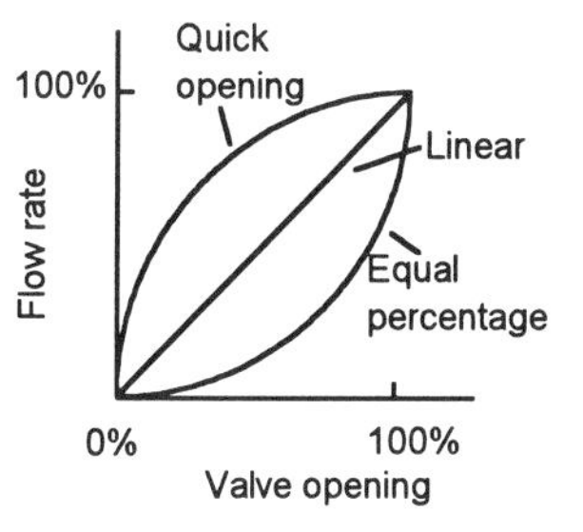

Figure 3.26 *Valve characteristics*

The shape of the plug determines how the movement of the throttle screw affects the rate of flow. Figure 3.26 shows the basic form of valve characteristics. With the linear form of plug the flow rate is proportional to the movement of the throttle screw. With the quick opening form of plug, the flow is more rapidly changed by the initial movement of the screw opening the orifice than later movements of the screw. With the equal percentage form of plug, equal percentage changes in the valve orifice produce equal percentage changes in the flow rate.

3.5.1 Speed control

Two basic methods are used with flow control valves to control the speed of a piston in a cylinder, the methods being termed *meter-in* and *meter-out* speed control. With meter-in control a flow valve is used to restrict the air flow of compressed air into a cylinder; with meter-out the restriction is of the air exhausted from a cylinder.

Figure 3.27 shows how a flow control valve in the meter-out position can be used to control the speed of movement of a piston in a cylinder. When the push-button on the first direction valve is pressed, compressed air flows into the cylinder and moves the piston to the right. The air on the other side of the piston has to escape and does so through the flow control valve. This valve is a one-way flow control valve and placed in such a direction that air escaping from the cylinder is controlled. The result is that the air is restricted so that it cannot escape quickly. This back pressure in the cylinder slows down the movement of the piston to the right. When the push-button of the second direction valve is pressed, the air is able to pass quickly through the flow control valve because of the bypass path. As there is no flow restriction on the air leaving from the left-hand side of the piston, it is able to move quickly to the left.

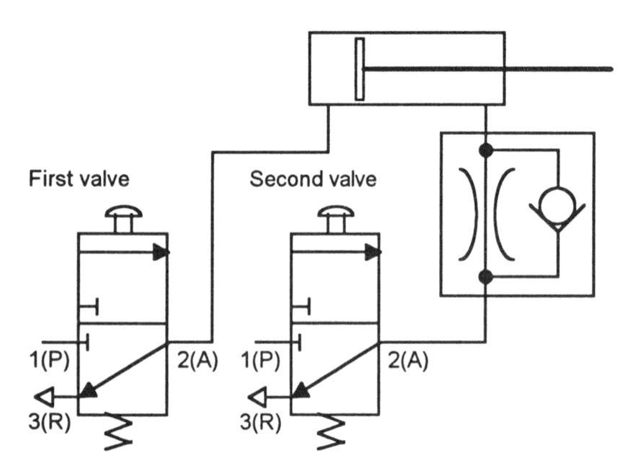

Figure 3.27 *Meter-out control*

Figure 3.28 shows a flow control valve used for meter-in control of the speed of a piston in a cylinder. When the push-button of the first direction valve is pressed, compressed air is applied via the flow valve to the left-hand side of the piston. The flow valve restricts the speed at which the compressed air is able to enter the cylinder and so slows down the speed of the piston movement. Meter-out control has advantages when compared with meter-in control in that it enables the speed of the piston to be better controlled and the back pressure provides end cushioning.

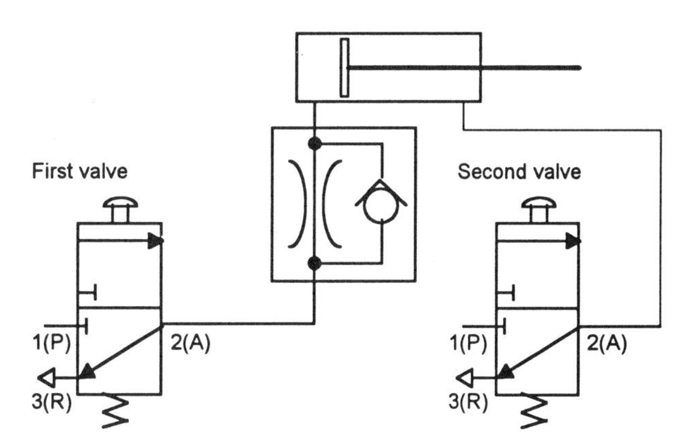

Figure 3.28 *Meter-in control*

3.5.2 Time delay valve

A time delay valve produces a delay between the application of a pressure and the valve operation. It can be produced by combining the functions of a one-way flow control valve, a reservoir and a 3/2 directional control valve.

Figure 3.29 shows the form of a time delay valve that is normally closed and its symbol. Air is supplied to port 12 and flows through a one-way flow control valve into the air reservoir. The throttling screw determines the rate at which air flows into the reservoir. The pressure in the air reservoir builds up with time. Eventually it reaches a value which is large enough for the spool of the 3/2 valve to be moved downwards against the spring and open the connection between ports 1 and 2 while disconnecting the connection between ports 2 and 3. The time delay between the pressure input to 12 and pressure being applied to port 2 is determined by the time taken for the pressure in the air reservoir to build up to the required value. This depends on the size of the reservoir volume and the size of the orifice in the flow valve through which the air enters the reservoir. A small orifice and a large reservoir mean a long time delay; a large orifice and a small reservoir a short time delay. The built-in one-way valve causes the reservoir space to vent quickly when the pressure at port 12 is removed.

Figure 3.30 shows a time delay valve that is normally open and its symbol. The valve is switched off by an input to port 10.

Example

How would you adjust the delay time for a normally open time delay valve?

This is done by adjusting the rate of flow of air into the reservoir by means of the throttling screw.

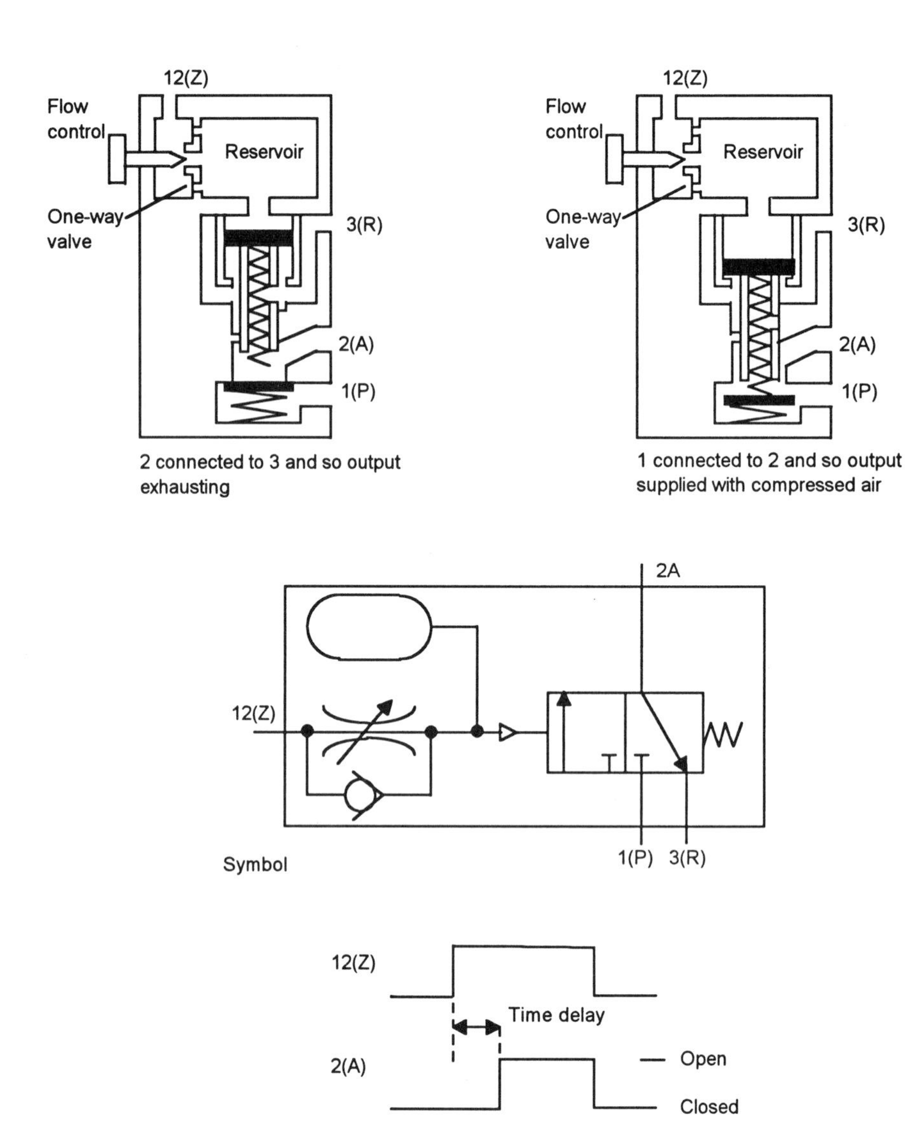

Figure 3.29 *Time delay valve: normally closed*

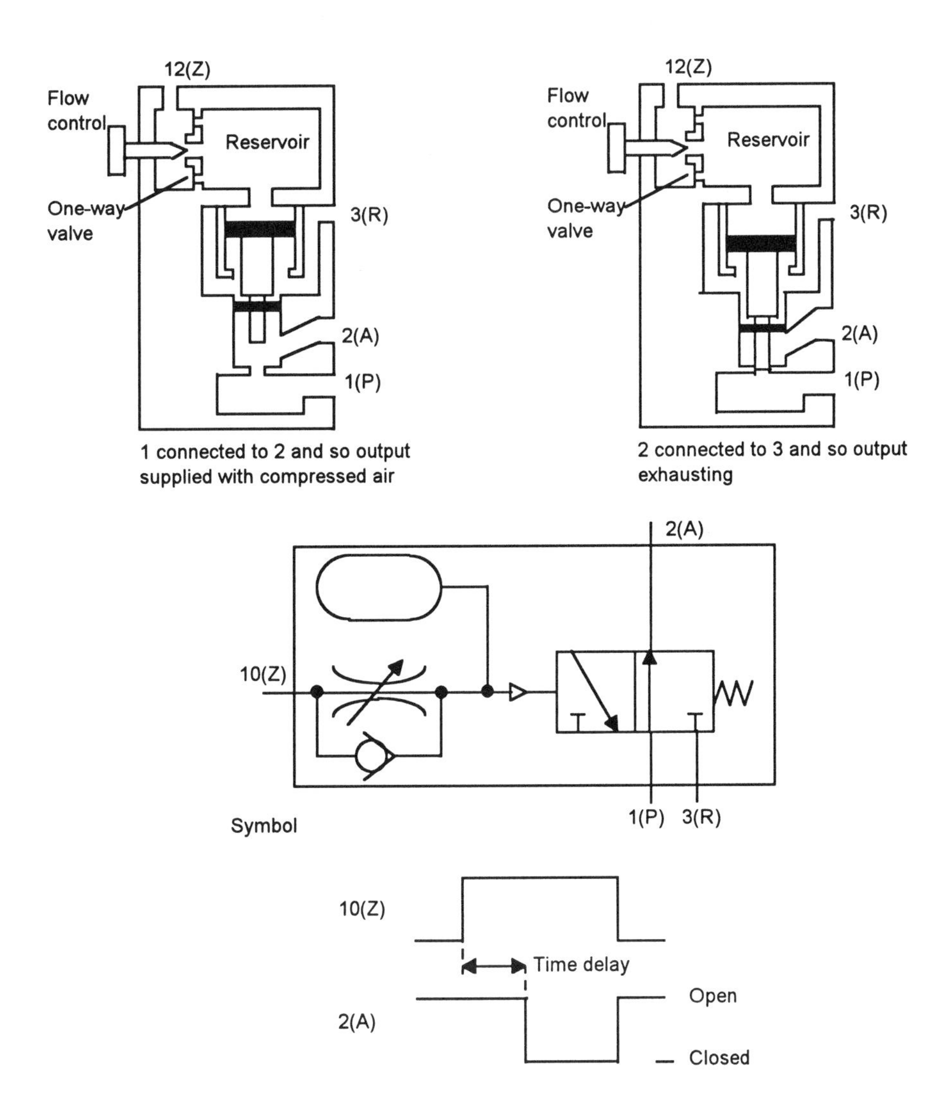

Figure 3.30 *Time delay valve: normally open*

Problems

1 For each of the valve symbols in Figure 3.31, state the method and outcomes of actuation.

2 Sketch the symbols for 2/2 valves that are:

(a) pedal actuated with spring return and normally open, closed by the actuation,

(b) push-button actuated with spring return and normally closed, opened by the actuation.

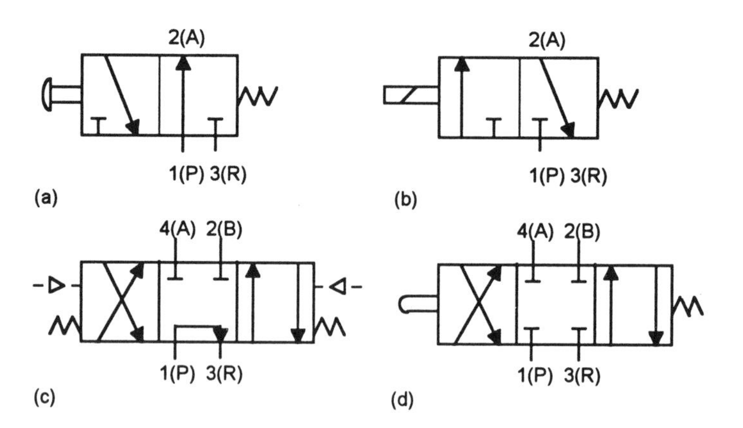

Figure 3.31 *Problem 1*

3 Sketch the symbols for 3/2 valves that are:

(a) solenoid actuated with solenoid return and normally open with pressure to output and exhaust closed, pressure closed and output to exhaust when solenoid actuated,

(b) actuated by a roller with spring return and normally with the pressure supply closed and output connected to exhaust, pressure to output and exhaust closed when pressure actuated.

4 State the outcomes of the pressing and then releasing of the push-buttons with regard to valves shown in Figure 3.32.

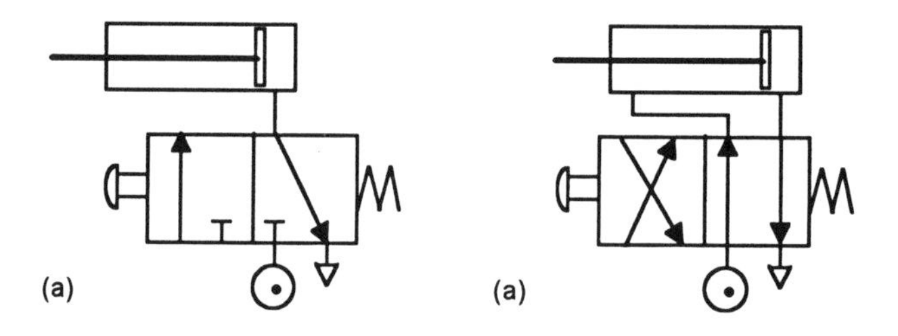

Figure 3.32 *Problem 4*

5 Sketch the symbol for the shuttle valve shown in Figure 3.33.

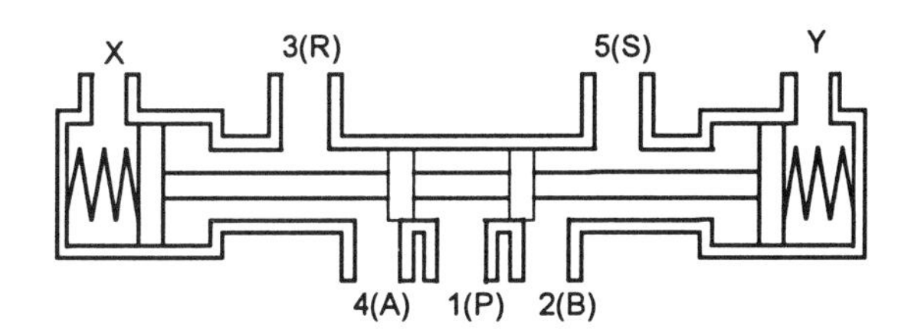

Figure 3.33 *Problem 5*

6 Explain the sequence of events that can follow for the valve shown in Figure 3.34 when the pressure X rises above the value needed to overcome the spring force for the sequence part of the valve.

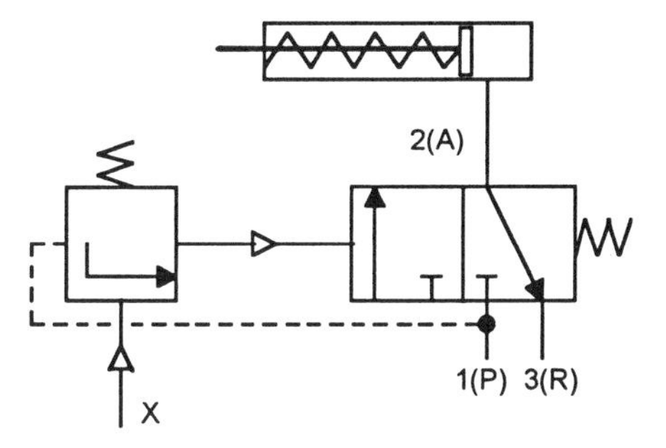

Figure 3.34 *Problem 6*

7 Compare the behaviour of the two pneumatic systems shown in Figure 3.35.

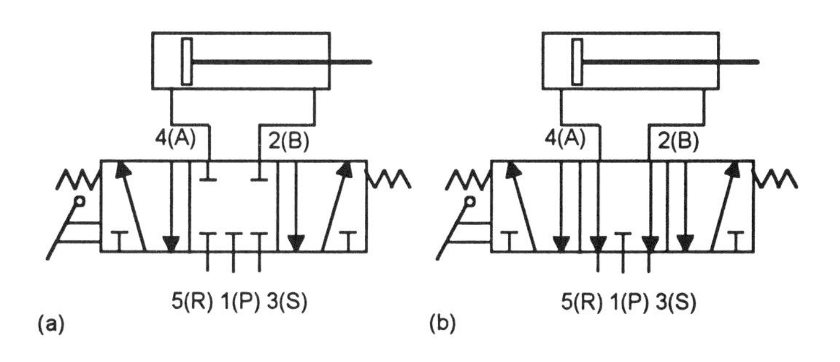

Figure 3.35 *Problem 7*

8 Explain how the piston will move in the cylinder in Figure 3.36 when the push-button on the directional valve is pressed and then released.

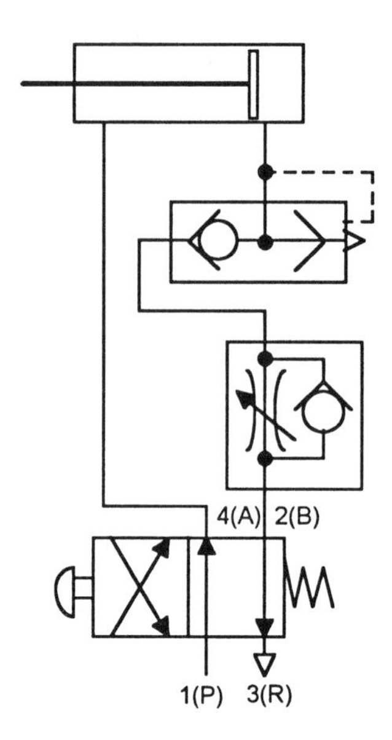

Figure 3.36 *Problem 8*

4 Actuators

4.1 Introduction

This chapter is a discussion of fluid power actuators. An *actuator* is a device that is used to apply a force to an object, and hence possible motion of that object.

Fluid power actuators can be classified in two groups:

1. *Linear actuators* are used to move an object or apply a force in a straight line. The basic linear actuator is a cylinder of the form shown in Figure 4.1. Linear actuators can be divided into two types, *single-acting cylinders* and *double-acting cylinders*. A single-acting cylinder is powered by fluid for the movement of the piston in one direction with it being returned in the other direction by an internal spring or an external force; a double-acting cylinder is powered by fluid in both directions.

2. *Rotary actuators* are used to move an object in a circular path. Rotary actuators are the fluid power equivalent of an electric motor.

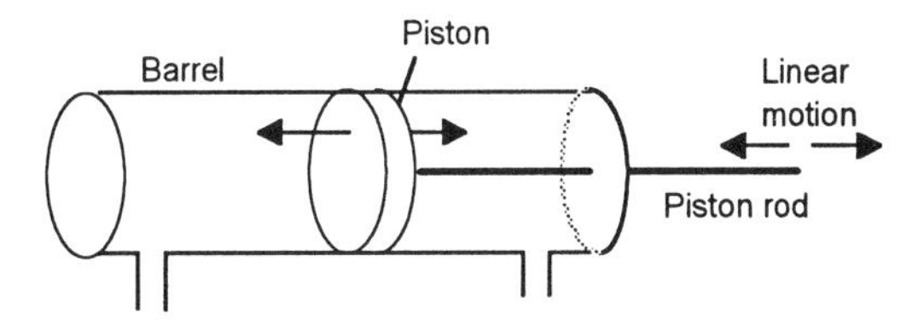

Figure 4.1 *The basic cylinder*

4.2 Single-acting cylinders

A *single-acting cylinder* is one powered by fluid being applied to one side of the piston to give movement of the piston in one direction, it being returned in the other direction by an internal spring or some external force. The other side of the piston is open to the atmosphere. Figure 4.2 shows one with a spring return.

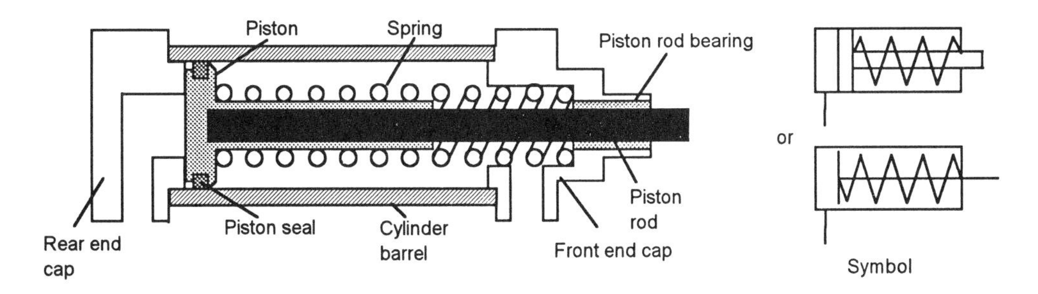

Figure 4.2 *Single-acting actuator*

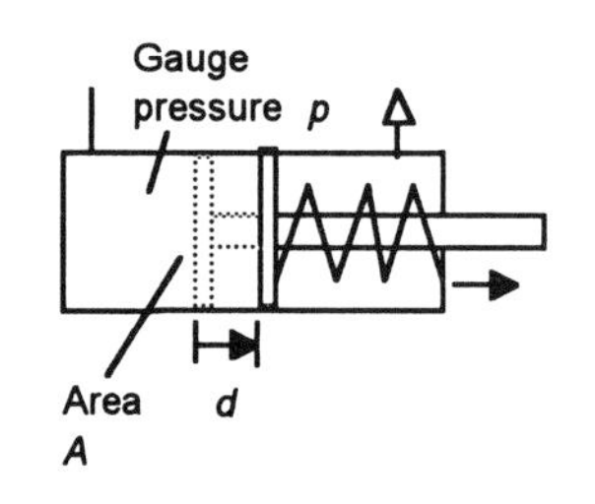

Figure 4.3 *Single-acting cylinder*

The fluid, applied to one side of the piston at a gauge pressure p with the other side being at atmospheric pressure (Figure 4.3), produces a force on the piston of:

$$\text{force arising from pressure} = pA \qquad [1]$$

where A is the area of the piston. To determine the actual force acting on the piston rod we also have to take account of friction. Because of friction we have:

$$pA = \text{force } F \text{ on piston rod} + \text{force 'lost' overcoming friction} \qquad [2]$$

The *efficiency* η of a cylinder is defined as:

$$\eta = \frac{\text{actual force } F \text{ on piston rod}}{\text{applied force on piston}} \qquad [3]$$

$$F = \eta pA \qquad [4]$$

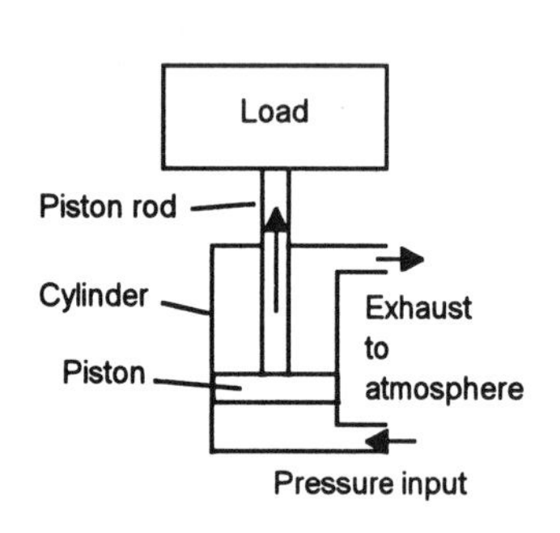

Figure 4.4 *Piston moved upwards by pressure, returned by load*

With an internal spring return, the force available from the pressurised fluid is reduced by the opposing spring force. If the cylinder is mounted in a vertical direction to move a load upwards, this return force may be provided by the load (Figure 4.4).

If the piston moves at speed v so that a distance d is covered in a time t, as in Figure 4.3, then:

$$\text{volume rate of flow of fluid into the cylinder} = Ad/t = Av \qquad [5]$$

Single-acting cylinders have stroke lengths up to about 100 mm, the stroke length being the maximum distance through which the piston is moved.

Example

A pneumatic single-acting cylinder is to be used to clamp work pieces in a machine tool. The piston has a diameter of 125 mm and the required clamping force is 6 kN. Determine the system pressure that has to be applied to the cylinder to achieve this force, neglecting any forces due to any in-built spring and (a) any frictional effects, (b) taking frictional losses to be 5%.

This example is concerned with what can be termed a static force calculation since we are not concerned with forces during the motion of the piston.

(a) If we assume that any frictional effects can be neglected, then:

$$\text{pressure} = \frac{F}{A} = \frac{6 \times 10^3}{\frac{1}{4}\pi 0.125^2} = 488.9 \text{ kPa}$$

(b) For frictional losses of 5% the efficiency is 95% and so, using equation [4], i.e. $F = \eta pA$:

$$p = \frac{F}{\eta A} = \frac{6 \times 10^3}{0.95 \times \frac{1}{4}\pi 0.125^2} = 514.7 \text{ kPa}$$

Example

Determine the volume rate of fluid flow required for a single-acting cylinder if the system gauge pressure is 5 bar, the piston has a diameter of 80 mm, the stroke length is 400 mm and the cylinder completes 10 strokes and returns per minute.

The volume of air required for one stroke is $\frac{1}{4}\pi \times 0.080^2 \times 0.400 = 0.002 \text{ m}^3$. This volume is required 10 times every minute and thus the rate at which air has to be supplied is 0.02 m^3/min. Volume rates of flow are, however, always specified in terms of free air and the above is the volume rate of flow required for air at a pressure of 5 bar. Using Boyle's law:

$$1V = (5 + 1) \times 0.02$$

Thus the volume rate of flow required of free air is 0.12 m^3/min.

4.2.1 Single-acting diaphragm cylinder

The single-acting diaphragm cylinder is a cylinder with a very short stroke and a large piston area. The return stroke is provided by an inbuilt spring or by the diaphragm being pre-tensioned. Figure 4.5 shows the basic form of such a cylinder. Stroke lengths are typically between 1 and 10 mm. Friction is negligible.

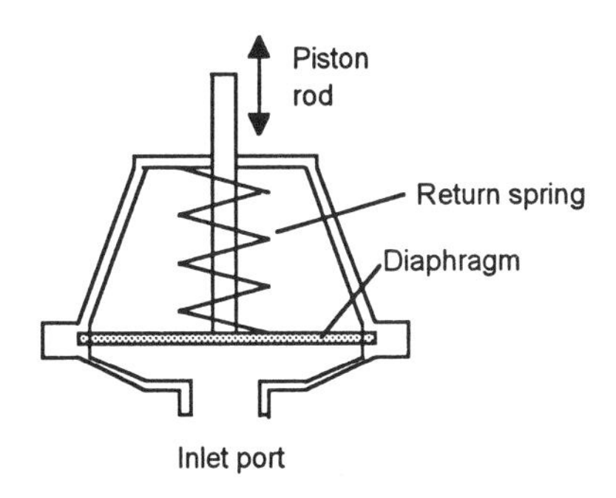

Figure 4.5 *Single-acting diaphragm cylinder*

4.2.2 Direct control of a single-acting cylinder

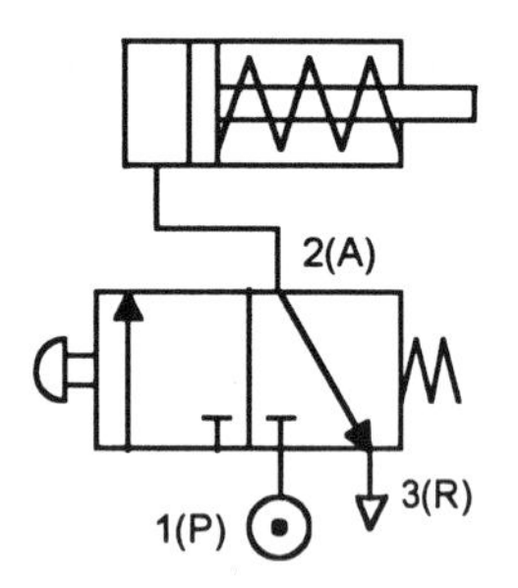

Figure 4.6 *Direct control*

Figure 4.6 shows how a single-acting cylinder can be directly controlled by using a normally closed 3/2 valve. When the push-button is pressed, compressed air is admitted to the left-hand end of the cylinder and forces the piston to the right. When the button is released, the air to the left of the piston is exhausted and so the piston, under the action of the return spring, returns to its initial position.

4.2.3 Indirect control of a single-acting cylinder

With indirect control, a 3/2 valve is used to control the cylinder and is controlled by pilot pressure from a push-button-operated 3/2 valve. This overcomes any problems that might occur as a consequence of pressure drops in the pipe runs connecting the push-button set/spring reset valve to the cylinder, it being possible to mount the second valve close to the cylinder and so keep pressure drops to a low value. Figure 4.7 shows such an arrangement when the cylinder is controlled by a normally closed valve.

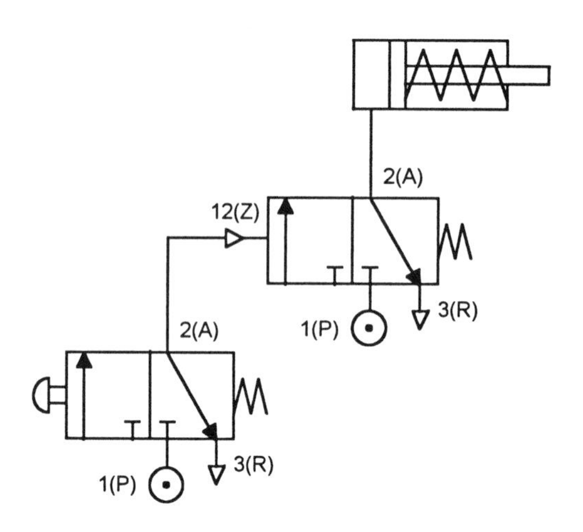

Figure 4.7 *Indirect control*

4.3 Double-acting cylinders

A *double-acting cylinder* is one powered by fluid for the movement of the piston in both the extend and return directions. With such a cylinder, two ports are used alternatively as supply and exhaust ports, so that pressure is used to give both the extend and return strokes. Thus, unlike the single-acting cylinder, no spring is required for the return stroke. Because of this, the double-acting cylinder is able to do work on both the extend and return strokes, the single-acting cylinder only being able to do work on the extend stroke. Figure 4.8 shows the basic form of such a cylinder and the symbols that are used.

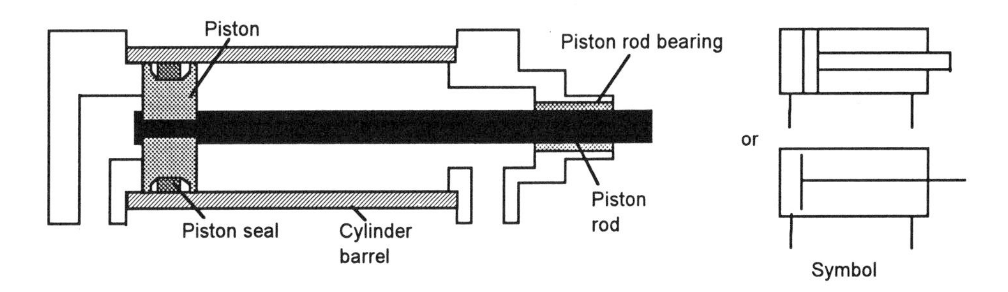

Figure 4.8 *Double-acting cylinder with single piston rod*

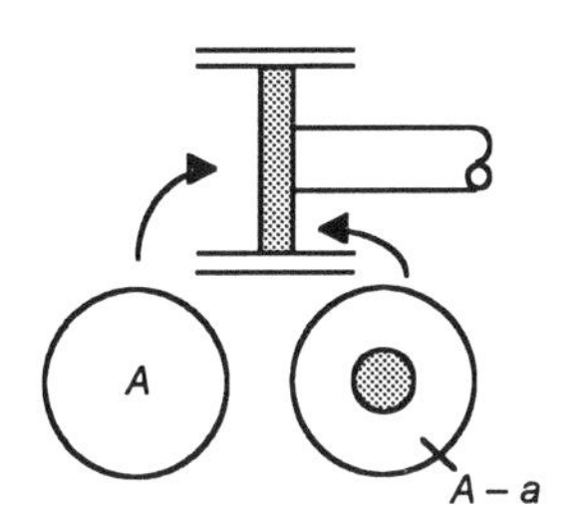

Figure 4.9 *The two sides of the piston*

With the arrangement shown in Figure 4.8 where there is just one piston rod, the area of the piston on which the pressure acts is different for the two sides of the piston. For a piston of area A and a piston rod of cross-sectional area a (Figure 4.9), if the pressure p is applied to the left-hand side of the piston and the right-hand side exhausted, then the force due to the pressure is pA. On the return stroke, when the pressure is applied to the right-hand side, the force on the piston due to the pressure is $p(A - a)$. If we have a mechanical efficiency η then the extend force is $pA\eta$ and the return force $p(A - a)\eta$. The extend stroke force is thus greater than the return force stroke. Typically a 20 mm diameter piston has a piston rod of diameter 8 mm. These give a piston area of 3.1 cm^2 and an annular area, i.e. $(A - a)$, on the rod side of 2.6 cm^2. As a result, the extend force with a pressure of 6 bar and an efficiency of 88% is 164 N and the return force 137 N. Tables are supplied by the manufacturers of cylinders giving such cylinder forces. Table 4.1 is an example of a segment of such a chart.

Table 4.1 *Effective cylinder forces*

Cylinder diameter (mm)	Piston rod diameter (mm)	Effective force at pressures					
		4 bar		6 bar		8 bar	
		Ext.	Ret.	Ext.	Ret.	Ext.	Ret.
10	4	28	21	42	35	56	46
20	8	109	92	164	137	218	183
32	12	282	243	422	364	563	486
40	16	444	373	665	560	887	746
50	20	690	581	1035	871	1380	1162

Example

A pneumatic double-acting single rod cylinder is to be used to clamp work pieces in a machine tool. The required clamping force is 3 kN and the supply pressure is 5 bar. Determine the size of the cylinder required. Assume an efficiency of 95%.

The effective piston area required is:

$$A = \frac{F}{p\eta} = \frac{3 \times 10^3}{5 \times 10^5 \times 0.95} = 0.0063 \text{ m}^2$$

This is a piston, and hence cylinder, diameter of 90 mm.

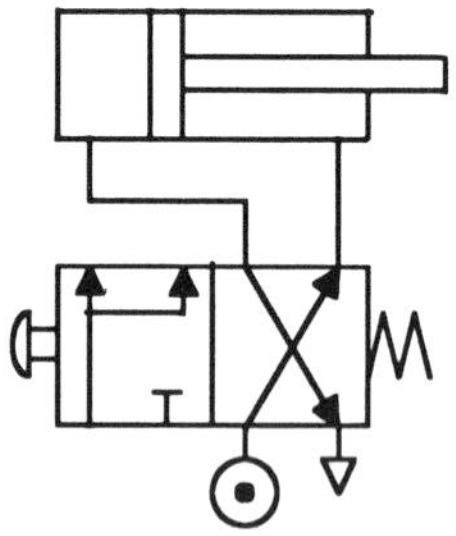

Figure 4.10 *Differential cylinder*

4.3.1 Differential cylinder

A cylinder which utilises the different areas for the two sides of the piston is termed a *differential cylinder*. If the same gauge pressure p is applied to both sides, as in Figure 4.10, the maximum extend force is $p(A - a)$, A being the piston area and a the piston rod area. The maximum return force, when the pressure is applied to just one side, is pa. Equal extend and return forces are given when $A - a = a$, i.e. $A/a = 2$. The different areas on each side do, however, mean the extend and return piston speeds are different.

Example

A double-acting, single piston cylinder has a piston of diameter 100 mm and a piston rod of diameter 25 mm. What is the ratio of the extend and return forces if a pressure of 6 bar is applied as shown in Figure 4.10?

The piston has an area of $\frac{1}{4}\pi 0.100^2 = 0.00785 \text{ m}^2$ and the annular area is $\frac{1}{4}\pi(0.100^2 - 0.025^2) = 0.00736 \text{ m}^2$. The ratio of the forces is the ratio of these areas and thus 1.07.

4.3.2 Double-acting cylinder with double-ended piston rods

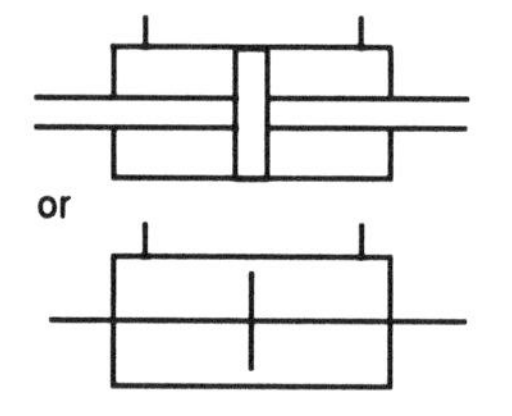

Figure 4.11 *Symbols*

Double-acting cylinders with piston rods protruding from both actuator ends give equal forces on both the extend and return strokes. Also, since the volumes in the cylinder on both the extend and return strokes are now the same, the resulting piston speeds are the same on both strokes. Better piston guidance is also provided since the rods run on bearings at each end of the cylinder. Figure 4.11 shows the symbols used.

4.3.3 Control of a double-acting cylinder

Figure 4.12 shows how a double-acting, single rod cylinder could be controlled by the use of a lever set/spring reset valve. In (a) the valve directly controls the cylinder, in (b) it indirectly controls it with the lever/reset valve used to provide pilot signals to a second valve. This overcomes any problems that might occur as a consequence of pressure drops in long pipe runs connecting the lever set/reset valve to the cylinder, since the second valve can be mounted close to the cylinder to keep pressure drops to a low value.

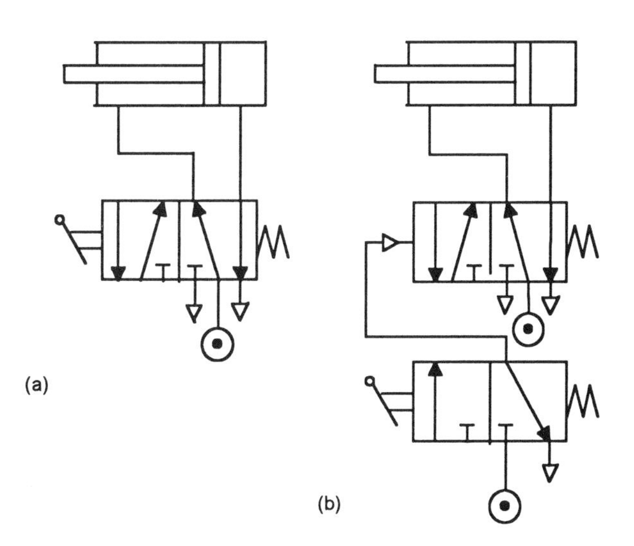

Figure 4.12 *(a) Direct control, (b) indirect control*

4.4 Cylinder construction and mounting

Basically a cylinder consists of two end caps with port connections with one end cap being a bearing cap in the case of a single rod cylinder, a cylinder barrel, a piston and the rod. The inner surface of the barrel has to be very smooth to prevent wear and leakage. It is generally a seamless drawn steel tube, though for special applications aluminium, a brass or a steel tube with a hard-chromed surface might be used. Pistons are usually made of steel. They have to transmit force to the rod, act as a sliding bearing and provide a seal between the high- and low-pressure sides of the cylinder.

Seals are fitted between static components and between the piston and the barrel to minimise leakage. *O-ring seals* are generally used as static seals. One form of seal used between the piston and the barrel is the *U-ring seal* (Figure 4.13(a)). The pressure forces the two lips of the U apart to give a positive seal. A variation is the *composite seal* (Figure 4.13(b)), this working on the same principle as the U-ring seal. In pneumatic systems Perbunan is commonly used for the seal material for temperatures between –20°C and 80°C, Viton between –20°C and 190°C and Teflon between –80°C and 200°C.

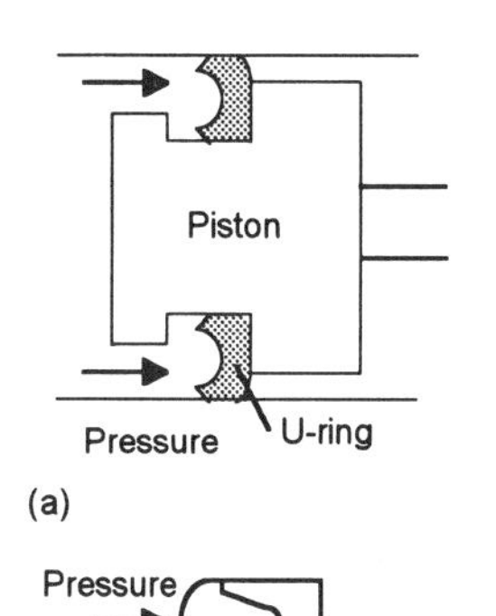

Figure 4.13 *Seals*

4.4.1 Cushioning

When large masses are being moved by a cylinder, *pneumatic cushioning* is used to absorb the shocks on the end caps when the piston is brought to a halt. Such cushioning slows the piston down gradually. Figure 4.14 shows the basic principle and the symbols used.

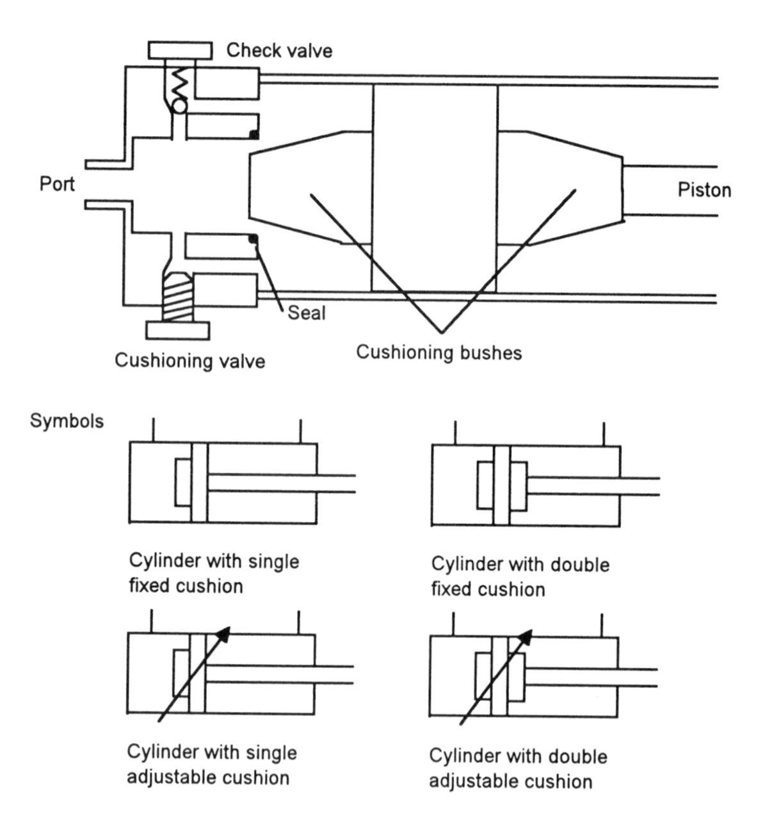

Figure 4.12 *Pneumatic cushioning*

Before reaching the end position, the direct flow path of the air being exhausted is reduced by a tapered bush entering the air exit tube. This throttling causes an initial speed reduction. Finally the bush closes the air exit tube and a cushion of air is trapped around the bush. This becomes compressed to a high pressure and supplies a resistive force which decelerates the piston and acts like a cushion.

4.4.2 The impact cylinder

Some applications of cylinders, e.g. metal forming, punching, stamping, require short duration and large forces to be applied to some target. Cylinders designed for this purpose are termed *impact cylinders*. Figure 4.13 shows the basic form of such a cylinder, it being vertical with the piston initially at the top.

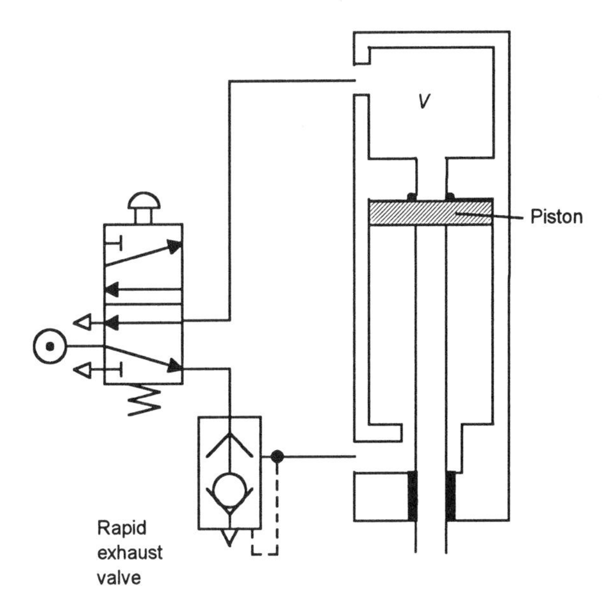

Figure 4.13 *Impact cylinder*

Initially, compressed air acts on the underside of the piston with the top side of the piston being connected to the exhaust port of the controlling valve. When the valve switches the compressed air from the lower to the upper part of the cylinder, the pressure begins to build up in the chamber, volume V, above the piston and decline in the region below the piston. The piston will start to move downwards when the force on its upper surface is greater than the force on its lower surface. But initially the piston is held against the upper end cap and the area exposed to the pressure is much less than the underneath area of the piston. Thus if the exposed upper surface is 1/9th the area of the lower surface, the piston will thus not start to move downwards until the pressure below the piston has fallen to 1/9th of the pressure in volume V. As soon as the pressure in the upper chamber reaches this value, the piston starts to move downwards. As soon as it does, the area of the piston exposed to the pressure increases by a factor of 9. This results in the force suddenly being increased by a factor of 9. The result is a sudden increase in force and downward acceleration of the piston. Because there is a large volume V of gas stored above the piston, it accelerates rapidly. Speeds of the order of 10 m/s can be attained. Since the piston and piston rod of an impact have a large mass, more than the normal cylinder, the kinetic energy of the piston when it reaches the end of its stroke is high and so enables significant work to be done. With a cylinder with bore size of 50 mm and a pressure supply of about 5 bar, an energy of about 25 J might be involved; with a bore of 150 mm about 260 J.

4.4.3 Cylinder mounting

Deciding the form of mounting to be used with a cylinder involves a consideration of the type of operation to be carried out and the nature of the load. Standard cylinders are not designed to absorb piston rod side loading and thus need mounting in such a way that the load moves precisely parallel to and in alignment with the actuator centreline. Figure 4.14 shows some of the forms of mount used. Where the load follows a straight line path with little or no deviation, a rigid mount such as end flange or foot mounting can be used. Where the load turns in one plane then a clevis or trunnion mounting might be appropriate. These allow the actuator to swivel around the mount as the direction of the load changes.

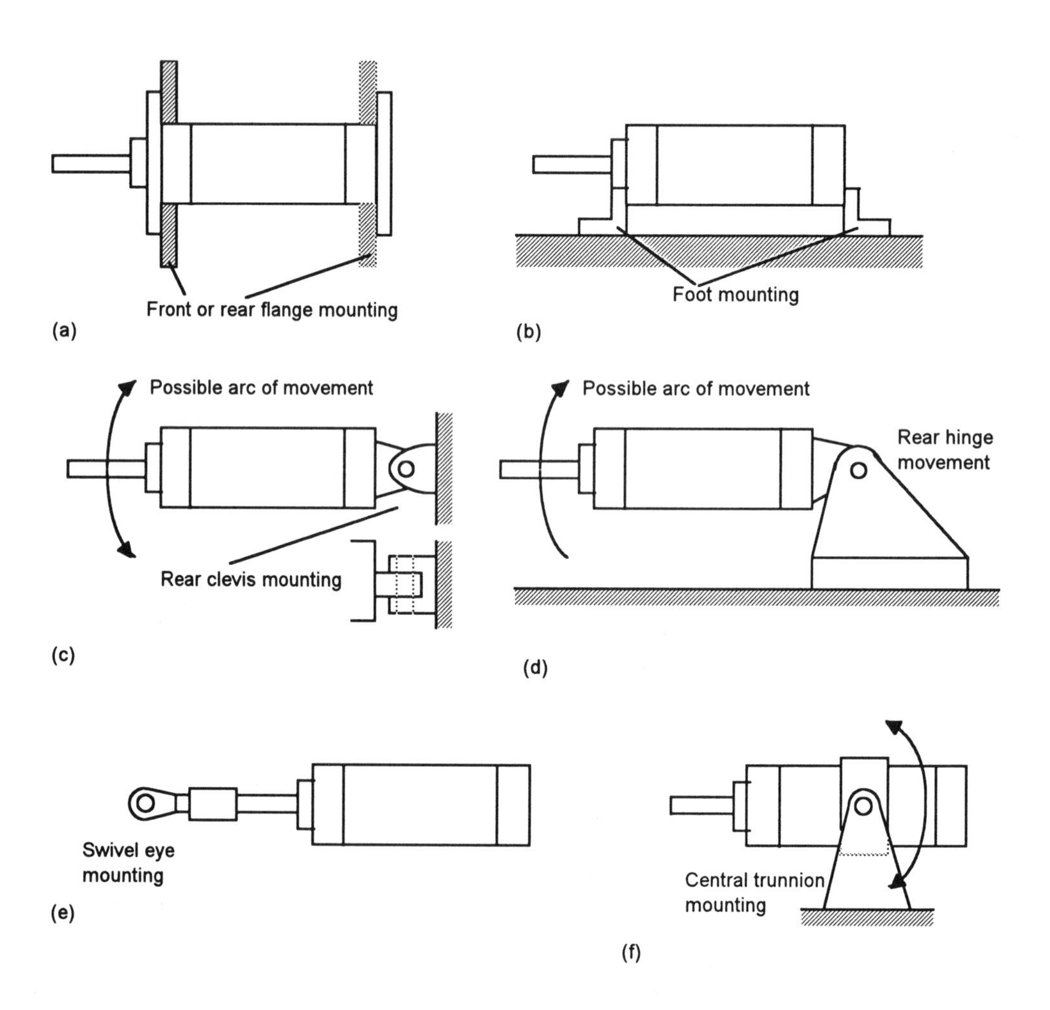

Figure 4.14 *(a) Front or rear flange mounting, (b) foot mounting, (c) rear clevis mounting, (d) rear hinged mounting, (e) swivel eye mounting, (f) central trunnion mounting*

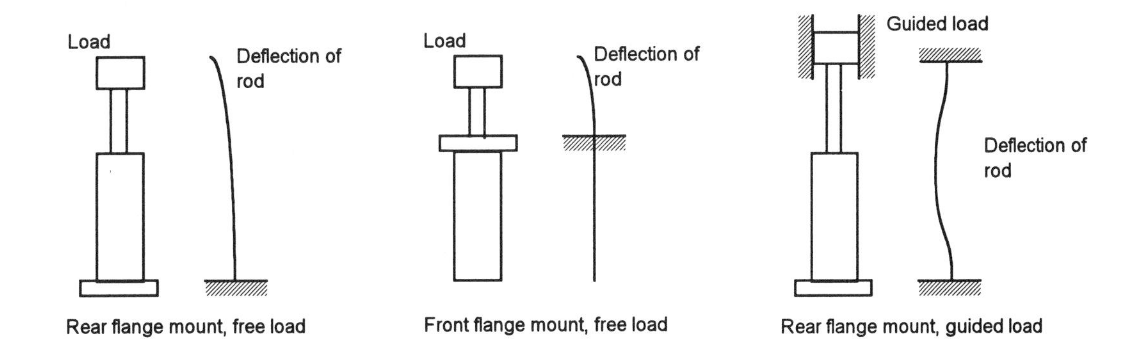

Figure 4.15 *Buckling*

The possibility of piston rod buckling has to be considered with long stroke lengths and high loads. Figure 4.15 shows some possible buckling modes. Euler's theory of buckling can be used to determine the maximum permissible buckling length of rods. However, to minimise calculations, cylinder manufactures produce charts and tables which relate piston rod diameters and operating pressures to the maximum permissible buckling lengths of rods. Figure 4.16 shows an example of a segment of such a chart.

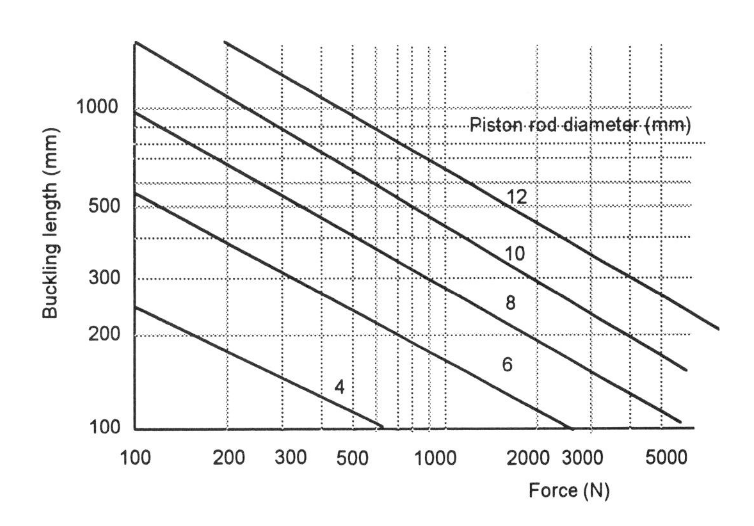

Figure 4.16 *Buckling length chart*

4.5 Rodless cylinders

Rodless cylinders are, as the name implies, cylinders without piston rods. The motion of the piston is communicated to an external guided carriage sleeve in a number of ways. The main types of rodless cylinder are:

1 *Magnetic cylinders*
Magnetic coupling is used to transmit the force from the piston to the external guided carriage.

2 *Band cylinders*
These use a cable or band to connect either side of the piston to the external guided carriage.

3 *Slot-type cylinders*
These involve the piston sliding along a barrel slot as the mechanism by which force is transmitted from the piston to the external guided carriage.

Figure 4.17 shows the basic form of a *magnetic cylinder*. The piston is fitted with a set of annular permanent magnets, as is the external carriage sleeve. Magnetic forces exist between the two sets of magnet, the magnetic field passing through the walls of the fully closed cylinder tube. Movement of the piston results in the magnetic forces pulling the carriage sleeve along without any direct mechanical contact between the two.

Magnetic rodless cylinders are available with pistons up to about 40 mm diameter and stroke lengths from 50 mm to 4 m. They can operate at high speeds, up to about 3 m/s. They have the advantages over conventional cylinders of almost no air leakage, not being subject to contamination-related wear and the carriage orientation can be rotated around the cylinder. However, the carriage can become separated from the piston, though recoupling is a simple process.

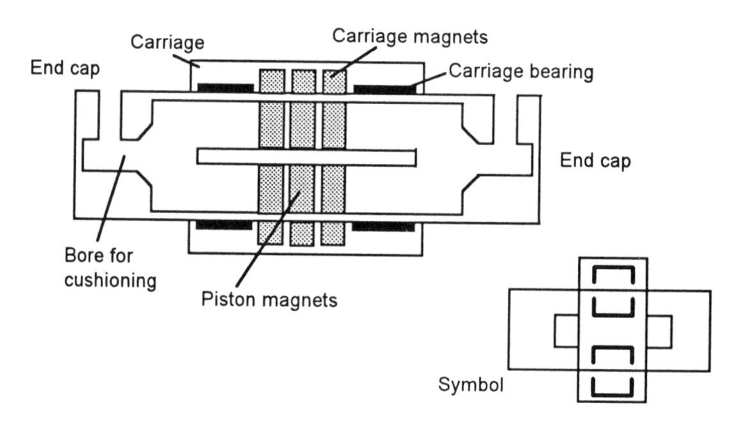

Figure 4.17 *Magnetic form of rodless cylinder*

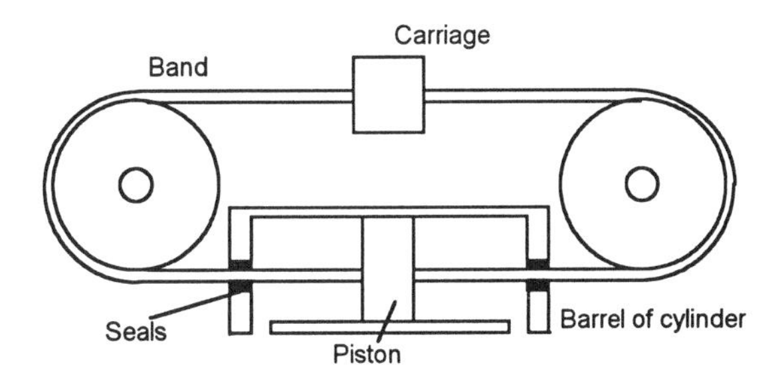

Figure 4.18 *Band form of rodless cylinder*

Figure 4.18 shows the basic form of the *band rodless cylinder*. This uses a cable or band to transmit force from the piston to the external carriage. When the piston moves it causes the band to move and so results in motion of the carriage. This type of cylinder is available with piston diameters up to about 60 mm and strokes of 7 m. It is capable, when used with lubricated air, of speeds up to about 2 m/s.

Figure 4.19 shows the basic form of the *slot type* of rodless cylinder. The barrel is slotted with the piston connected through the slot to the external carriage. The rodless piston moves inside the slotted cylinder and is sealed with two seal strips to prevent air leakage and dirt ingression. As it moves the seal strips separate to allow the tongue connecting the piston and carriage to pass, returning to their at-rest position after the piston has passed. Such a cylinder can be made very compact with cylinder bores from about 16 mm to 80 mm and strokes up to 10 m. Air pressures up to about 10 bar can be used. A common application of this form of cylinder is with sliding doors in railway carriages and buses.

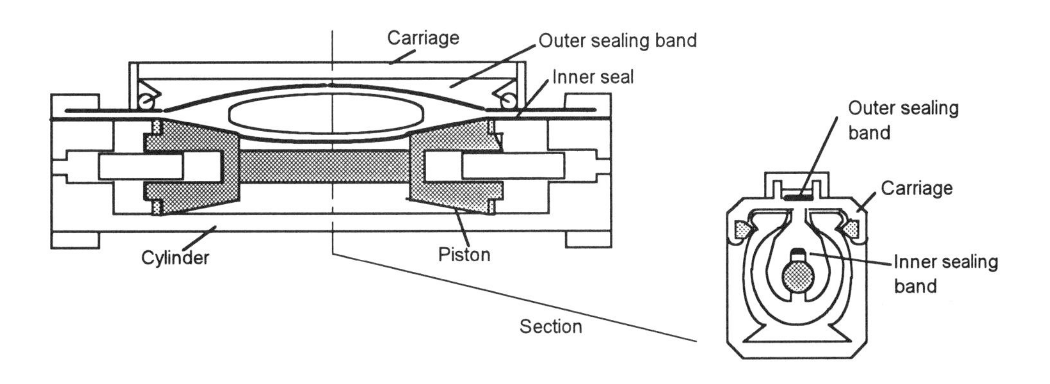

Figure 4.19 *Slot form of rodless cylinder*

4.6 Cylinder dynamics

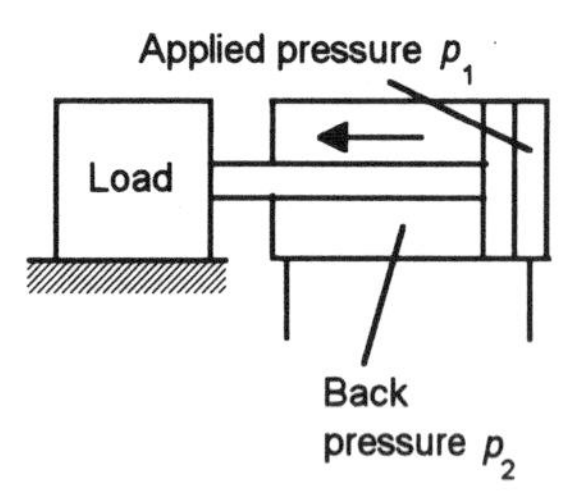

Figure 4.20 *Double-acting cylinder*

Consider a single piston in a double-acting cylinder being used to move a load (Figure 4.20) as a result of the input pressures changing. The supply pressure is connected to the right-hand side of the piston and begin to build up, likewise the pressure on the left-hand side will begins to fall when it is connected to the exhaust. Initially the force provided by this pressure difference across the piston will be less than the force resisting the motion and there will be no motion. As the force provided by the increasing pressure difference rises, a point is reached when it becomes greater than the resistive forces and there is then a net force and so acceleration. The time taken for this point to be reached is termed the *response time* and the applied pressure at that point is termed the *breakaway pressure*. As the piston accelerates, air is driven out of the exhaust side of the piston. However, air only leaves as long as the pressure is above the exhaust pressure. Then the air becomes trapped and a back pressure builds up. Consequently the force due to the pressure decreases until eventually it again becomes equal to the resistive forces. There is then no net force acting on the piston and so no acceleration and it has a constant velocity. If the cylinder has cushioning, the flow of the exhausting air is restricted and the back pressure is increased even more. This rise in back pressure results in a decelerating force which bring the piston to rest.

Figure 4.21 shows the pressures and motion of the piston as a function of time. The applied pressure is about 10% more than the breakaway pressure and the back pressure during the constant velocity part of the motion is about 40% of the breakaway pressure. Thus the net pressure used to drive the load during the constant velocity motion, allowing some slight margin for friction, is often taken as 50% of the force that could be supplied by the supply pressure alone acting on the piston face.

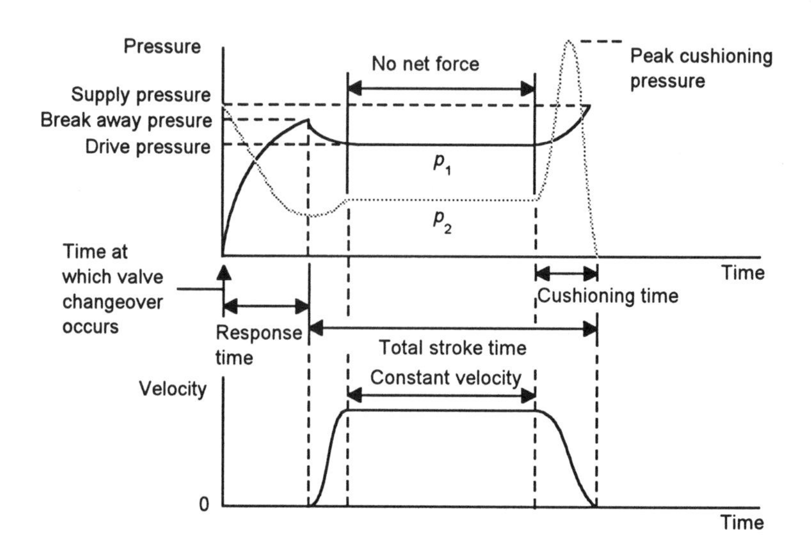

Figure 4.21 *Pressure and motion changes with time*

Example

A pneumatic cylinder is required to move a load of 1000 kg along a horizontal surface, the coefficient of friction μ between the load and the surface being 0.15. The acceleration of the load is to occur within a distance of 40 mm and the constant velocity part of the motion is to be at 600 mm/s. The system pressure is 6 bar. Internal frictional losses in the cylinder are equivalent to 10% of the total applied force. Determine the actuator size required.

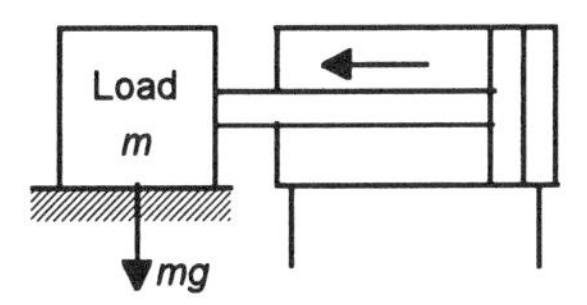

Figure 4.22 *Example*

Figure 4.22 shows the arrangement. The frictional force is $\mu mg = 0.15 \times 1000 \times 9.81 = 1471.5$ N. The force required to accelerate the load to the velocity of 0.6 m/s in 0.040 m is ma, with the acceleration a being given by $v^2 = u^2 + 2as$ and so $a = v^2/2s = 0.6^2/(2 \times 0.040) = 4.5$ m/s^2, and so the force is $1000 \times 4.5 = 4500$ N. Thus the total external force that has to be applied to the load is $1471.5 + 4500 = 5971.5$ N. Internal frictional losses in the cylinder amount to 10% of $5971.5 = 597.15$ N and thus the total force the cylinder has to supply is 6568.65 N. The piston area thus required for this $F/p = 6568.65/6 \times 10^5 = 0.0109$ m^2. This is a piston diameter of 118 mm.

Example

A pneumatic cylinder is required to move loads of 120 kg a distance of 600 mm up a ramp inclined at 50° to the horizontal, the coefficient of friction being 0.10. The acceleration of the load is to occur within a distance of 30 mm and the load is to attain a steady velocity thereafter of 500 mm/s. The supply pressure is 6 bar. Internal frictional losses in the cylinder are equivalent to 10% of the total applied force. Determine the actuator size required.

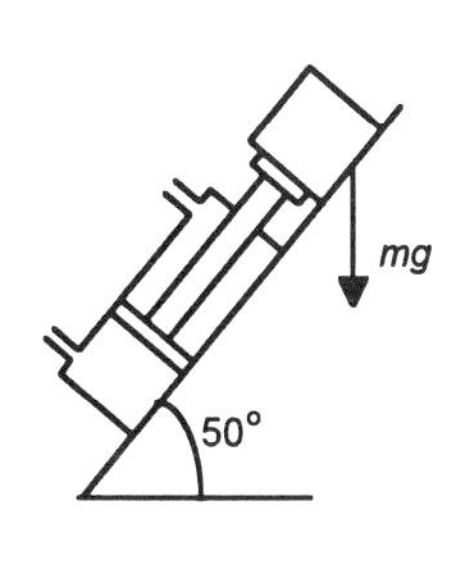

Figure 4.23 *Example*

Figure 4.23 shows the arrangement. The component of the load weight mg acting down the plane is $mg \sin \theta = 120 \times 9.81 \sin 50° = 901.8$ N. The frictional force is $\mu mg \cos \theta = 0.10 \times 120 \times 9.81 \cos 50° = 75.7$ N. The acceleration required up to the plane is given by $v^2 = u^2 + 2as$ as $a = v^2/2s = 0.500^2/(2 \times 0.030) = 4.17$ m/s^2 and thus the force required to give this acceleration is $ma = 120 \times 4.17 = 500$ N. Thus the total force required is $901.8 + 75.7 + 500 = 1477.5$ N. Internal frictional losses in the cylinder amount to 10% of $1477.5 = 147.75$ N and thus the total force the cylinder has to supply is 1625.25 N. The piston area thus required for this $F/p = 1625.25/6 \times 10^5 = 0.0027$ m^2. This is a piston diameter of 59 mm.

Example

A double-acting pneumatic cylinder is required to exert a dynamic force of 1.2 kN on the extend stroke. Determine the size of cylinder required if the dynamic force is assumed to be 50% of the static force and the supply pressure is 5 bar.

The static force is 2.4 kN and thus the piston area required is $F/p = 2.4 \times 10^3/5 \times 10^5 = 0.048$ m^2 and the effective piston diameter is 78 mm.

4.7 Semi-rotary actuator

Where rotary movement through some angle less than 360° is required, it is possible to use linear actuators, i.e. cylinders, and levers. Figure 4.24 illustrates this. However, such a method tends to be restricted to relatively small angles, be rather bulky and give an output torque, i.e. turning moment, which varies throughout the stroke of the piston.

A better alternative is a *semi-rotary actuator*. Figure 4.25 shows the symbol for such an actuator.

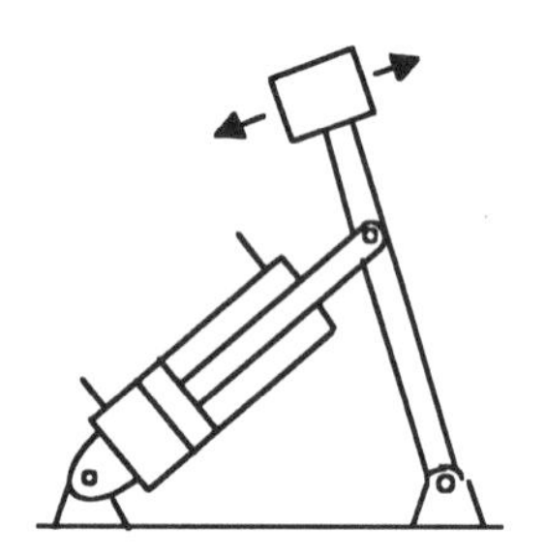

Figure 4.24 *Producing rotation*

4.7.1 Types of semi-rotary actuator

There are two basic types of semi-rotary actuator, the piston type and the vane type. The *piston type* (Figure 4.26) has a piston rod in the form of a rack with a piston at each end and a pinion, which carries the output shaft, meshing with the rack. The *vane type* (Figure 4.27) has a movable vane which is rotated by a pressure difference occurring between the two ports.

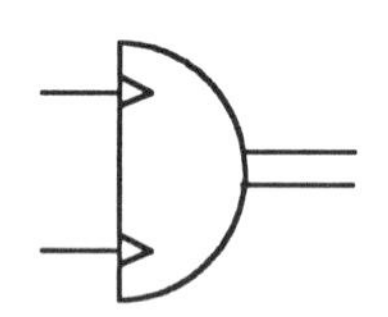

Figure 4.25 *Symbol for semi-rotary actuator*

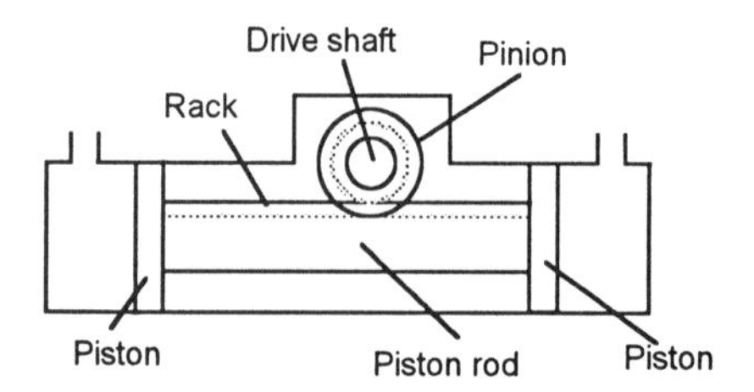

Figure 4.26 *Piston-type semi-rotary actuator*

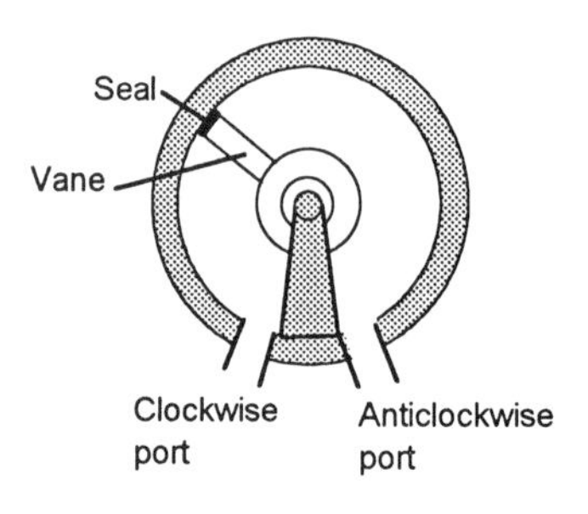

Figure 4.27 *Vane-type semi-rotary actuator*

4.8 Air motors

While the use of rack and pinion drives or levers can be used to obtain a rotary movement from a linear actuator, i.e. cylinder, they are only able to give rotation through an angle which is less than 360°. Air motors, however, can give a continuous rotation and so unlimited angular rotation. Such motors have the following advantages:

1 A high power to weight ratio,

2 Easy to reverse,

3 Simple speed control,

4 Can be stalled without damage,

5 Torque (i.e. the turning moment) remains reasonably constant over a wide range of speeds,

6 Capable of high speeds,

7 Robust, can be used in hazardous areas,

8 Simple to install,

9 Have minimal maintenance.

Fixed capacity with one direction of flow

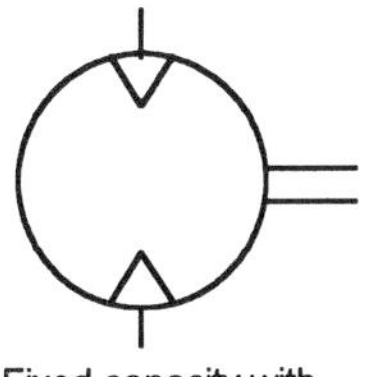

Fixed capacity with two directions of flow

Figure 4.28 *Symbols, an arrow across the symbol indicating variable capacity*

Commonly used air motors are vane motors, gear motors, piston motors and turbine motors. Compressors and motors are very similar, the graphical symbols used for the two reflecting this. Figure 4.28 shows symbols for air motors.

4.8.1 Vane motors

A vane motor (Figure 4.29) is very similar to a vane pump (see Section 2.2.2) and is very widely used. An eccentric rotor has slots in which vanes are forced outwards against the walls of the cylinder by the rotation. The vanes divide the chamber into separate compartments, each a different size. Compressed air enters one such compartment as its volume is getting bigger. The air pressure on the vanes creates a force and hence a torque which causes the rotor to rotate, the compressed air being released from the exhaust port. Such a motor can be made to reverse its direction of rotation by using a different inlet port for the compressed air, one on the opposite side of the cylinder to the other inlet port.

Vane motors are available to supply powers from less than 1 kW up to about 20 kW. Maximum speed tends to be about 30 000 rev/min. The maximum operating pressure is about 8 bar.

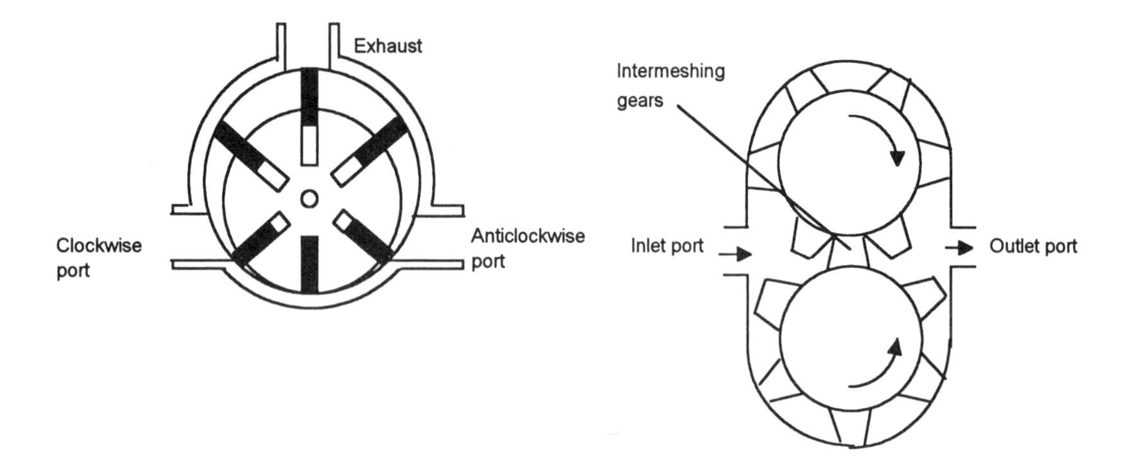

Figure 4.29 *Vane motor*

Figure 4.30 *Gear motor*

4.8.2 Gear motors

Figure 4.30 shows the basic form of gear motors. The fluid entering the upper chamber produces a pressure difference between the upper and lower chambers which results in the gears rotating. They typically have powers from about 0.5 kW to 5 kW, maximum speeds of about 15 000 rev/min and a maximum operating pressure of about 10 bar.

4.8.3 Piston motors

Piston motors can be used to provide more power than vane or gear motors, typically in the range 1.5 to 30 kW, and are more efficient since they suffer from less leakage. They have speeds of up to about 5000 rev/min and maximum operating pressures of 10 bar. They can easily be reversed by reversing the direction of air flow. However, they are much more bulky and have a more complex form of construction than vane or gear motors.

Figure 4.31 shows the basic form of a *radial piston motor*. It has a rotary valve which applies compressed air to above two of the pistons and exhausts air from above the other two pistons. The result is movement of the pistons which causes the crankshaft to rotate. In doing so the rotary valve rotates and the air pressure and exhausts are switched round the pistons in sequence, so maintaining the rotation.

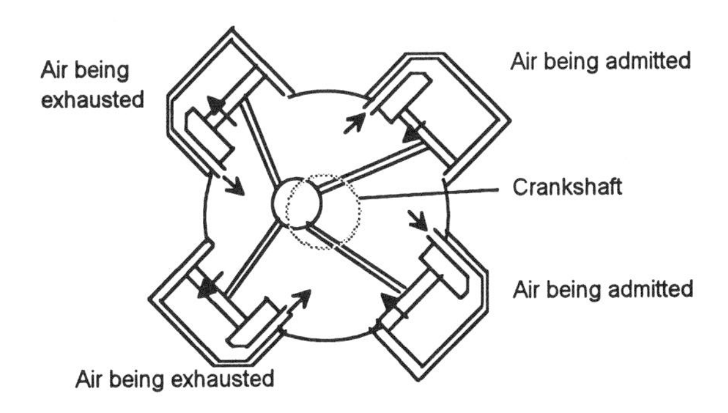

Figure 4.31 *Radial piston motor*

Another form of piston motor is the *axial piston motor*. This is of the same form as the axial piston pump with swash plate (see Figure 2.34).

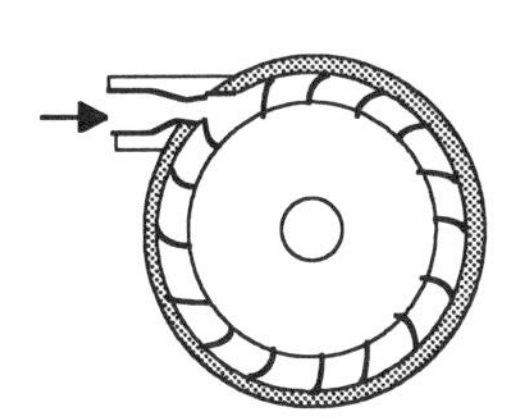

Figure 4.32 *Turbine motor*

4.8.4 Turbine motors

The turbine motor consists of a small turbine (Figure 4.32) which is caused to rotate by the kinetic energy of an air jet impinging on its blades. Such a motor can only be used where low power is required. The speed can be very high, up to 100 000 rev/min.

4.8.5 Power–speed–torque diagram

Figure 4.33 shows the typical form of power–speed–torque diagram that occurs with air motors such as vane and piston motors. As the speed increases:

1 The torque decreases linearly.

2 The power rises from zero to a maximum and then decreases back to zero again, the maximum power occurring at approximately half the maximum no-load speed.

3 The air consumption increases to become almost constant at high speeds.

Example

Determine the speed, torque and power required of a pneumatic motor if it is required to lift a load of 10 kg by means of a cable wrapped round a drum of radius 0.2 m at 1 m/s, frictional losses amounting to 25% of the applied torque.

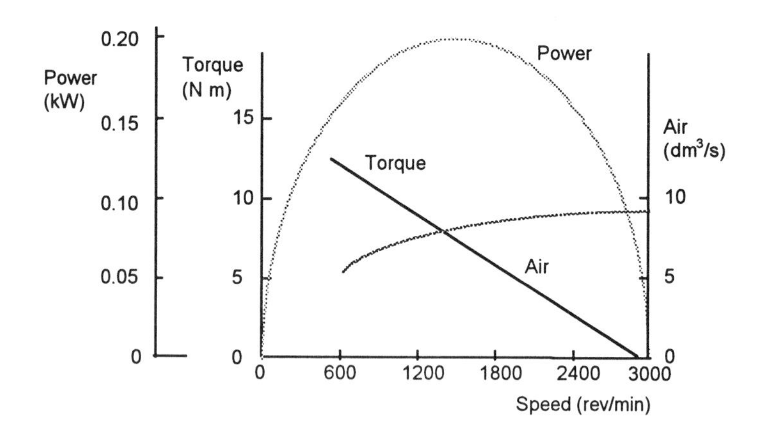

Figure 4.33 *Typical pneumatic motor characteristics*

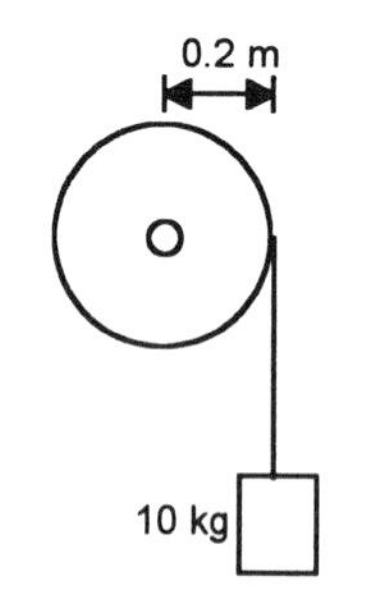

Figure 4.34 *Example*

Figure 4.34 shows the situation. The torque $T = Fr$, where F is the tangential force and r the radius of the cable drum. Thus:

$$T = mgr = 10 \times 9.81 \times 0.2 = 19.62 \text{ N m}$$

Frictional losses amount to 25% of 19.62 = 4.905 N m and thus the torque required of the motor is 24.5 N m.

If the load has to have a vertical velocity of v, then in 1 s the load has to be lifted a distance v. This means that this amount of cable must become wrapped round the drum. The circumference of the drum is $2\pi r$. If it has a rotational speed of n revolutions per second then in 1 s a point on its circumference moves through a distance of $2\pi rn$. Thus we must have $v = 2\pi rn$ and the required motor speed is:

$$n = \frac{v}{2\pi r} = \frac{1}{2\pi \times 0.20} = 0.796 \text{ rev/s} = 47.7 \text{ rev/min}$$

For a rotating shaft of radius r, the distance travelled by a point on its surface in one revolution is $2\pi r$ and if there are n revolutions per second then the distance travelled per second is $2\pi rn$. If the shaft is acted on by a torque T then, since $T = Fr$ with F being the tangential force, the work done per second by the rotating shaft, i.e. the power, is $F \times$ distance travelled in 1 s = $2\pi nrF = 2\pi nT$. Hence the required power is:

$$\text{power} = 2\pi nT = 2\pi \times 0.796 \times 24.5 = 123 \text{ W} = 0.123 \text{ kW}$$

4.9 Hydraulic motors

Figure 4.35 shows the symbols used for hydraulic motors. As the symbols suggest, there is considerable similarity between motors and pumps. Commonly used hydraulic motors are the gear motor, which is similar to

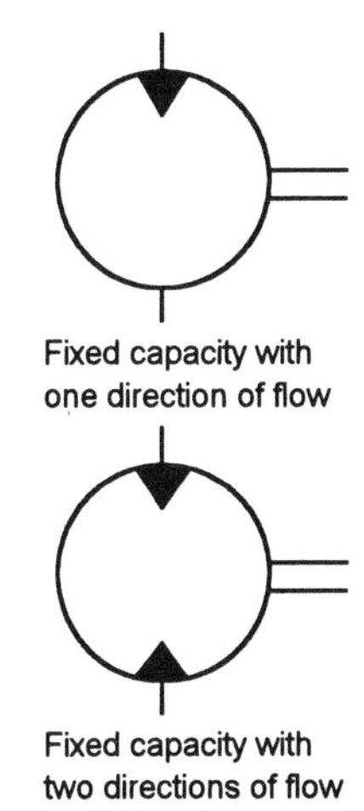

Figure 4.35 *Symbols, an arrow across the symbol indicating variable capacity*

the gear pump (see Figure 2.30), and the axial piston motor, which is similar to the axial piston pump (see Figure 2.34). The speed of the axial piston motor can be changed by changing the angle of the swash plate. Piston motors are generally the most efficient and give the highest torques, speeds and powers.

4.9.1 Hydraulic transmission

This consists of a pump, driven by a petrol engine, a diesel engine, an electric motor or some other prime mover, and one or more hydraulic motors and is used to transmit power from one rotating shaft to another at a distance from the first. Such a transmission is used in motor vehicles.

Figure 4.37 shows the basic form of such a transmission. The shaft of the pump is driven by the prime mover and pumps oil from the reservoir to the control valve. A pressure relief valve is included. When the upper lever of the control valve is used, the flow passes one way through the motor, when the other lever is used the flow passes in the opposite direction through the motor. The result is rotation of the drive shaft in one direction or the reverse, depending on which lever is used. Reversal is thus effected by a directional control valve.

The transmission system shown in Figure 4.37 is one where the exhausting fluid from the motor is discharged directly into the reservoir tank. Such a system is termed *open circuit*. Figure 4.38 shows an alternative where the discharge from the motor is fed directly back to the inlet side of the pump, no passage via the reservoir being required. This is termed a *closed circuit* system. The drive shaft can be reversed by reversing the pump, or by the use of a directional control valve (as in Figure 4.37).

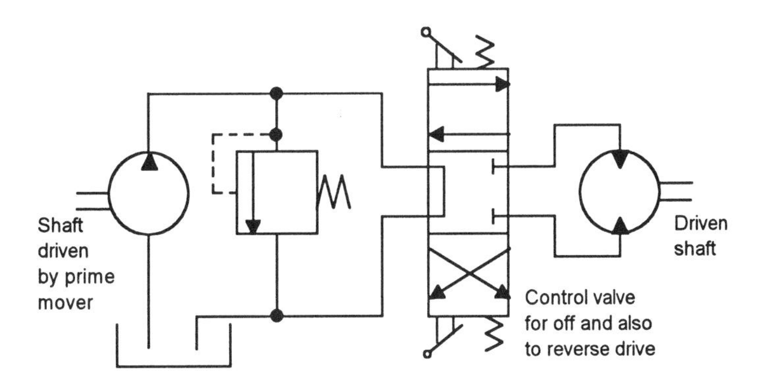

Figure 4.37 *Open-circuit hydraulic transmission*

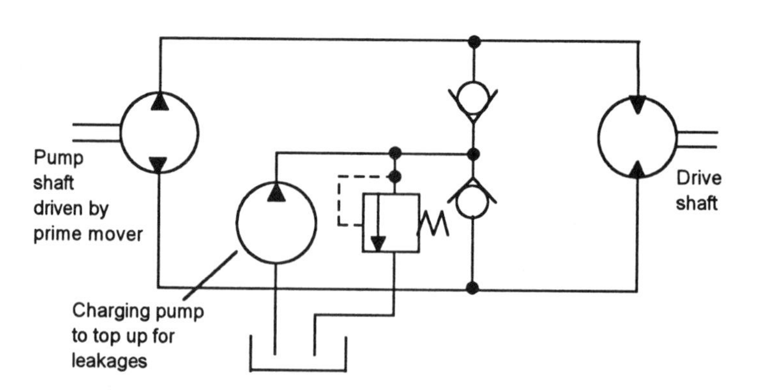

Figure 4.38 *Closed-circuit hydraulic transmission*

Problems

1 A pneumatic single-acting cylinder is to be used to clamp work pieces in a machine tool. The piston has a diameter of 80 mm and the required clamping force is 4 kN. Determine the system pressure that has to be applied to the cylinder to achieve this force. Neglect any forces due an in-built spring and take frictional losses to be 10%.

2 Determine the volume rate of flow required for a single-acting cylinder if the system gauge pressure is 6 bar, the piston has a diameter of 100 mm, the stroke length is 400 mm and the cylinder completes 5 strokes and returns per minute.

3 Devise a circuit for the direct speed control of both strokes of a single-acting cylinder.

4 A pneumatic double-acting, single-rod cylinder is to be used to clamp work pieces in a machine tool. The required clamping force is 4 kN and the supply pressure is 6 bar. Determine the size of the cylinder required. Assume an efficiency of 95%.

5 Devise a circuit which has one push-button to extend the piston in a double-acting, single-rod cylinder and one push-button to return it. Assume that the push-buttons have to be located some distance from the cylinder.

6 A single-rod, double-acting cylinder has a piston of diameter 40 mm and a piston rod of diameter 16 mm. Determine the extend and return forces if a gauge pressure of 7 bar is first applied to one side of the piston and then the other, the other side in each case being connected to the exhaust line. Neglect any consideration of frictional forces.

7 A single-rod, double-acting differential cylinder, connected as in Figure 4.10, has a piston of diameter 80 mm and a piston rod of diameter 50 mm. What will be the ratio of the maximum extend to return forces?

8 A pneumatic cylinder is required to move a load of 1000 kg along a horizontal surface, the coefficient of friction between the load and the surface being 0.12. The acceleration of the load is to occur within a distance of 30 mm and the constant velocity part of the motion is to be at 500 mm/s. The system pressure is 6 bar. Internal frictional losses in the cylinder are equivalent to 10% of the total applied force. Determine the actuator size required.

9 A pneumatic cylinder is required to move loads of 200 kg a distance of 600 mm up a ramp inclined at 60° to the horizontal, the coefficient of friction being 0.15. The acceleration of the load is to occur within a distance of 30 mm and the load is to attain a steady velocity thereafter of 600 mm/s. The supply pressure is 6 bar. Internal frictional losses in the cylinder are equivalent to 10% of the total applied force. Determine the actuator size required.

10 A double-acting pneumatic cylinder is required to exert a dynamic force of 1 kN on the extend stroke. Determine the size of cylinder required if the dynamic force is assumed to be 50% of the static force and the supply pressure is 6 bar.

11 Determine the speed, torque and power required of a pneumatic motor if it is required to lift a load of 2 kg, by means of a cable wrapped round a drum of radius 0.1 m, at a velocity of 0.5 m/s, frictional losses amounting to 20% of the applied torque.

5 Sensors

5.1 Introduction

The term *sensor* is used with instrumentation and control systems to describe the element which takes information about the variable being measured or monitored and changes it into some form which enables the rest of the measurement or control system to utilise it. This chapter is a brief consideration of the sensors that are commonly used with pneumatics and hydraulic systems for position sensing and how they might be used. With machine control they take information about the positions of, say, some machine element and transform it into a suitable form for the pneumatic or hydraulic system.

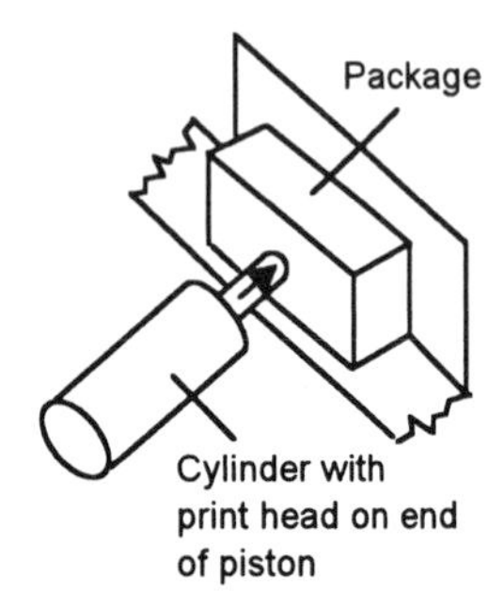

Figure 5.1 *Package printing machine*

For example, suppose a pneumatic machine is required to push a printing head onto each package produced from a packaging machine and print the name and address of the company (Figure 5.1). The packages may be of different sizes and not spaced regularly on the delivery belt. What is required is a means of determining when a package is in the correct position for a cylinder actuator to be actuated and push the printing head onto the face of the package. A position sensor is required. This might be some device which senses the package by coming into contact with it, for example it might move a roller or plunger or operate an electrical switch, or it might be some device which is non-contact and detects the presence of the package by changing the escape of air from a jet or operating a reed switch or other electrical non-contact sensor. The sensor might then be used to mechanically, or by a pneumatic pilot signal, or electrically, actuate a valve.

Position sensors can be grouped into two basic types:

1 *Contact sensors*
The object at its position comes in contact with the sensor.

2 *Non-contact sensors*
There is no contact between the object and the sensor, the term *proximity sensor* often being used.

The following sections give common mechanical, pneumatic and electrical examples of these types of position sensors.

5.2 Mechanical contact sensors

Mechanical contact sensors are very widely used for position sensing, common forms involving roller and plunger valves (Figure 5.2). The object being detected impinges on the end of the roller or plunger and results in its movement, so switching the valve from one position to another. Precise end-position control is possible with the valve remaining switched into its new position as long as the object is in contact with the roller or plunger.

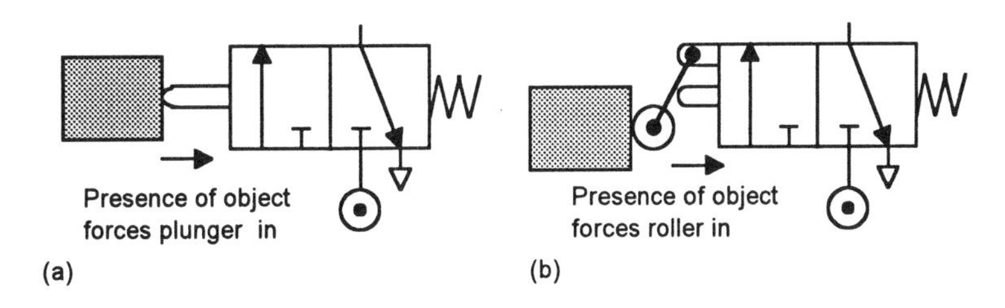

Figure 5.2 *Roller and plunger position sensors*

Figure 5.3 gives an example of a *one-way trip valve*. Such a valve uses a roller but the roller is so pivoted that it is only actuated when the projection from the object moves in just one direction past the roller. The valve is actuated and switches position for only as long as the projection depresses the roller by the right amount. Thus the trip valve can be used to give a pulse output for a period of time as the object moves past the roller.

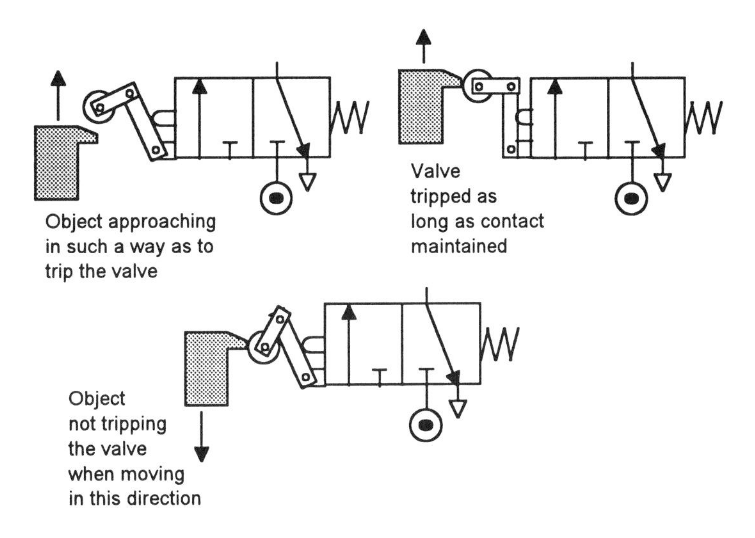

Figure 5.3 *One-way trip valve*

5.2.1 Back pressure sensor

Figure 5.4 shows the basic form of such a sensor. This form of sensor supplies no pressure signal when the object is not in contact with its protruding plunger and a pressure signal when the plunger is depressed. Such pressure signals can be used as pilot signals for valves.

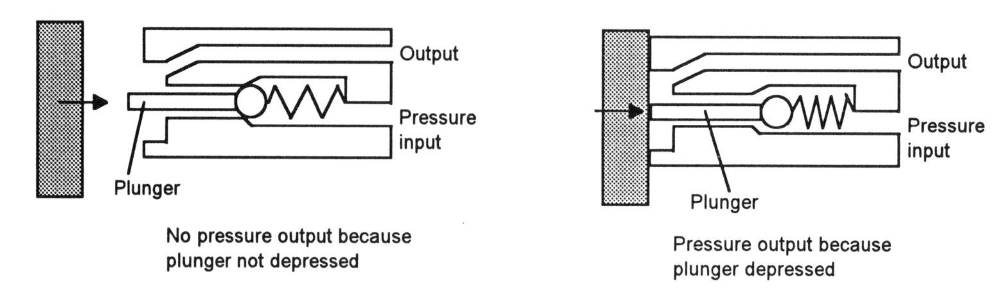

Figure 5.4 *Back pressure sensor*

5.3 Non-contact pneumatic sensor

Figure 5.5 shows the basic principle of a non-contact pneumatic sensor. Low-pressure air is allowed to escape through a port in the front of the sensor. This escaping air, in the absence of any close-by object, escapes and in doing so also reduces the pressure in the nearby sensor output port. However, if there is a close-by object, the air cannot so readily escape and the result is that the pressure increases in the sensor output port. The output pressure from the sensor thus depends on the proximity of objects.

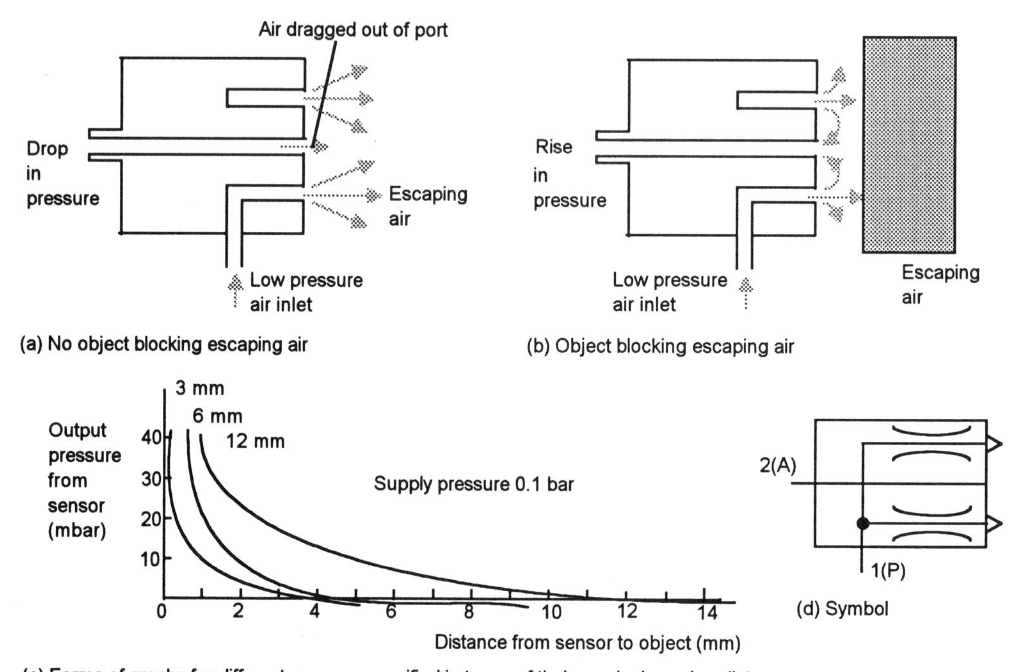

Figure 5.5 *Pneumatic proximity sensor*

Increasing the supply pressure to a sensor changes the output pressure relationship with sensing distance, increasing the sensing distance over which it can be used.

Other forms of this type of sensor are the *gap sensor* (Figure 5.6) and the *air barrier sensor* (Figure 5.7). With the gap sensor, the supplied compressed air can escape from two vent ports. However, in the line to one of these ports, and the output port 2(A), there is a constriction. As a result of the air passing through the constriction there is a pressure drop at the outlet port. However, there is also air entering the vent port from the other vent and this builds up a back pressure. Thus the pressure at the output is the sum of these two pressures, the back pressure dominating. If an object interrupts the path of the air, the back pressure is prevented and the output pressure drops. The air barrier sensor works on a similar principle. However, instead of the object interrupting the air path between the two vents, the path is interrupted by air from a distant vent.

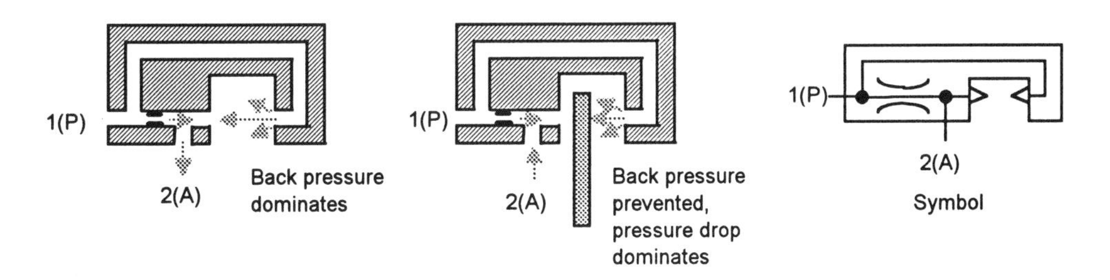

Figure 5.6 *Gap sensor*

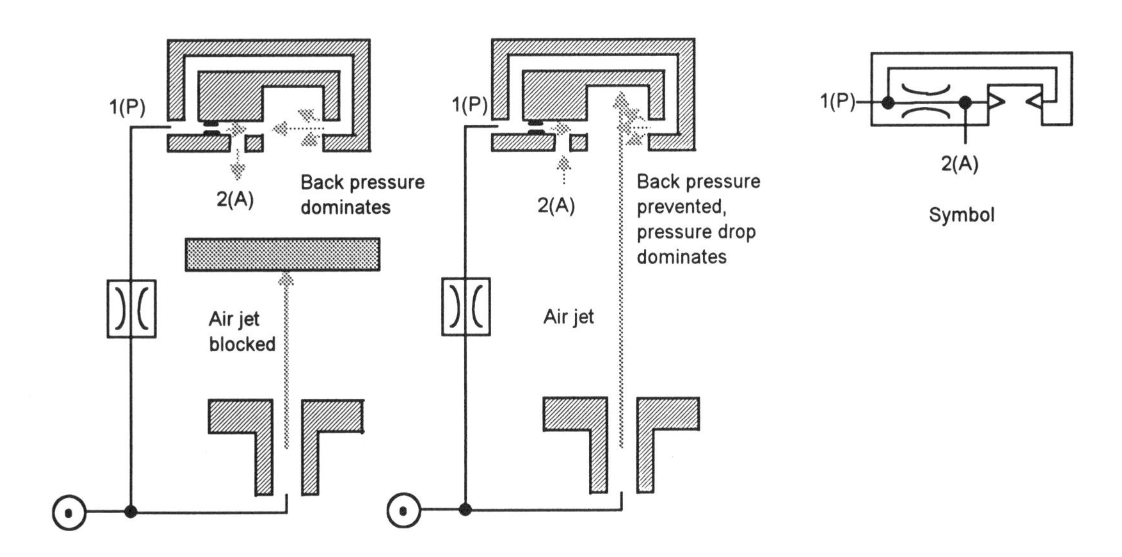

Figure 5.7 *Air barrier sensor*

5.3.1 Amplifiers

The proximity sensor described above, and other forms of pneumatic sensors, gives a low pressure signal which needs amplification before it can be used as the pilot signal for a directional control valve. At least 2 to 4 bar are required for such a signal and the sensor typically gives only about tens of millibars.

Figure 5.8 shows one basic form of an *amplifier valve*. When there is no input air pressure signal, the input diaphragm is undeflected and the output 2(A) is connect to the exhaust 3(R), the input supply pressure port 1(P) being shut off by the ball valve. When there is an input signal, the diaphragm deflects and closes off the connection between the output port and the exhaust. The output diaphragm is also deflected and opens the connection between the supply pressure and the output. Thus the input signal gives rise to a much larger change in the output pressure.

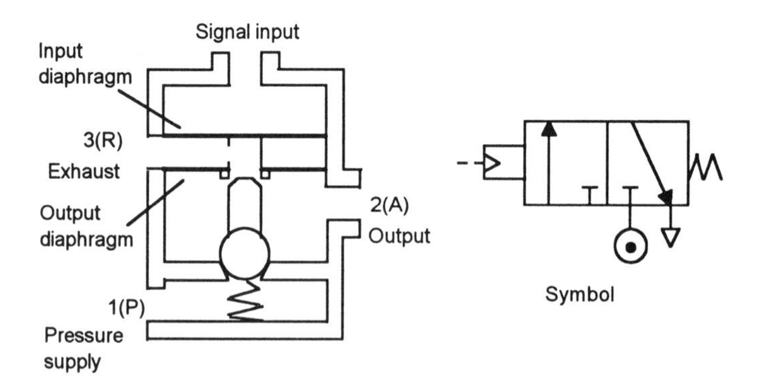

Figure 5.8 *Amplifier valve*

Figure 5.9 shows how an amplifier might be used with a pneumatic proximity sensor to switch the extend pressure off and the retract pressure on when a piston reaches the end of its stroke with a cylinder. Initially, with the piston position as shown and not entering the sensor, the sensor is supplying a high-pressure signal which has the piston maintained in position. The piston movement operation is started by pressing the start push-button. This causes the piston to extend. When the end of the piston enters the gap sensor the output pressure from the sensor drops and, after amplification, the pilot signal is cancelled. This switches the pressure to cause the piston to retract.

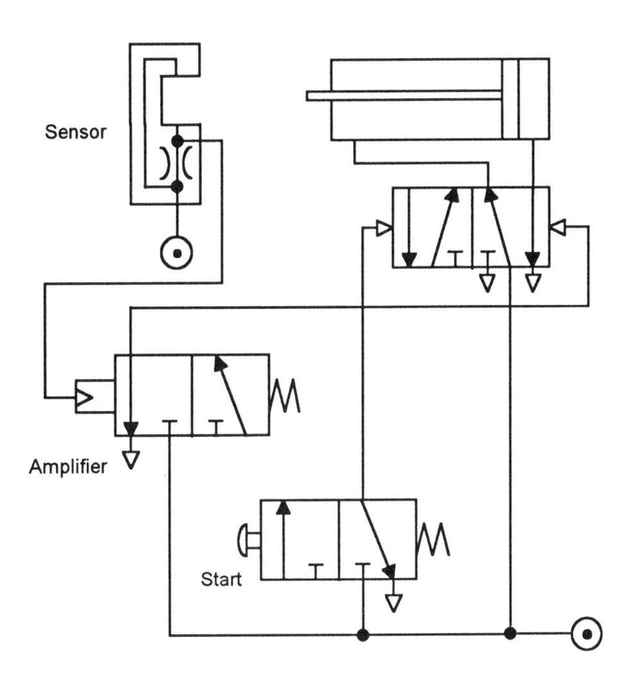

Figure 5.9 *Example of a gap sensor used with an amplifier*

5.4 Electrical switches

Electrical switches are a form of sensor in which the activation of the sensor produces an electrical current. Mechanical forms of switches are specified in terms of the number of poles in the switch, this being the number of separate circuits that can be switched simultaneously, and the number of ways, this being the number of positions to which each pole may be switched. As an illustration of these terms, Figure 5.10(a) shows a single-pole, single-throw (SPST) switch, Figure 5.10(b) a single-pole, double-throw (SPDT) switch and Figure 5.10(c) a double-pole, double-throw (DPDT) switch. The switch action may be a momentary action which resets the switch when the operating force is removed, e.g. the push-button door bell with the circuit only being made when the button is pressed and ceasing when the finger is removed from the button, or latched with the switch remaining in the operated position even when the operating force is removed. To reset the switch a reset force must be applied.

Switches can be operated in a number of ways, e.g. by push-button, lever, roller, rotating cam (Figure 5.11). They are available with normally open (NO) or normally closed (NC) contacts or can be configured as either by the choice of relevant contacts. Figure 5.12 illustrates this for a lever-operated switch.

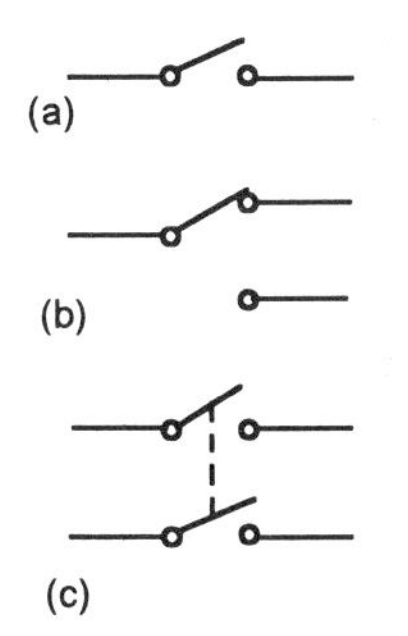

Figure 5.10 *(a) SPST, (b) SPDT, (c) DPDT*

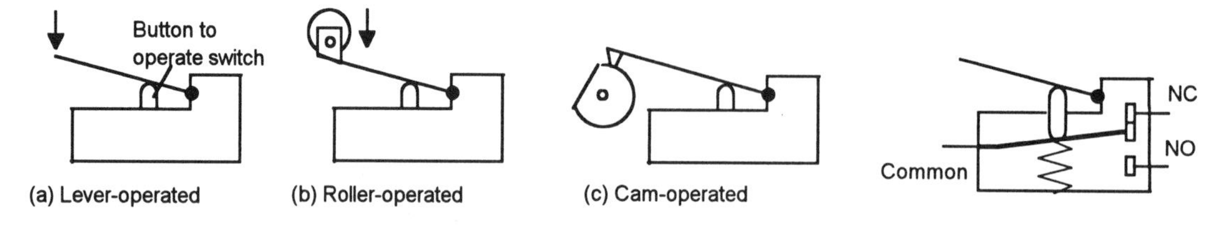

Figure 5.11 *Forms of switch actuation*

Figure 5.12 *NC/NO switch*

5.4.1 Non-contacting switches

There are a number of different types of non-contacting switches, some of which are only suitable for switching as a result of the proximity of metal objects while others work with both metallic and non-metallic objects.

The *eddy current* type of proximity switch has a coil which is energised by a constant alternating current. This current produces an alternating magnetic field. When a metallic object is close to it, eddy currents are induced in it. The magnetic field due to these eddy currents induces an e.m.f. back in the coil with the result that the voltage amplitude needed to maintain the constant current changes. The voltage amplitude is thus a measure of the proximity of metallic objects. The voltage can be used to actuate an electronic switch circuit and so give an on–off device. The range over which this sensor can operate is typically about 0.5–20 mm.

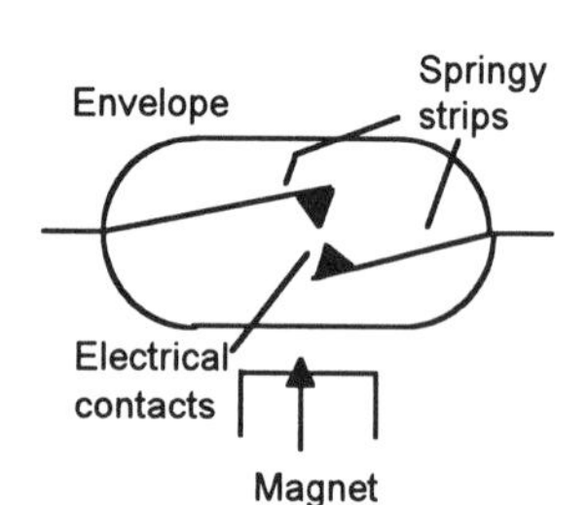

Figure 5.13 *Reed switch*

A very commonly used proximity sensor is the *reed switch*. This consists of two overlapping, but not touching, strips of springy ferromagnetic material sealed in a glass or plastic envelope (Figure 5.13). When a magnet or current-carrying coil is brought close to the switch, the strips become magnetised and attract each other, so closing the contacts. Typically the contacts close when the magnet is about 1 mm from the switch.

A proximity switch that can be used with metallic and non-metallic objects is the *capacitive proximity switch*. For a parallel plate capacitor, the capacitance depends on the separation of the plates and the dielectric between them. Thus the approach of a metallic object to the plate of the capacitive sensor changes its capacitance and also the approach of a non-metallic object changes its capacitance because of a change of dielectric. The change in capacitance can be used to activate an electronic switch circuit and so give an on–off switch. Capacitive proximity switches can be used to detect objects between about 4 mm and 60 mm from the sensor head.

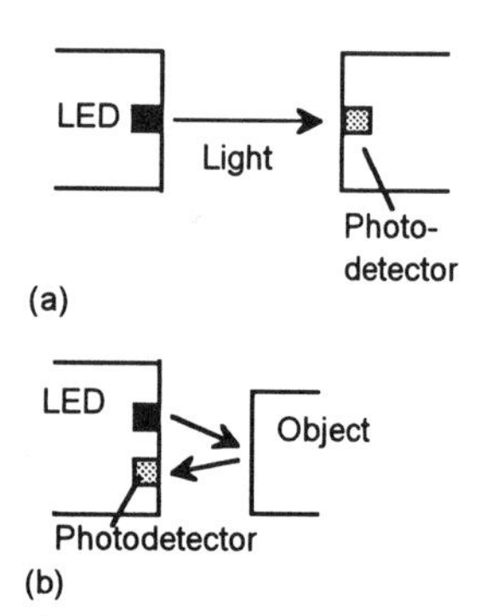

Figure 5.14 *Photoelectric sensors*

Another form of non-contact proximity sensor involves *photoelectric devices*, i.e. light-emitting diodes and photodetectors. With the transmissive form (Figure 5.14(a)), the object being detected prevents the beam of light from the light-emitting diode (LED) reaching the photodiode and so changes its output current. With the reflective form (Figure 5.14(b)), the presence of the object reflects light onto the detector and so changes its output current.

Problems

1 Describe a non-contact pneumatic sensor that could be used to detect the presence of an object.

2 A mechanical contact sensor is required to operate a valve when the piston of a cylinder reaches the end of its stroke. Suggest a suitable sensor.

3 Sketch a possible circuit that could be used with a pneumatic gap sensor to switch off the extend stroke of a cylinder when the piston reaches the end of its stroke.

4 Suggest a mechanical sensor that could be used to switch on the current to a solenoid valve for a controlled fixed period of time.

5 Give an example of a (a) contacting, (b) non-contacting electrical sensor for the position of an object.

6 Electro-pneumatics/ hydraulics

6.1 Introduction

The electrical control of pneumatic/hydraulic systems has the great advantage that the speed of response is very much greater than with pressure-based systems. Also, programmable logic controllers can be used (see Chapter 9 for more details) and readily programmed to implement control strategies, considerably simplifying the process of designing control systems when more than a few valves and cylinders are involved.

The electrical control is based on the use of solenoid valves and this chapter discusses such valves and their control, particularly control by the use of relays.

6.2 Solenoid valves

With solenoid valves, a current through a solenoid is used to implement switching between positions. Figure 6.1 shows the basic principle. When a current flows through the solenoid, a magnetic field is produced which pulls the laminated steel core into the coil, the force depending on the current through the solenoid. The larger the current the greater the force exerted. Figure 6.2 shows the symbol used for a 3/2 valve with a single direct-acting solenoid, return being by means of a spring.

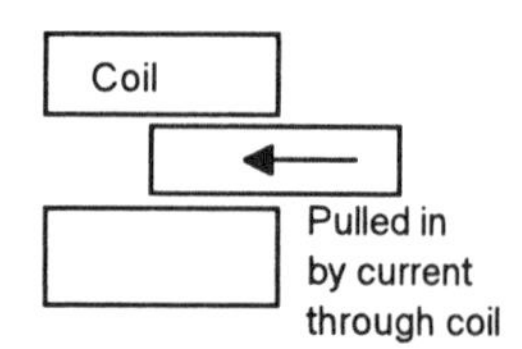

Figure 6.1 *Solenoid valve*

With a solenoid valve, and if hydraulic fluids are involved, high forces have to be exerted to move the spool against its seals and possibly fluid pressure over a relatively long stroke. Thus rather than use high currents with the solenoid to give a *direct-acting valve*, it is common to use the solenoid to operate a small pilot valve which in turns sends a pressure signal to operate the valve spool of the main valve. This is termed *indirect acting*.

Figure 6.3(a) shows the basic form of an indirect-acting, single-solenoid valve with a current applied to the solenoid of the pilot valve. The spool of this valve moves to the left as a result of the solenoid core being drawn into the solenoid. As a consequence it applies pilot pressure to the left-hand end of the main valve, this then shifting its spool to the right and so connecting the supply pressure, port 1(P), to output port 2(B). Figure 6.3(b) shows the separate symbols for the pilot and main valves and their interconnections. Although pilot operation can be achieved with separate valves, it is more usual to use a single assembly containing both the pilot and main valves. The assembly can then be represented by the simplified symbol shown in Figure 6.3(c).

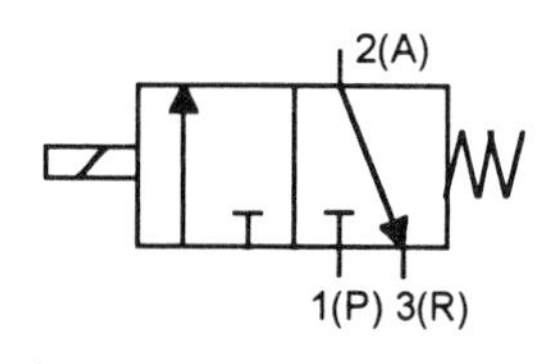

Figure 6.2 *Symbol for a single direct-acting solenoid valve*

Solenoids used with valves tend to have operating voltages of 12 V d.c., 24 V d.c., 110 V d.c. or 110 V a.c.

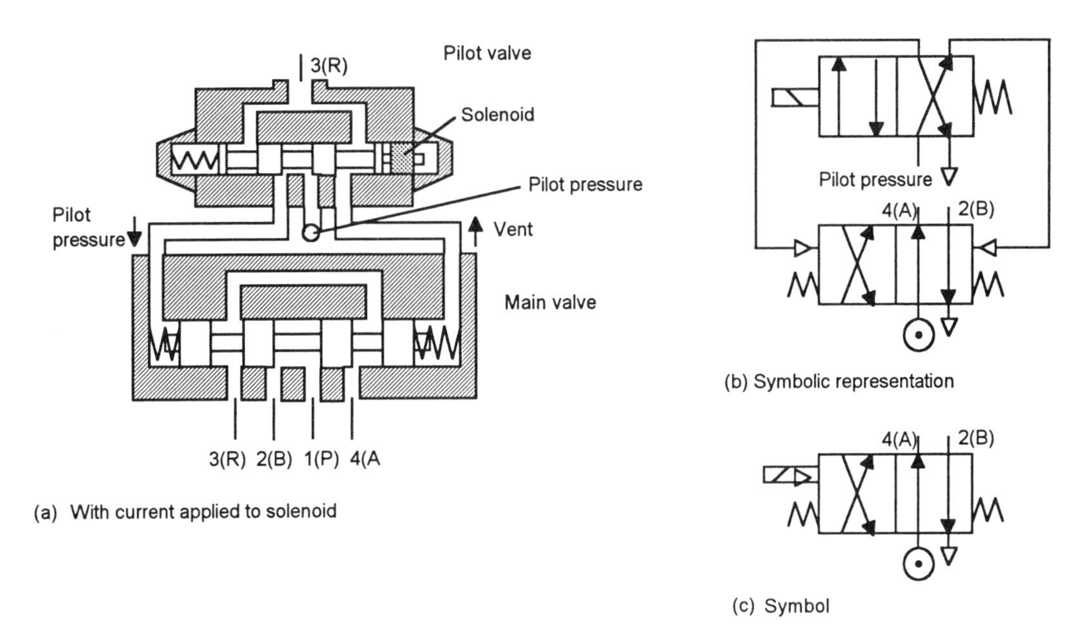

Figure 6.3 *Single-solenoid indirect-acting valve*

6.2.1 Single-solenoid valves

These have a single solenoid with return generally being by means of a spring when the solenoid is de-energised. The valve shown in Figure 6.3 is of this type. Figure 6.4 shows an example of a simple circuit that might be used with a single-solenoid valve to electrically actuate a single-acting cylinder. When the switch is closed and a current passes through the solenoid, the valve switches position and pressure is applied to extend the piston in the cylinder.

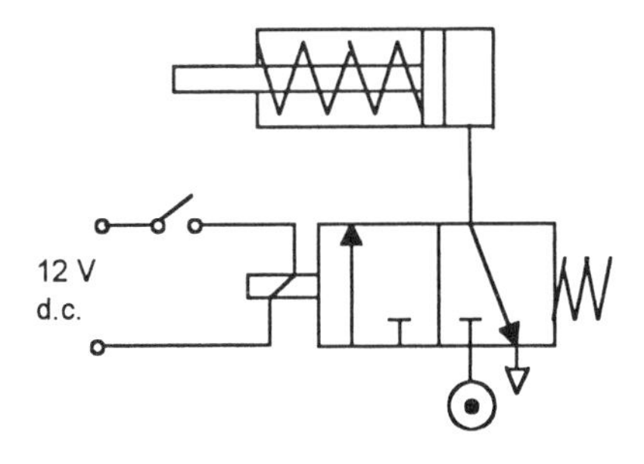

Figure 6.4 *Electrical control of a single-acting cylinder*

Care has to be taken with the use of such a single-solenoid valve since in the event of a solenoid current failure the valve will reset and this might cause an actuator to operate in a dangerous manner.

6.2.2 Double-solenoid valves

Figure 6.5 shows a double-solenoid valve. This is set in one position by the operation of one solenoid and in the other position by the second solenoid, remaining in each position until it receives a signal to change its position.

As an illustration of the use of a double-solenoid valve, Figure 6.6 shows how such a valve might be used to control a double-acting cylinder. Momentary closing switch S1 causes a current to flow through the solenoid at the left-hand end of the valve and so result in the piston extending. On opening S1 the valve remains in this extended position until a signal is received by the closure of switch S2 to activate the right-hand solenoid and return the piston.

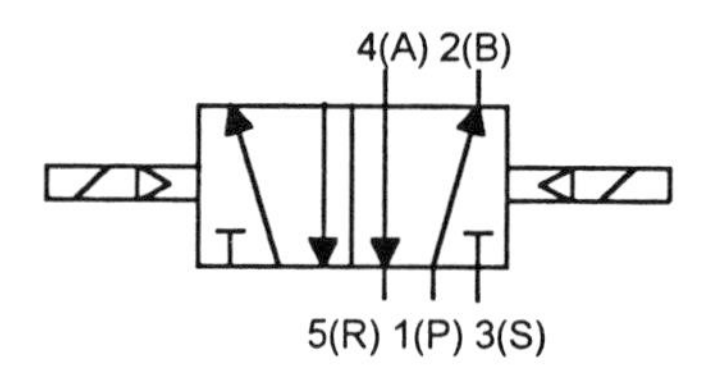

Figure 6.5 *Example of a double-solenoid valve*

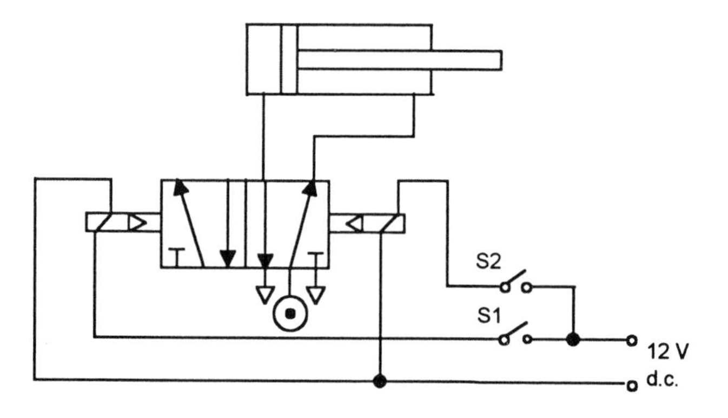

Figure 6.6 *Control of a double-acting cylinder*

6.2.3 Suppression

When the current through a solenoid coil is abruptly switched off, a large back e.m.f. can be produced as a result of the rapid change in the magnetic field. This large back e.m.f. can produce arcing across switch contacts, possibly damaging the switch. A simple method that is used with d.c. solenoid coils is to connect a diode across them (Figure 6.7). This is in such a direction as to offer very high resistance when the coil is being driven but low resistance for the back e.m.f. The result is a path for the induced current which safeguards the switch.

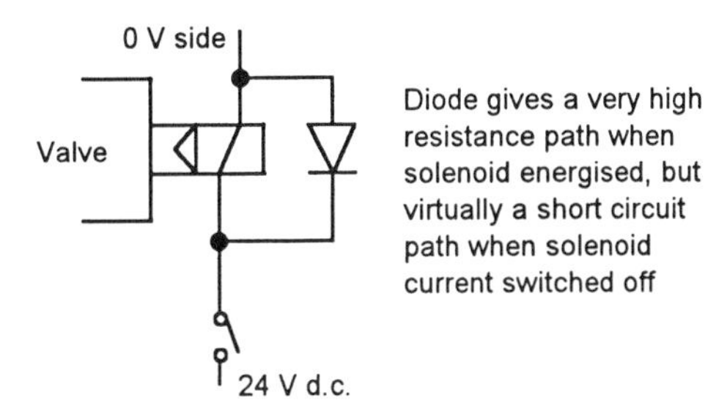

Figure 6.7 *Diode protection*

6.3 Electrical relays

Electrical relays (Figure 6.8) are electrically operated switches in which changing a current in one circuit can switch a current in another circuit. When there is a current flowing through the relay solenoid a magnetic field is produced which attracts the iron armature, so bringing the switch contacts of a normally open switch into contact and making the circuit. Thus switching on the current through the solenoid can be used to switch a current on in another circuit. Relays can have normally open and/or normally closed contacts. With normally open, a current through the solenoid closes the relay contacts; with normally closed it opens the relay contacts.

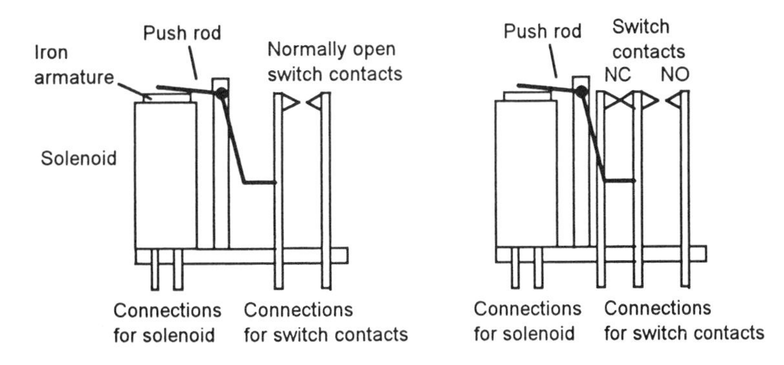

Figure 6.8 *Relay*

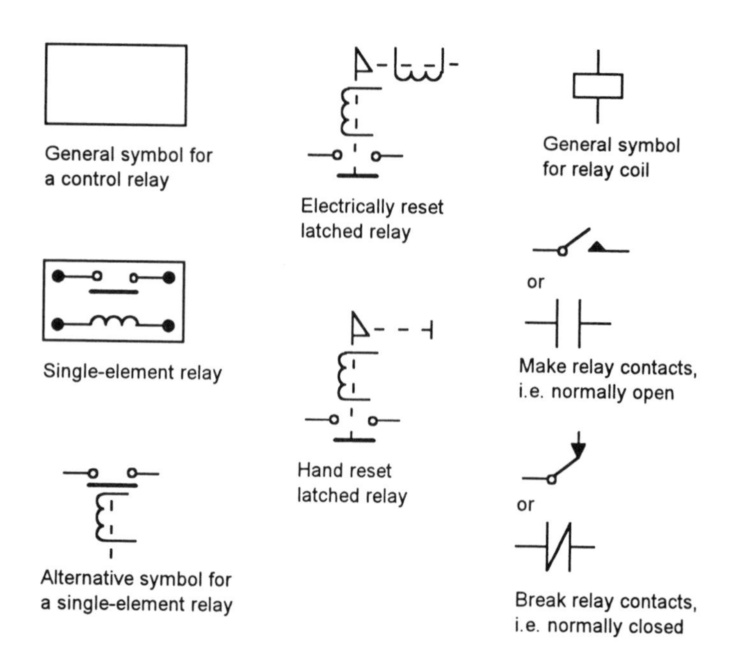

Figure 6.9 *Relay symbols*

The symbols used with relays are shown in Figure 6.9. Latched relays are ones that remain on, even when the current to the solenoid has ceased (see Section 6.3.1).

Example

State what happens when, for the circuit shown in Figure 6.10, the switch S1 is closed.

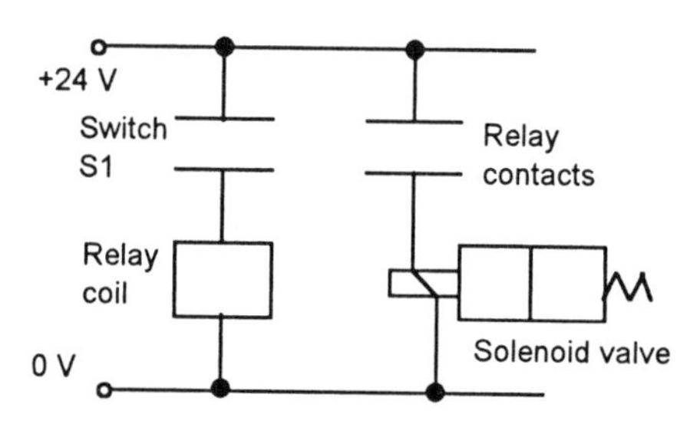

Figure 6.10 *Example*

When S1 is closed the relay coil is energised. This closes the relay contacts and energises the solenoid of the solenoid valve, so making it switch positions.

6.3.1 Latching relay circuits

It is often the case that a relay needs to remain energised and its contacts activated, even when the current to the solenoid has ceased. A latching circuit is required, such a circuit being a self-maintaining circuit that, after being energised, maintains that state until another input is received. Figure 6.11 shows such a circuit.

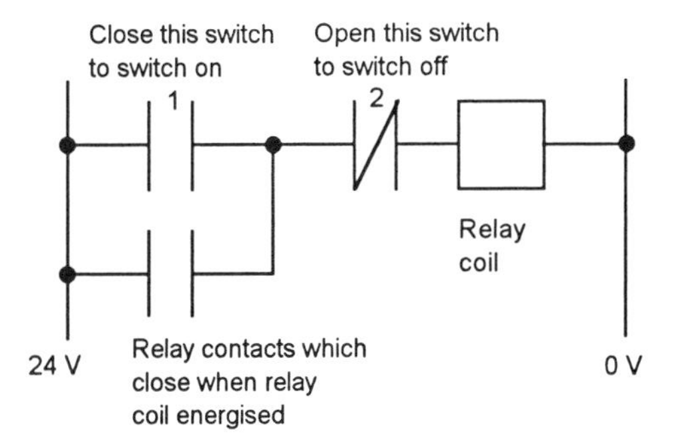

Figure 6.11 *Latching circuit*

When switch 1 is closed, the relay coil is energised and its contacts close. Thus, even when switch 1 opens, there is still connection to the supply to energise the relay coil and so the circuit remains on. It is only switched off when switch 2 is opened, remaining off even if switch 2 is then closed again.

Example

Devise a system which will extend a single-solenoid cylinder when one switch is closed and latch it so that it remains extended when the switch opens and will retract it when another switch is opened, it remaining retracted even when the switch closes.

Figure 6.12 shows a possible circuit. When switch S1 is closed, the solenoid of the solenoid valve is energised and the valve switches the air supply to the cylinder to give extension of the piston. When the piston is extended it closes the limit switch S2. This enables the current to remain connected to the solenoid, even when switch S1 is opened. Thus the piston remains extended. It requires switch S3 to be operated before the solenoid ceases to be energised and the piston returns.

Figure 6.13 shows a circuit involving a relay. When switch S1 is activated, a current passes through the relay coil and its normally open contacts close. This switches on the current to the solenoid of the solenoid valve. The relay contacts close and hence the solenoid of the valve is energised. When the cylinder extends, limit switch S2 closes and latches the solenoid on. Only when switch S3 is activated will the relay contacts open and the solenoid be no longer energised.

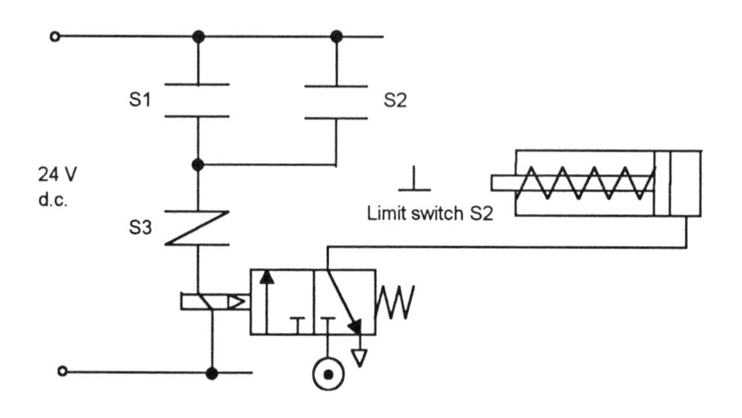

Figure 6.12 *Example*

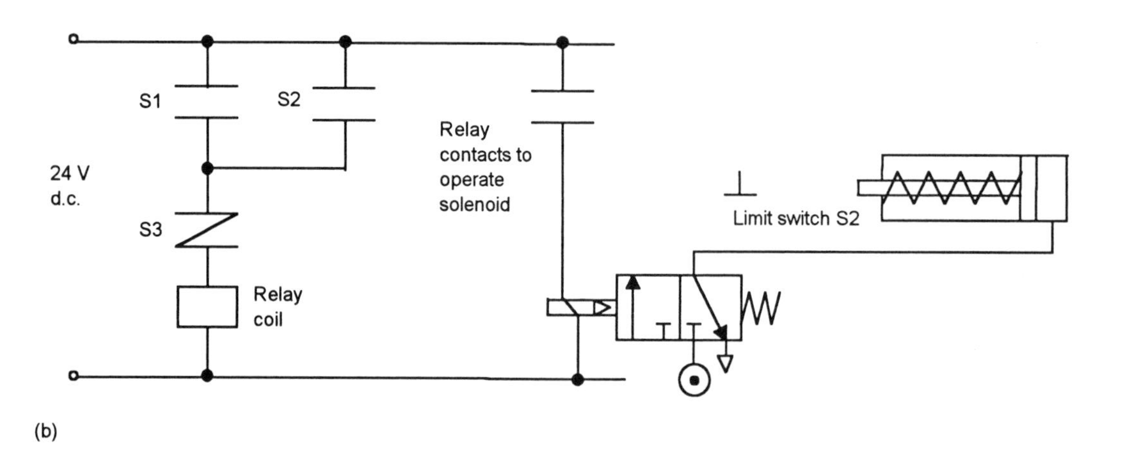

Figure 6.13 *Example*

Problems

1 Explain the difference between direct- and indirect-acting solenoid valves.

2 Explain the difference between the solenoid valves described by the symbols in Figure 6.14 and the reason for the difference.

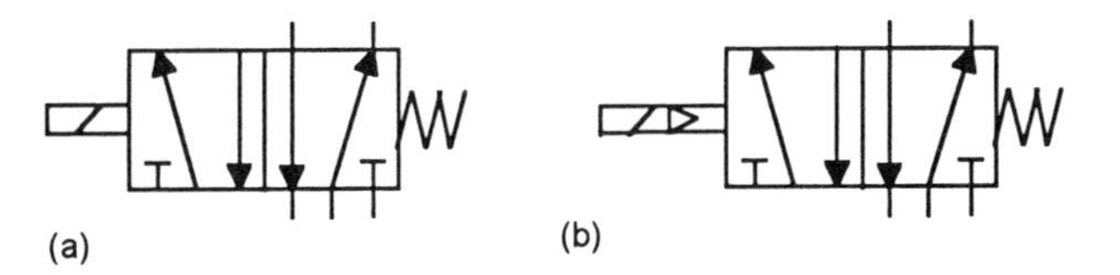

Figure 6.14 *Problem 2*

3 For the valve shown in Figure 6.15, explain how the valve can be made to apply the system pressure to output 2(B) and exhaust output 1(A).

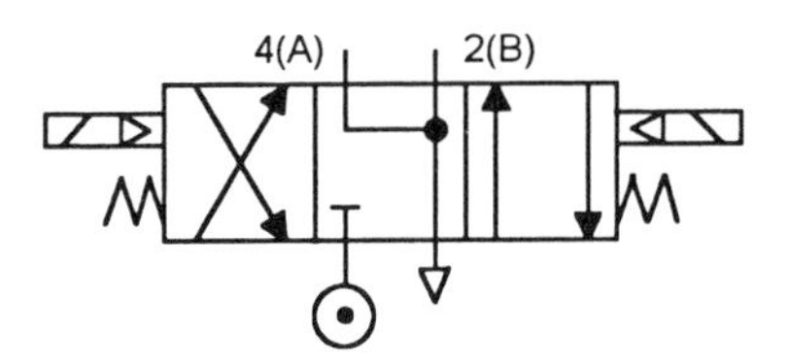

Figure 6.15 *Problem 3*

4 Describe what happens to the piston in the cylinder for the circuit shown in Figure 6.16 when the switch S1 is closed.

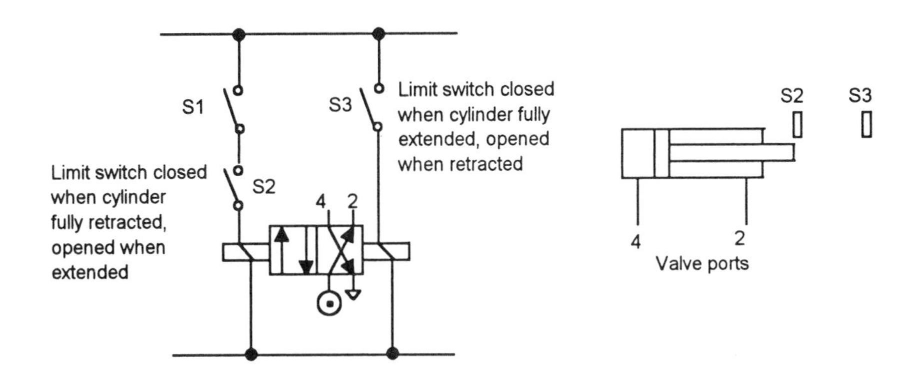

Figure 6.16 *Problem 4*

5 For the relay-controlled solenoid valve circuit shown in Figure 6.17, explain which switches have to be activated for the solenoid to be energised.

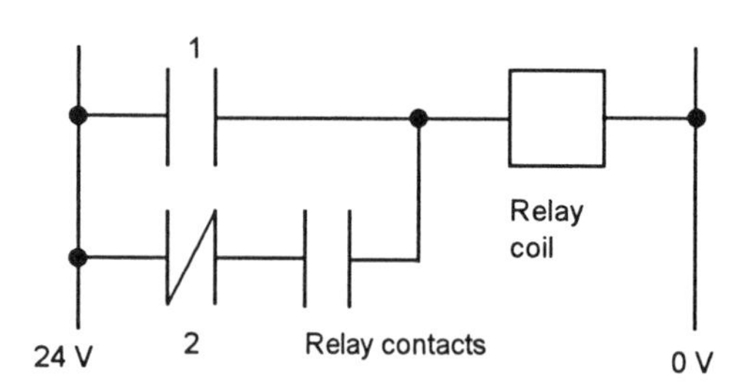

Figure 6.17 *Problem 5*

6 Explain how a relay can be used to latch a solenoid valve so that the solenoid remains energised even when the switch used to switch on the current to the solenoid is no longer closed and supplying current.

7 Cylinder control

7.1 Introduction

This chapter brings together the principles of control valves from Chapter 3, actuators from Chapter 4 and sensors from Chapter 5 in consideration of the control of cylinders. This involves directional control, speed control, pilot operation, sequential control and cascade control, as well as hydro-pneumatic systems.

As an illustration of the types of problems that this chapter deals with, consider the problem of the circuits required for automatic machines which are required to perform a number of sequential actions such as positioning objects, operating clamps and then operating some machine tool (Figure 7.1). This requires the switching in sequence of a number of cylinders, the movements of the cylinder pistons being the mechanisms by which the actions are initiated.

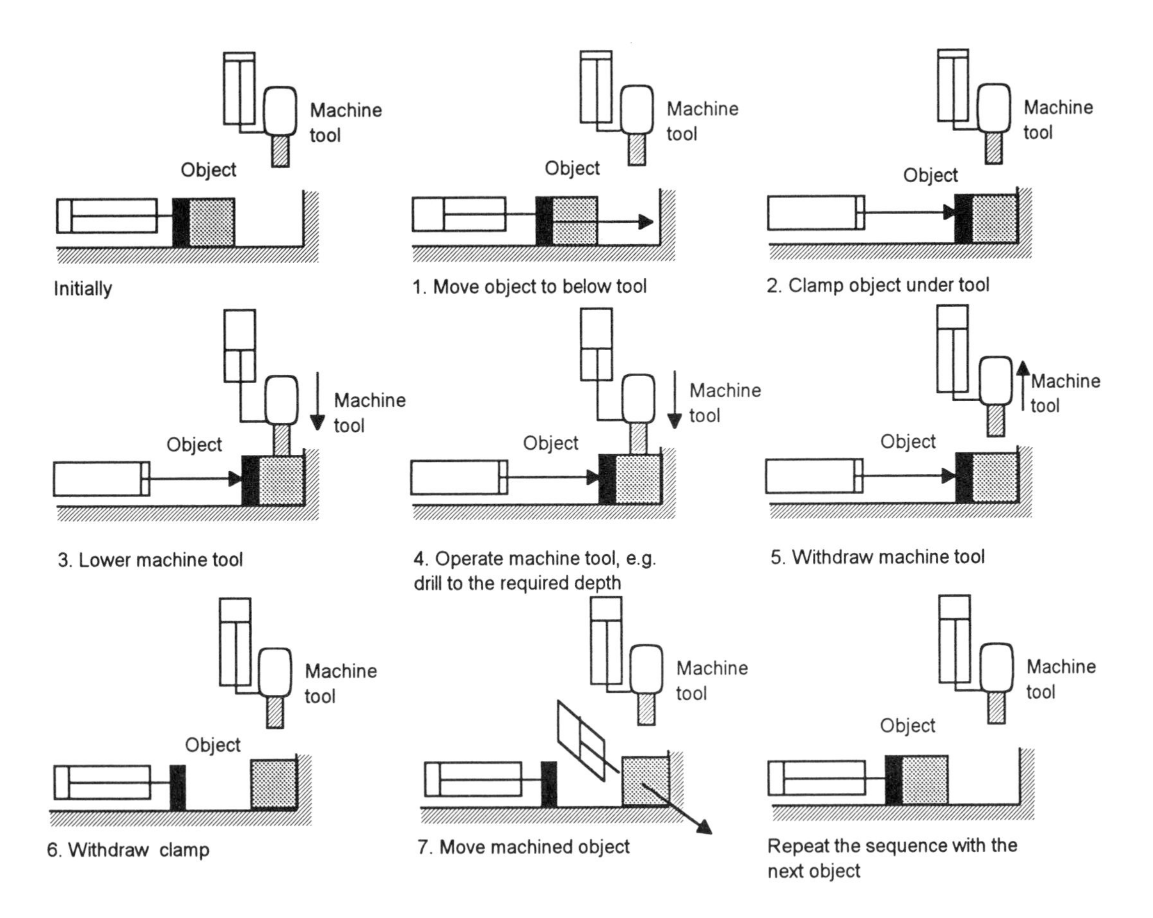

Figure 7.1 *Operations required for an automatic machine operation*

7.2 Directional control

The directional control of a piston rod in a cylinder, i.e. extending or retracting it, is by directing pressurised fluid to one side of the piston and exhausting the other side. Figure 7.2 illustrates this with control of a single-acting cylinder by the use of a 3/2 valve. Figures 7.3, 7.4 and 7.5 illustrate this with control of a double-acting cylinder. While two 3/2 valves could be used, as in Figure 7.3, it is more common to use a 5/2 valve (Figure 7.4) or a 5/3 valve (Figure 7.5).

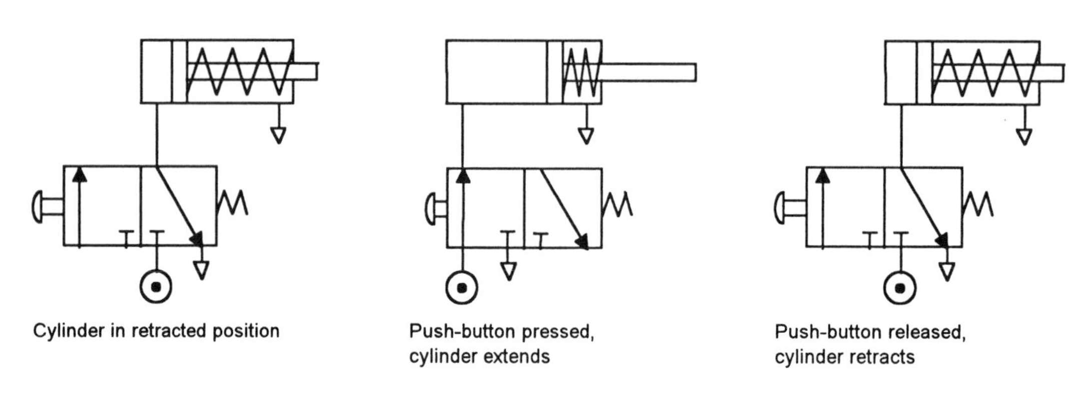

Figure 7.2 *Control of a single-acting cylinder*

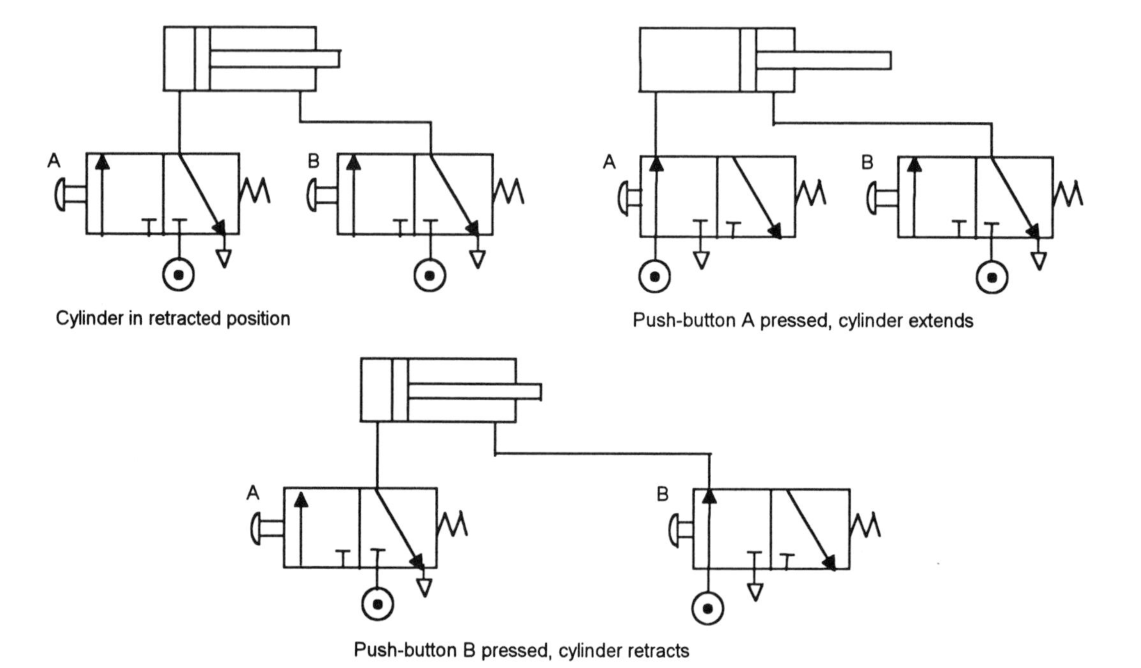

Figure 7.3 *Control of a double-acting cylinder by two 3/2 valves*

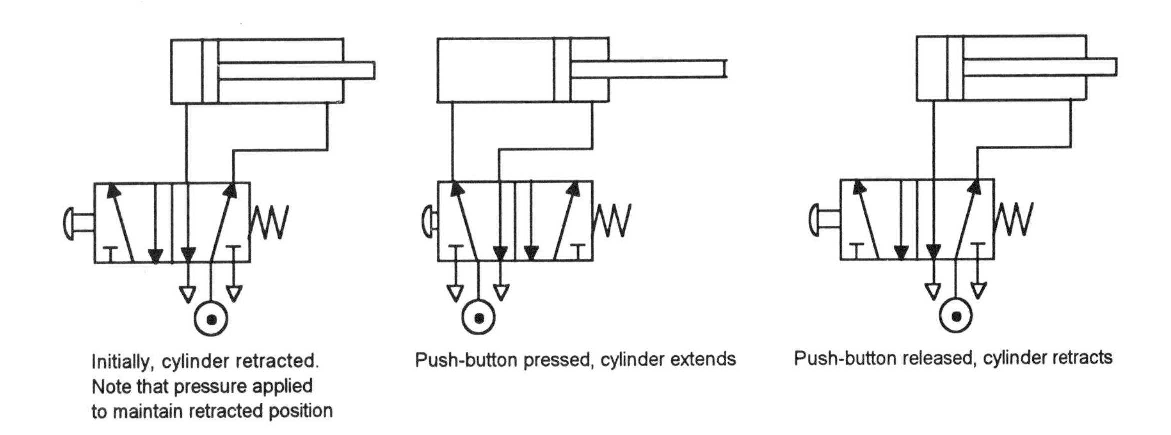

Figure 7.4 *Control of a double-acting cylinder by a 5/2 valve*

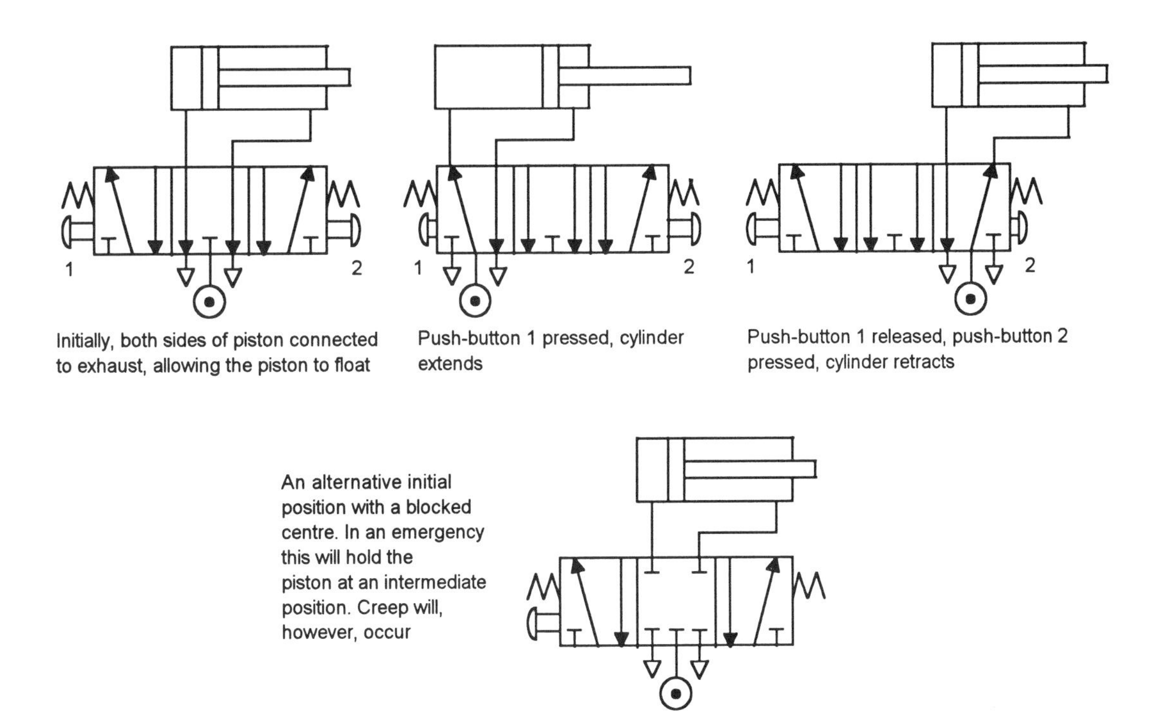

Figure 7.5 *Control of a double-acting cylinder by a 5/3 valve*

7.3 Speed control

The speed with which a piston moves in a cylinder depends on the rate at which fluid enters and leaves the cylinder. Thus we can control the speed by controlling the flow of fluid entering the cylinder, the so-termed *meter-in* situation (Figure 7.6(a)), or by controlling the rate at which fluid is allowed to leave the cylinder, the so-termed *meter-out* situation (Figure 7.6(b)) (also see Section 3.5.1). Meter-out control can be achieved by using a flow control valve and non-return valve in the line between the cylinder and the direction control valve or by using a flow control valve fitted to the exhaust port of the direction control valve.

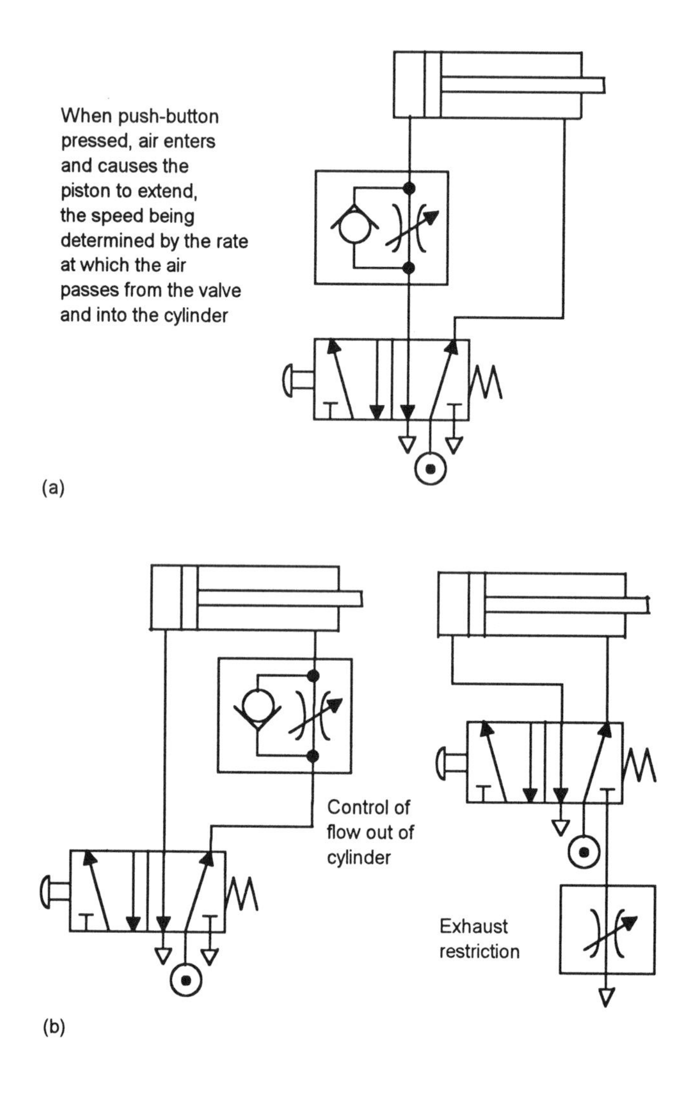

Figure 7.6 *(a) Meter-in control, (b) two alternatives for meter-out control*

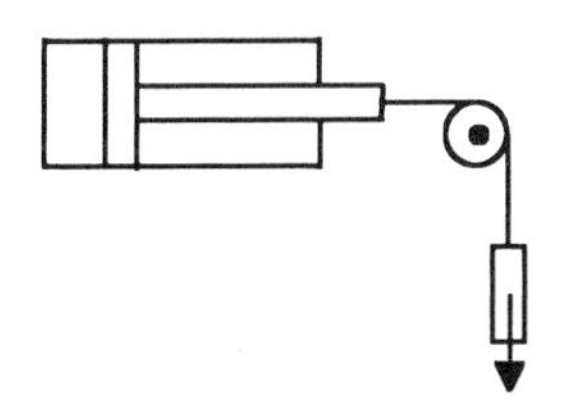

Figure 7.7 *Overhauling load*

Meter-out control has advantages when compared with meter-in control in that it enables the speed of the piston to be better controlled and the back pressure produced by restricting the flow of air out of the cylinder provides end cushioning. Meter-in control presents problems if the load can run away from the actuator, e.g. as in Figure 7.7, and in such situations meter-out should be used since the back pressure produced by restricting the flow of air out of the cylinder provides cushioning to stop such run-away occurring. With meter-in there is no such cushion.

With pneumatic systems, since air is compressible it is difficult to produce a very slow, steady movement of a piston in a cylinder using either meter-in or meter-out speed control. An alternative system which will give such control involves a hydro-pneumatic system. This is discussed in Section 7.3.1.

With hydraulic systems, meter-in and meter-out control can be used in the same way as shown in Figure 7.6. In addition, speed control can be exercised by *bleed-off*. A bleed-off valve allows a variable amount of the incoming fluid to be extracted and so control the amount entering the cylinder (Figure 7.8).

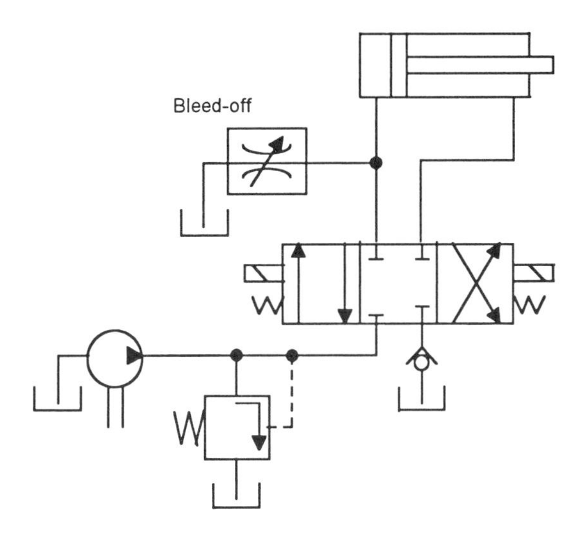

Figure 7.8 *Bleed-off speed control*

7.4 Hydro-pneumatic systems

Slow, smooth speed control of pneumatic cylinders is not achieved with a simple pneumatic actuator because of the compressibility of air. This can present problems where cylinders are used for a machining operation, since slow, smooth control is required. This problem can be overcome if compressed air is used to pressurise oil, the oil being almost incompressible.

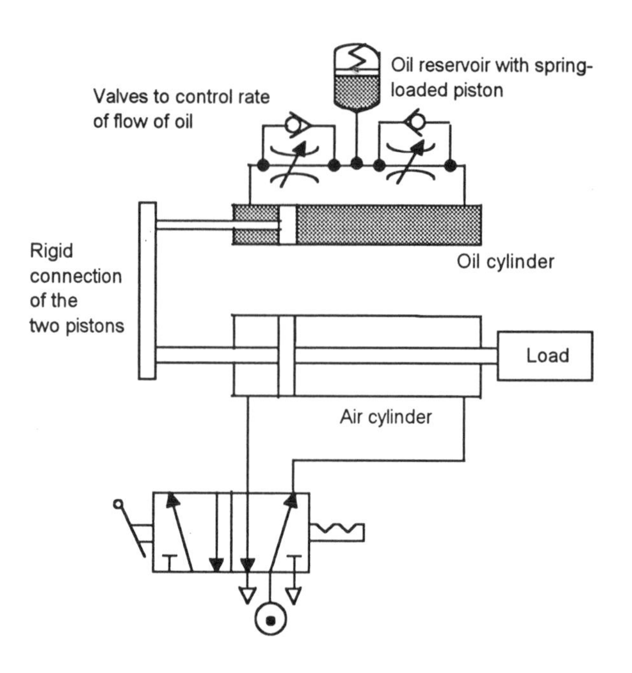

Figure 7.9 *Air–oil system for speed control*

Figure 7.9 shows an *air–oil system* for speed control of a pneumatic cylinder, this often being termed a *hydraulic-check system*. The piston in the air cylinder is directly connected to the piston in the oil cylinder so that they both move together. The movement causes oil to be displaced through the bypass from one side of the oil cylinder to the other. A variable restrictor valve is fitted in this bypass to control the rate at which oil can move between the two sides of the cylinder. This valve may be inward or outward checking or both. Such a system enables slow, smooth control to be achieved. The oil reservoir makes up differing oil volumes displaced by the hydraulic piston and makes up for oil leakages.

An alternative way of achieving control uses *air–oil reservoirs*. Figure 7.10 shows how such a reservoir can be used with an air–oil system that might be used in a garage to lift a car for servicing. The load acts on the piston in an oil cylinder. The oil output from the cylinder is controlled by a lock valve. When the valve is closed, no oil can escape from the cylinder and so the load is locked in position. When this valve is open, the oil feeds into an air–oil reservoir, displacing air which is exhausted through a pneumatic valve. Thus lowering the load is achieved by just switching the lock valve open. If the pneumatic valve is then switched, air pressure is applied to the air–oil reservoir and forces oil into the oil cylinder and hence lifts the load. Thus lifting the load requires the lock valve to be open and also the pneumatic valve to be switched.

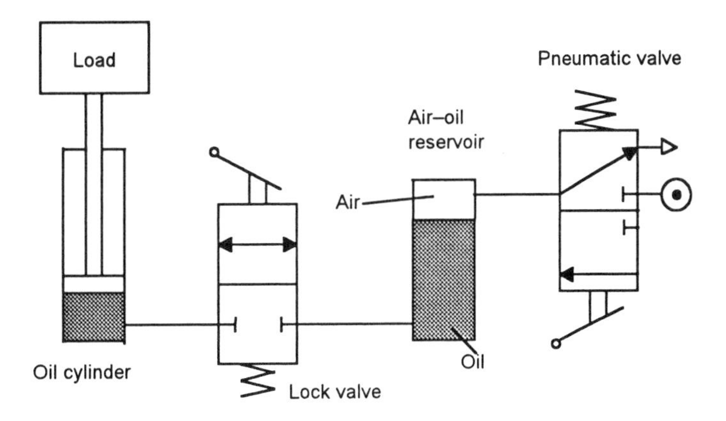

Figure 7.10 *Air–oil reservoir system*

The system in Figure 7.10 can be speed controlled by using variable restrictions in the feed from the oil end of the air–oil reservoir to the oil cylinder. Figure 7.11 shows how this principle can be extended to the speed control of both the extend and retract strokes of a double-acting oil cylinder. This employs two air–oil reservoirs and flow controls for the air.

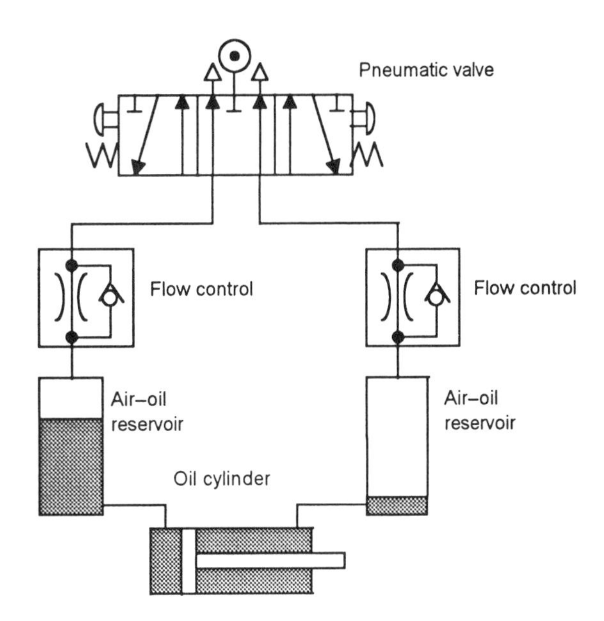

Figure 7.11 *Two air–oil reservoirs used to control a double-acting oil cylinder*

7.4.1 Air–oil intensifier

These involve a pneumatic pressure being used to produce a higher hydraulic pressure. Such a situation may be required to power single-acting, short stroke, hydraulic cylinders for machine operations such as clamping or bending. Figure 7.12 shows the basic principle of such a device, a large bore pneumatic cylinder being used to drive a smaller bore hydraulic cylinder. The force exerted on the piston by the pneumatic pressure p_a is p_aA_a, where A_a is the area of the pneumatic cylinder piston. This same force will be exerted on the hydraulic cylinder piston. Thus $p_aA_a = p_hA_h$, where p_h is the resulting hydraulic pressure and A_h the area of the hydraulic cylinder piston. Thus, since the pistons are circular:

$$\text{intensification ratio} = \frac{p_h}{p_a} = \frac{d_a^2}{d_h^2} \qquad [1]$$

where d_a is the diameter of the pneumatic cylinder piston and d_h the diameter of the hydraulic cylinder piston.

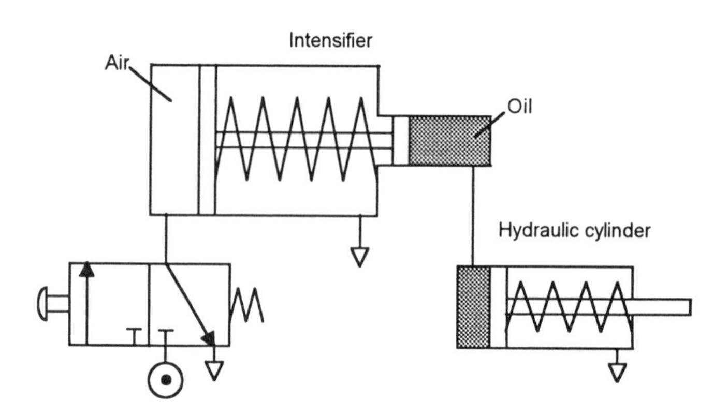

Figure 7.12 *An air–oil intensifier*

Example

An air–oil intensifier has an intensification ratio of 10:1. If the maximum air supply gauge pressure is 4 bar, what will be the maximum oil gauge pressure produced?

Using equation [1]:

$$\text{maximum oil pressure} = \text{intensification ratio} \times \text{pneumatic pressure}$$

$$= 10 \times 4 = 40 \text{ bar}$$

7.5 Pilot operation

If there is a long distance between a pneumatic valve and the cylinder it is controlling, then the air consumption per cycle can be large due to all the air in the connecting pipe work that has to be moved. A large air consumption means a large bore valve. This can be avoided if the valve is placed very close to the cylinder and pilot operation used to actuate the valve. The pilot signal can then be transmitted over the long distance.

Figure 7.13 illustrates this with a simple pneumatic circuit involving push-button control of a valve producing a pilot signal which is then transmitted over a distance to the directional control valve which is situated close to the cylinder. When the push-button is pressed, pilot pressure is produced which then actuates the directional control valve and switches pressure into the cylinder to cause it to extend. When the push-button is released, the pilot pressure signal ceases and the directional control valve reverts to its at rest position, so causing the cylinder to retract. Figure 7.14 shows a pneumatic circuit for double pilot control of a cylinder, the extension and the retraction being controlled by push-button operated valves producing pilot signals for extension and retraction. The directional control valve here remains in the last state it has been put by a pilot signal.

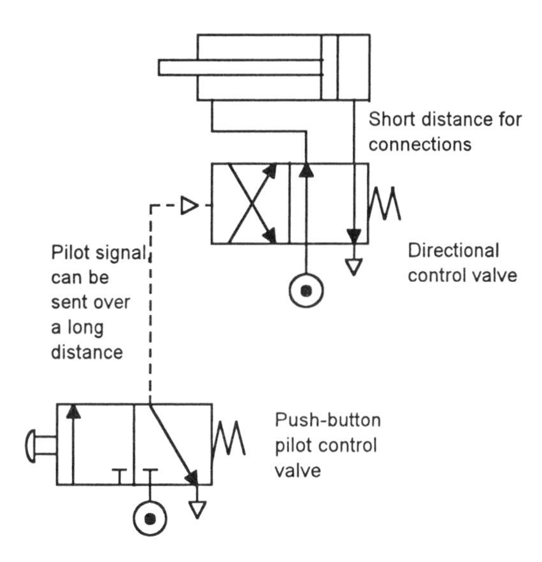

Figure 7.13 *Pilot control*

7.5.1 Trapped signals

Problems can occur with double pilot control when both pilot signals exist simultaneously on opposite ends of a directional valve. This can occur when pilot signal fluid is trapped and cannot escape, being given no exhaust connection. Figure 7.15 illustrates this. Pilot signals must be able to exhaust and thus 2/2 valves are not suitable for use with pilot signals.

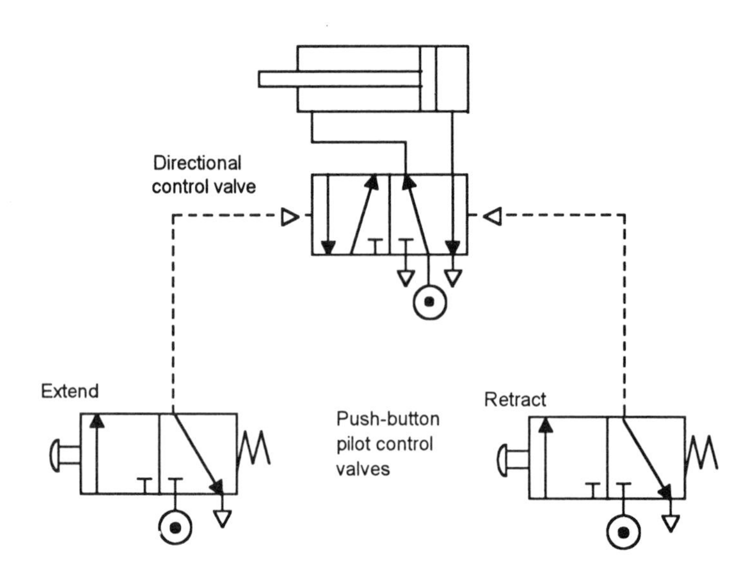

Figure 7.14 *Double pilot control*

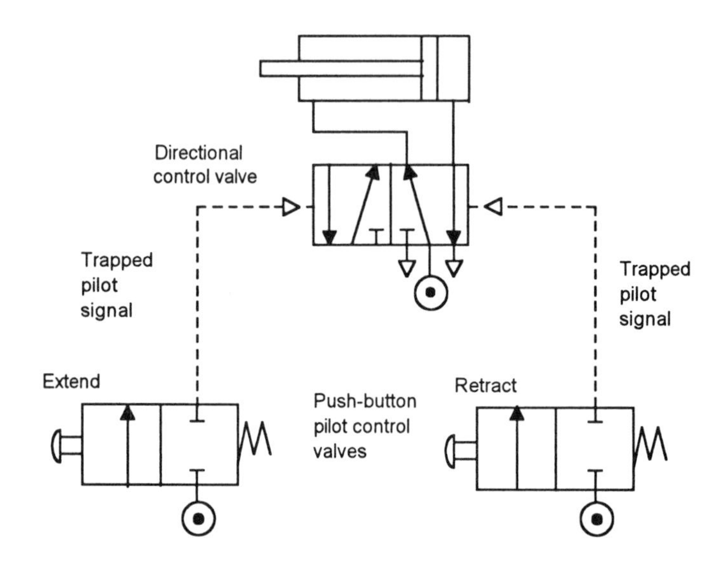

Figure 7.15 *Locked-in pilot signal fluid*

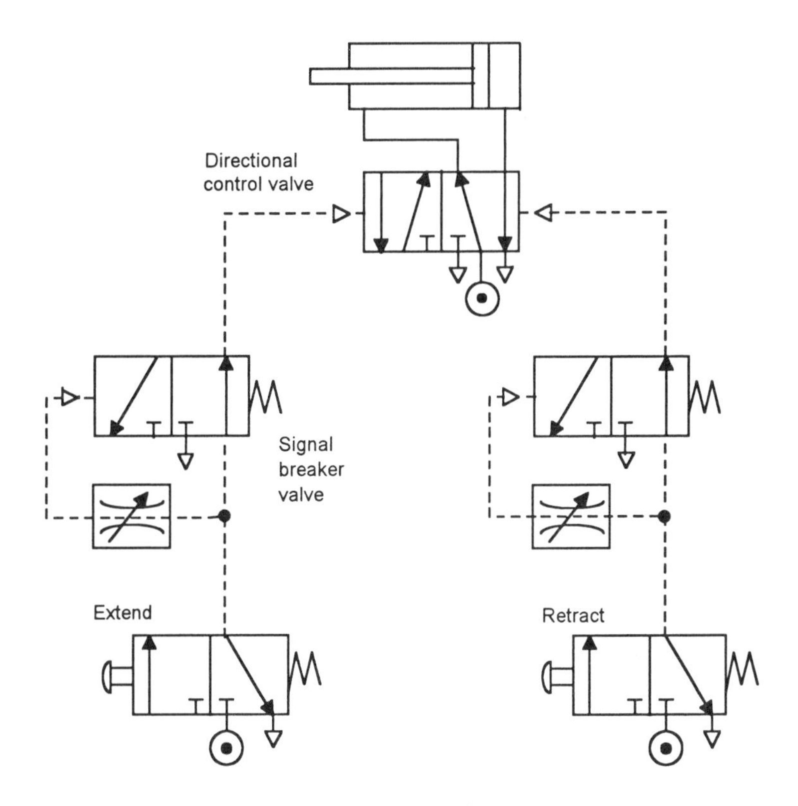

Figure 7.16 *Circuit with signal breaker valves*

One way of overcoming this problem of trapped pilot signals is by the use of a *signal breaker valve* (Figure 7.16). When the push-button is pressed on the pilot control valve for extend, the pilot signal operates the directional control valve and also, some time later because of the restriction, the signal breaker valve. The result is that the signal breaker valve vents the pilot signal after it has operated the directional control valve. The signal breaker remains in its switched position until the push-button on the extend valve is released. Trapped signals are thus prevented.

7.6 Automatic operation

Position sensors can be used to operate valves, and so automate processes, by initiating the next step of a sequence. Figure 7.17 shows a simple circuit for the automatic return of a cylinder when it reaches the end of its stroke. The extend valve is switched by pressing the push-button on the start valve and extension occurs. When the piston rod reaches the end of its stroke, it trips a roller limit switch which then switches the retract valve so that the piston then retracts.

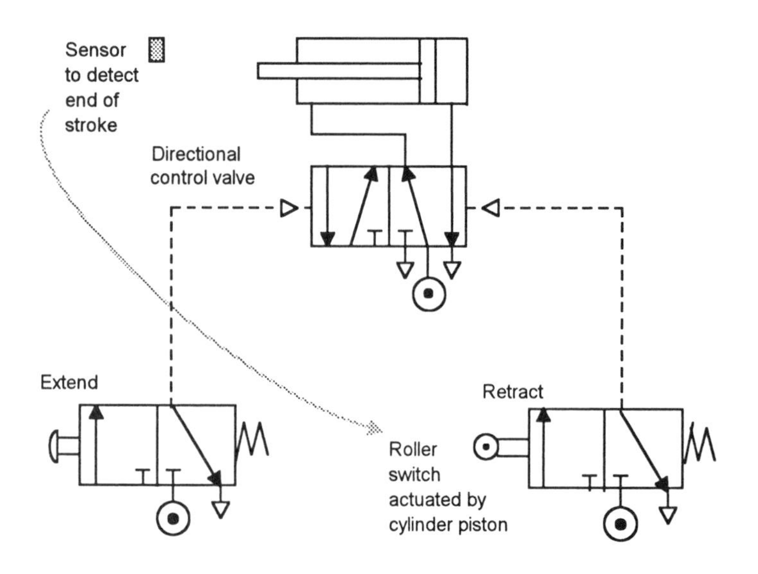

Figure 7.17 *Automatic return*

7.6.1 Time control

Figure 7.17 is an example of position-sensed control events happening at particular positions. Another situation that can be encountered is time control events happening after particular times. Time delay valves (see Section 3.5.2) can be used to delay pilot signals. Such valves can be built up from a flow control valve with free reverse flow and an air pilot-operated normally closed 3/2 valve. The flow control valve slows down the time taken for pressure to build up and hence switching action occurring. This time can be increased by the inclusion of a reservoir. Switching does not occur until the pressure has built up to a value which gives a force greater than the spring used with the valve.

Figure 7.18 shows a circuit, with the time delay circuit built up from its components (though they can be purchased as single entities), where the movement of the piston in the cylinder takes place after a time delay from when the push-button is pressed. After the push-button is pressed, the pilot pressure slowly builds up until it is big enough to switch the positions for the time delay valve. This then allows the system pressure to be applied, via the 4/2 valve to the cylinder.

Figure 7.19 shows how time delay can be combined with position sensing. The cylinder extends when the start valve is actuated and position sensor 1 is actuated. When position sensor 2 is actuated, a time delay occurs before the direction control valve switches position and retraction of the piston occurs. Thus the cylinder remains in the fully extended position for this time. This type of operation might be used to clamp an object for a fixed time.

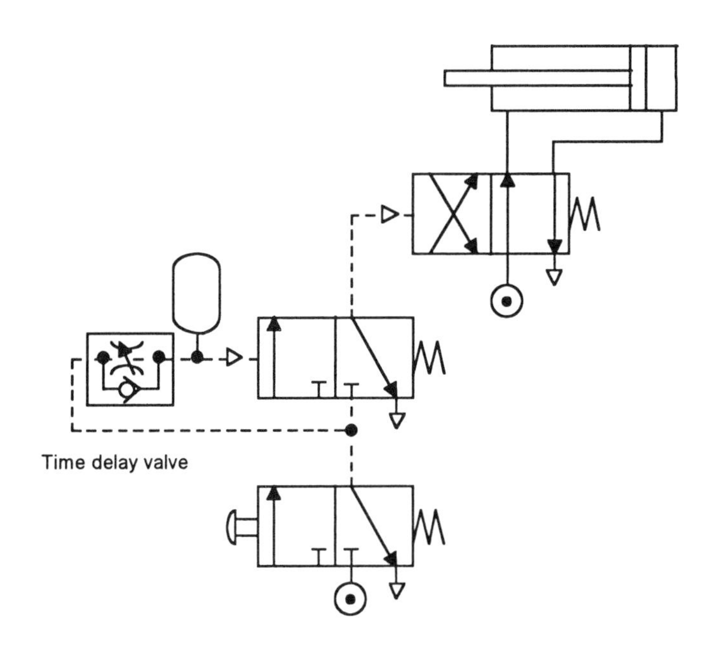

Figure 7.18 *Time delay*

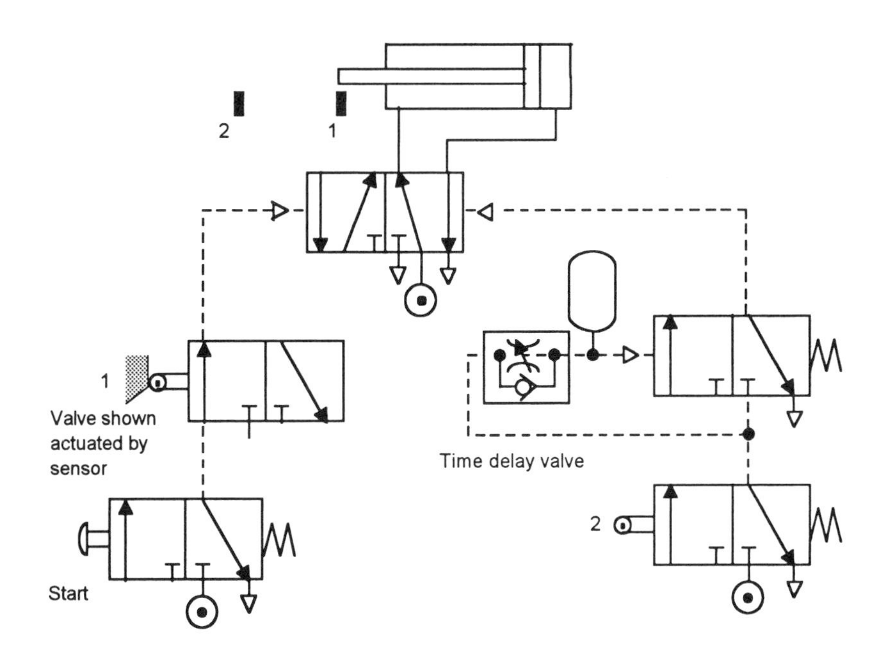

Figure 7.19 *Hold extension for a fixed time*

7.6.2 Pressure-sensed control

Pressure sensing is used where a specific pressure is required for a machine operation such as clamping. The pressure-sensing valve is basically a 3/2 valve which is normally closed (see Section 3.3 and Figure 3.18). The valve switches when the force due to the pilot pressure exceeds the spring force. Figure 7.20 illustrates this with a circuit where there is both positional and pressure control. When the start push-button is pressed, the pressure to the right of the piston rises and extension occurs. Also, the pilot pressure on the pressure-sensing valve increases. When the pressure in the cylinder reaches the required value, the pressure-sensing valve switches position. Then, when the position sensor is actuated, the pressure-sensing valve applies pilot pressure to the directional control valve and it switches and the extension stops.

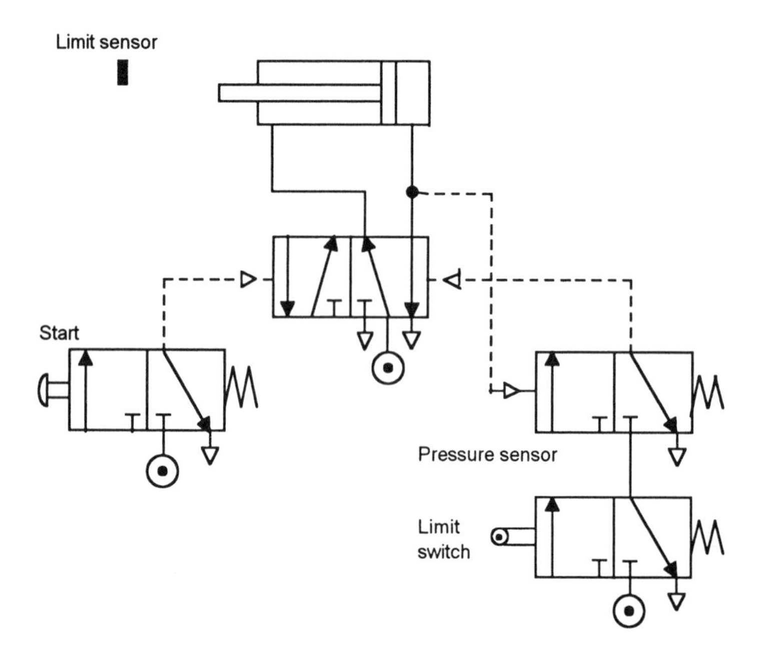

Figure 7.20 *Pressure and position control*

7.7 Shutdown systems

One way of stopping cylinder action is to exhaust the air supply or return the hydraulic fluid to its reservoir. However, this might leave a device in an unsafe position. A safer way of achieving shutdown which allows a cylinder to be parked in an extended or retracted position is shown in Figure 7.21. In the event of an emergency requiring shutdown, the push-button of the shutdown valve is pressed. This shuts off the pilot signal and the spring then switches the directional control valve to give, in the situation shown in the figure, an automatic retraction of the cylinder.

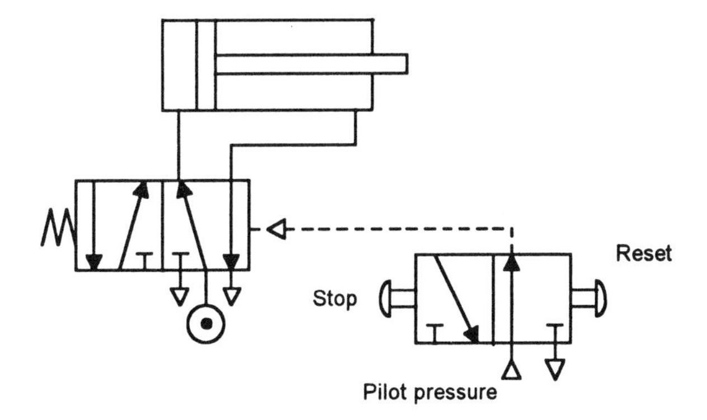

Figure 7.21 *Shutdown*

Figure 7.22 shows how a cylinder can be used to shut down a cylinder to a mid-stroke position. The emergency stop push-button is used to remove the pilot signals so that the 5/3 valve is switched to its mid-switching position by the springs.

Figure 7.23 shows a shutdown circuit involving a double pilot pressure-operated valve. When the stop push-button is pressed, the pilot signal to the 5/2 valve becomes trapped and so drives the cylinder to the fully retracted position. The figure shows the switching positions with the press-button not switched and then when switched.

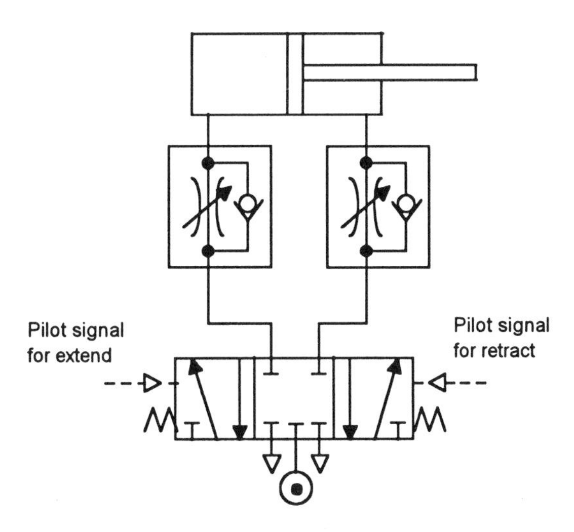

Figure 7.22 *Shutdown to mid position*

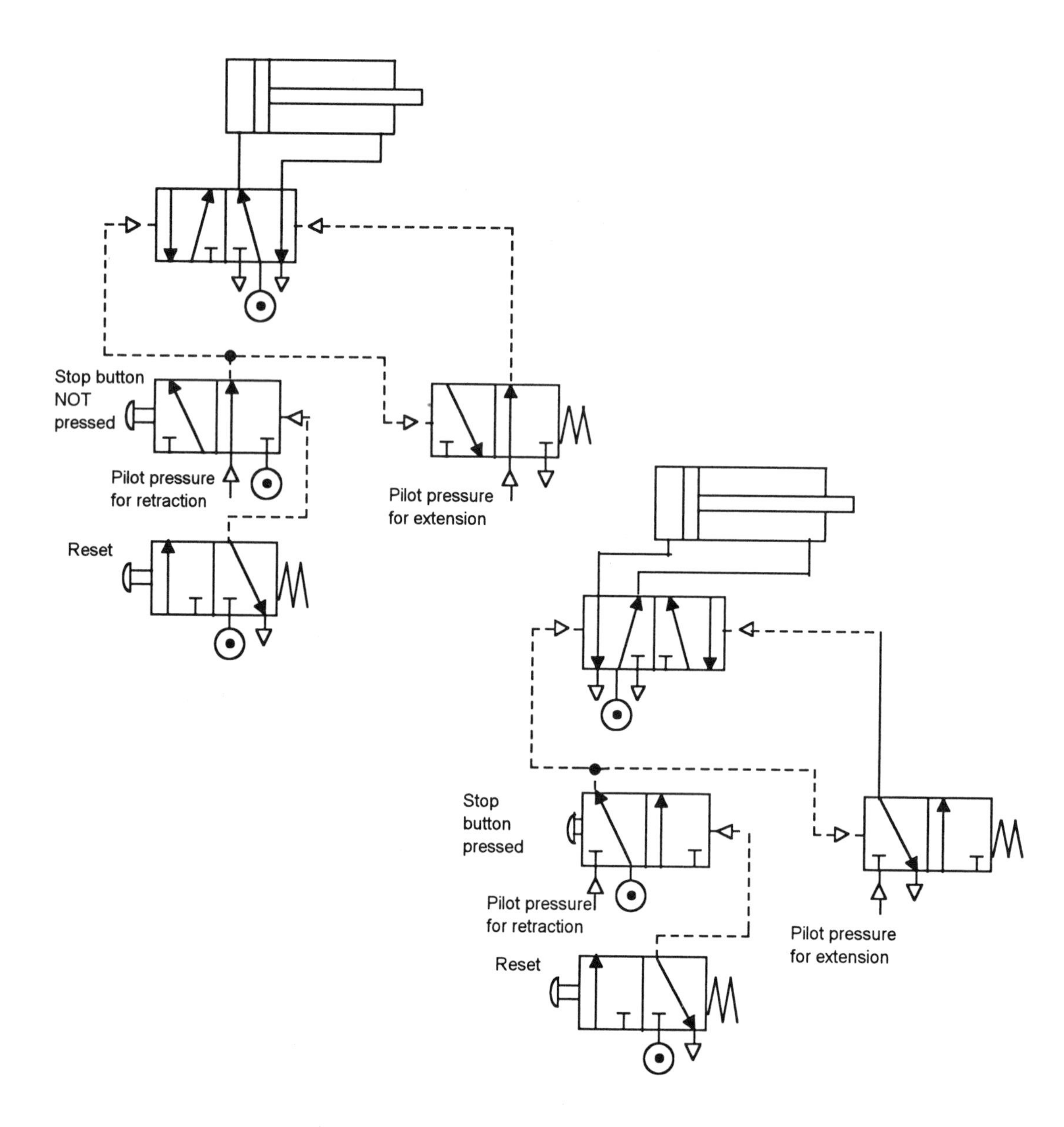

Figure 7.23 *Shutdown*

7.8 Sequential circuits

Situations often occur where it is necessary to activate a number of cylinders in some sequence. Thus event 2 might have to start when event 1 is completed, event 3 when event 2 has been completed. For example, we might have: only when cylinder A is fully extended (event 1) can cylinder B start extending (event 2), and cylinder A can only start retracting (event 3) when cylinder B has fully extended (event 2). In discussions of sequential control it is common practice to give each cylinder a reference letter A, B, C, D, etc., and to indicate the state of each cylinder by using a + sign if its extended or a – sign if retracted. Thus a sequence of operations might be shown as A+, B+, A–, B–. This indicates that the sequence of events is cylinder A extend, followed by cylinder B being extended, followed by cylinder A retracting, followed by cylinder B retracting. Figure 7.24 illustrates this with a displacement-step diagram. Figure 7.25 shows a circuit that could be used to generate this displacement-event diagram for two cylinders A and B.

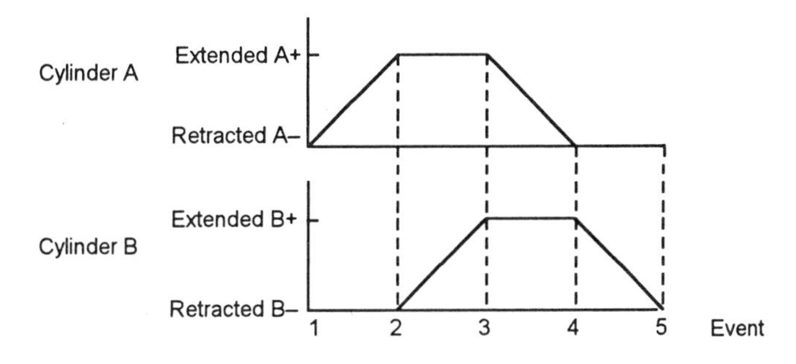

Figure 7.24 *Displacement-event diagram*

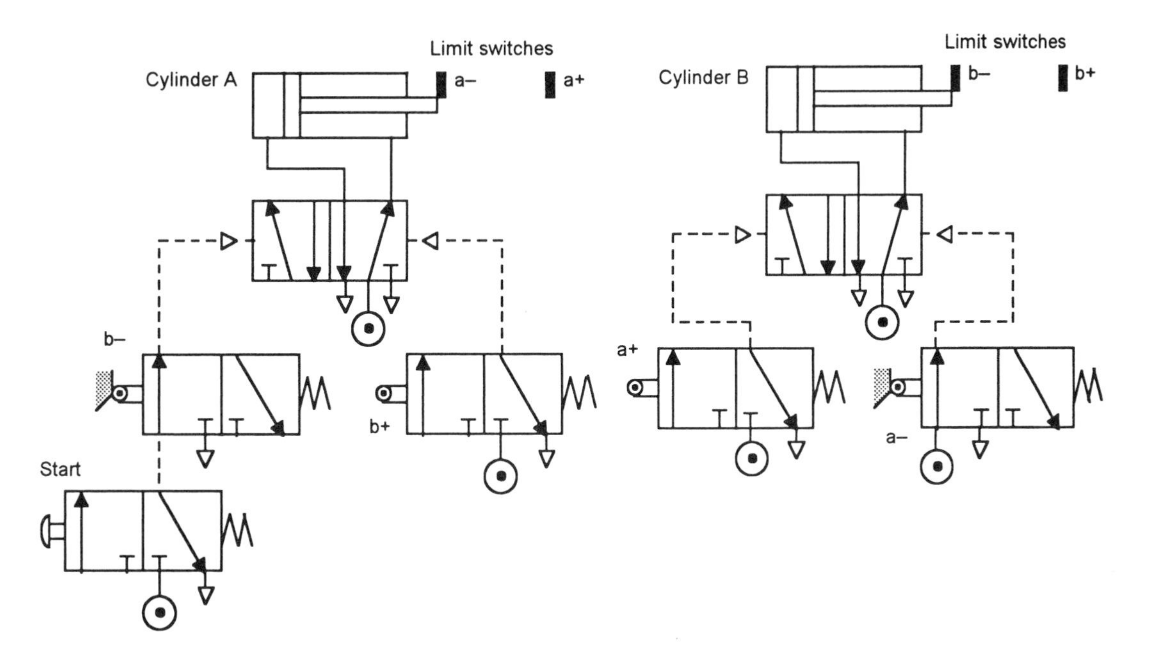

Figure 7.25 *Two-actuator sequential operation*

In order to generate the displacement-event diagram of Figure 7.24 the sequence of operations with Figure 7.25 is:

Event 1

1 Start push-button pressed.

2 Cylinder A extends, releasing limit switch a–.

Event 2

3 Cylinder A fully extended, limit switch a+ operated to start B extending.

4 Cylinder B extends, limit switch b– released.

Event 3

5 Cylinder B fully extended, limit switch b+ operated to start cylinder A retracting.

6 Cylinder A retracts, limit switch a+ released.

Event 4

7 Cylinder A fully retracted, limit switch a– operated to start cylinder B retracting.

8 Cylinder B retracts, limit switch b+ released.

Event 5

9 Cylinder B fully retracted, limit switch b– operated to complete the cycle.

The cycle can be started again by pushing the start button. If we wanted the system to run continuously then the last movement in the sequence would have to trigger the first movement.

Example

Design a sequential circuit for two actuators A and B to give the sequence A+, B+, B–, A–. This might be for a machine where A extends and clamps the component and B extends and stamps a design onto it, before B and then A retracting; the sequence then being clamp, lower stamp, raise stamp, unclamp.

The sequence required is:

1 Start push-button pressed.

2 A extends, limit switch a– released.

3 A fully extended, limit switch a+ operated to start B extending.

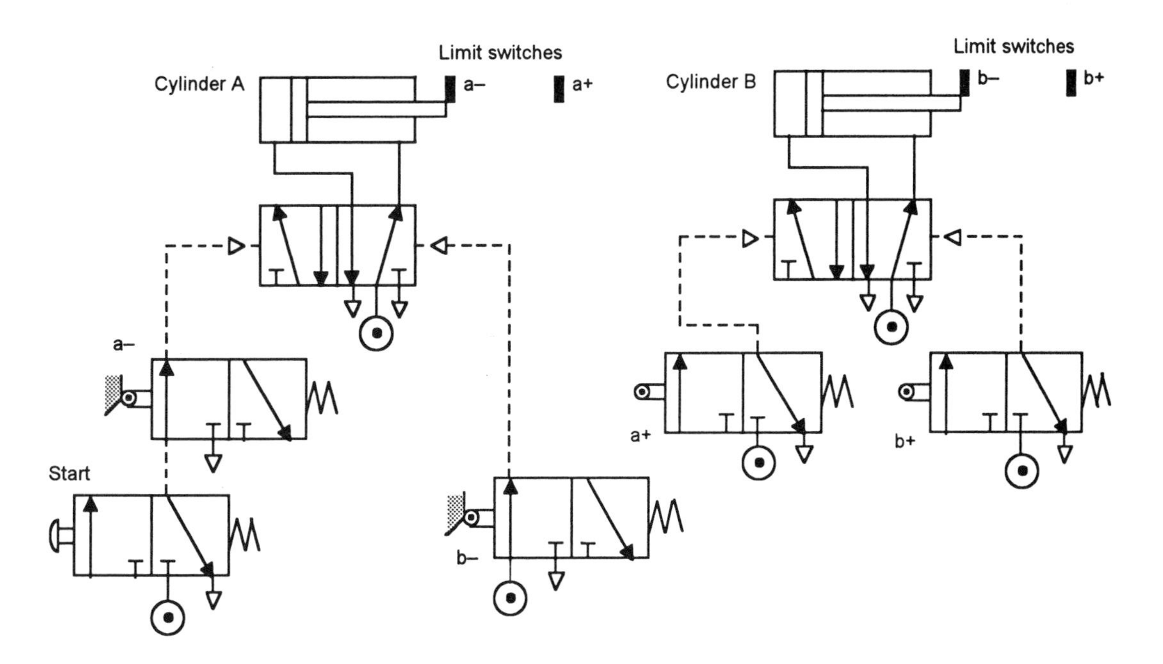

Figure 7.26 *Example*

4 B extends, limit switch b– released.

5 B fully extended, limit switch b+ operated to start B retracting.

6 B retracts, limit switch b+ released.

7 B fully retracted, limit switch b– operated to start A retracting.

8 A retracts, limit switch b+ released.

9 A fully retracted, limit switch b– operated to complete the cycle.

Figure 7.26 shows the resulting circuit. A problem with this circuit is that pilot signals are trapped. Thus either signal breakers need to be included, in the b– and the a+ lines, or an alternative circuit designed to avoid the problem. Figure 7.27 shows Figure 7.26 amended to include signal breakers.

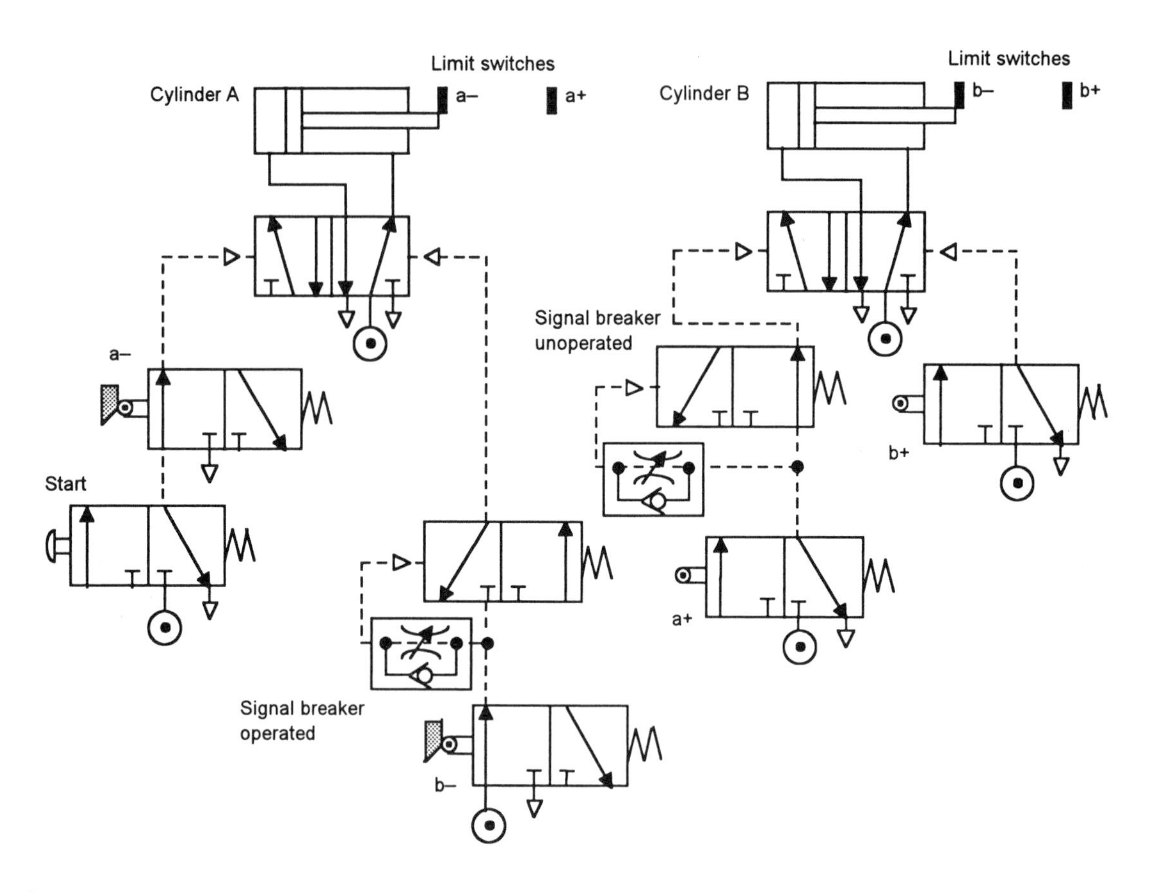

Figure 7.27 *Example*

7.9 Cascade control

Cascade control is a technique used to give sequential circuits by switching on and off the air supply to valves in groups. It avoids the problem of trapped signals. The procedure is to divide the sequence program into groups as follows, note that Roman numerals are used for the group numbers:

1 No cylinder letter must appear more than once in any group. Thus if we have the sequence A+, B+, B–, A– we can have groups:

A+, B+,	B–, A–
Group I	Group II

2 For continuous cycling, a cycle should be regarded as part of a continuous program and thus a cascade group may continue into the next cycle. Thus if we have the cycle sequence A+, B+, C+, B–, A– , C– we could have, with a continuous program, the groups:

A+, B+, C+, B–, A– , C– *or* A+, B+, C+, B–, A– , C–
Group I Group II Group I Group II Group I

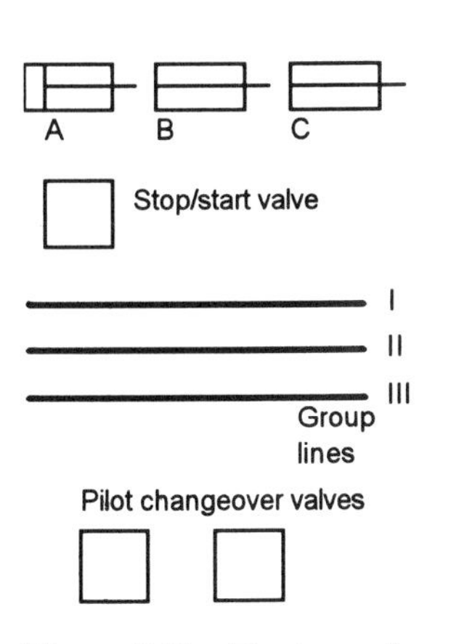

Figure 7.28 *The ingredients of a cascade circuit*

The aim should be to minimise the number of groups required.

3 A start/stop valve is needed. If the cycling is to be continuous, use the last operation to supply a signal to start the sequence over again. The start/stop valve should be in the line that selects group I.

4 Draw the pilot changeover valves for the groups at the bottom of the circuit (Figure 7.28). Draw group pressure lines above the cascade valves with initially the pilot air on group I, these lines being termed 'bus bars'.

5 Each group is activated in turn. Figure 7.29 shows with 5/2 cascade valves how they are connected to the bus bars and the selection signals.

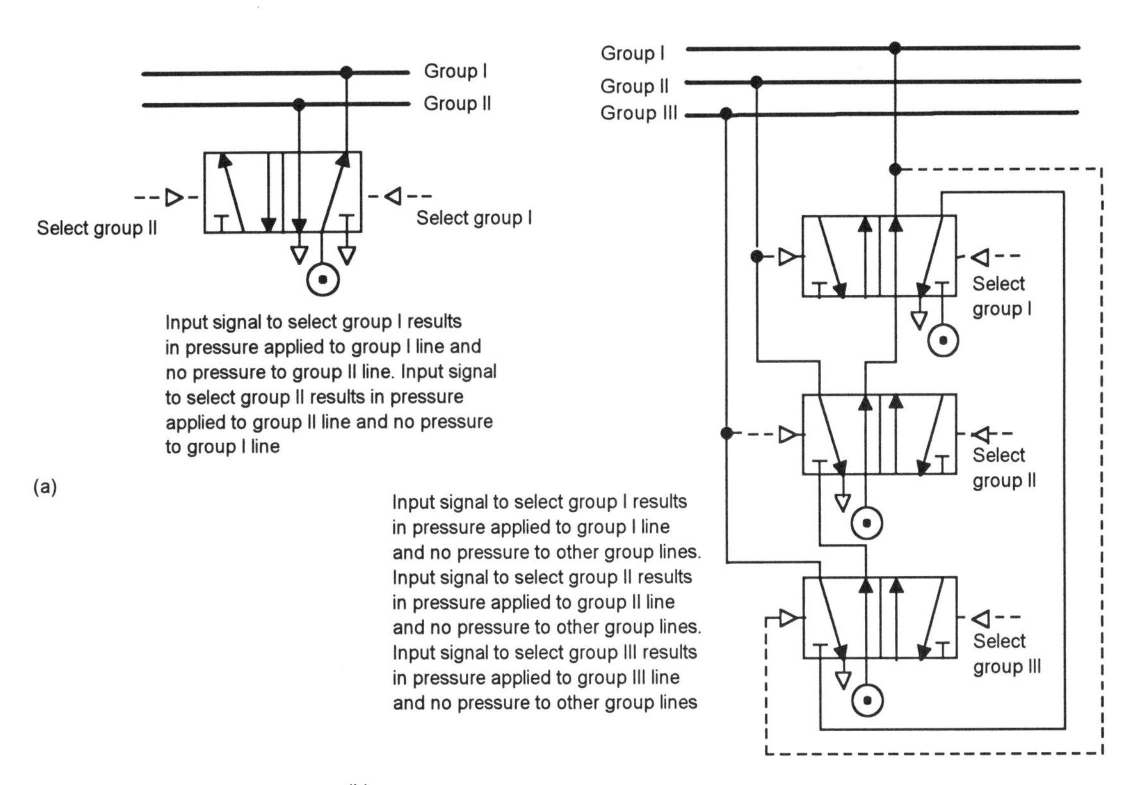

Figure 7.29 *Cascade valves: (a) two group, (b) three group*

6 The first function in each group is signalled directly by that group supply. The last trip valve to be operated in each group will by supplied with main air and cause the next group to be selected. The other trip valves in each group will be operated with air from their respective group and initiate the next function.

Figure 7.30 illustrates the above with a circuit to generate the cycle A+, B+, B–, A–, this having been divided into two groups – group I of A+, B+ and group II of B–, A–.

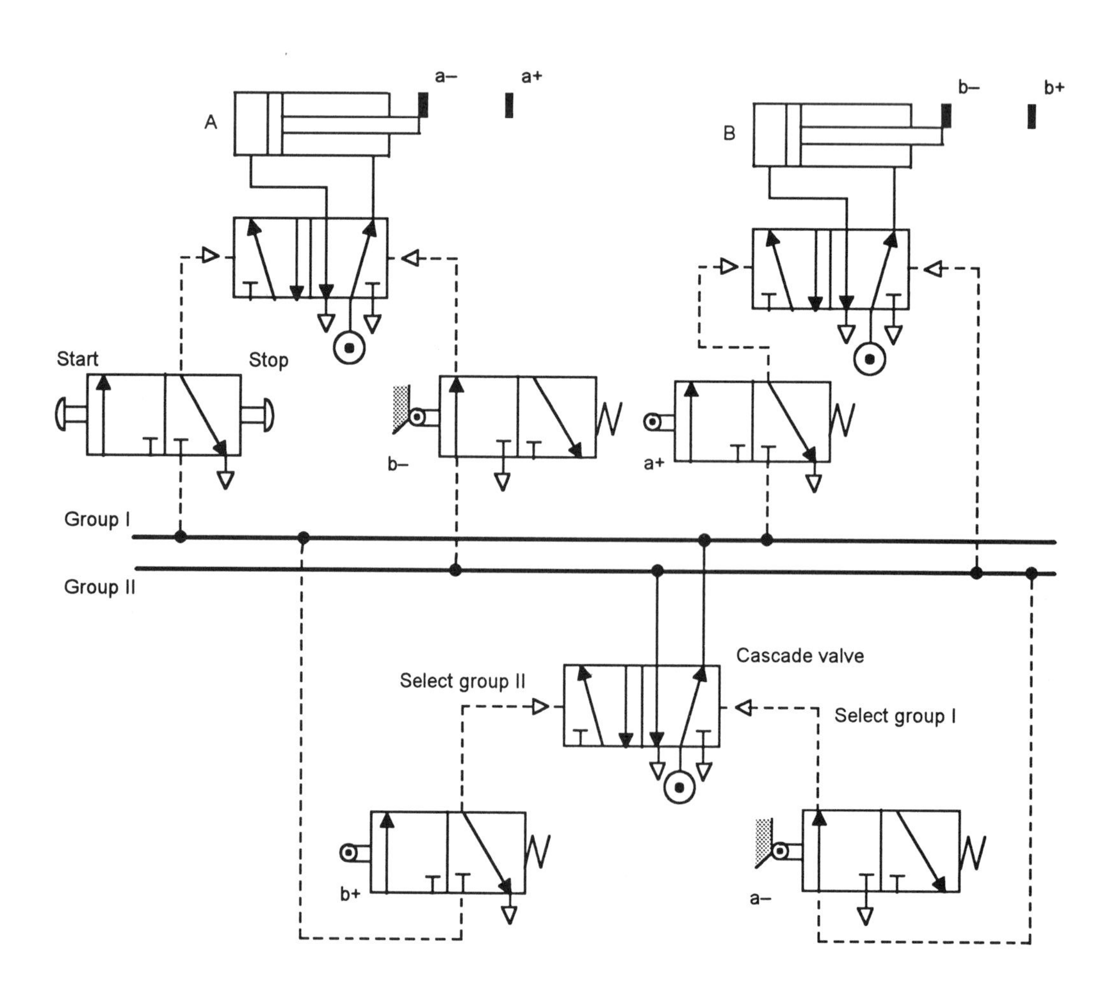

Figure 7.30 *A+, B+, B–, A–*

The circuit can be traced as:

1 Set start/stop valve to start and generate a command to select group I.

2 Group I gives the command for pressure to be applied to give A+ and cylinder A extends.

3 Limit sensor a+ operates and generates the command for group I pressure to be applied to give B+ and cylinder B extends.

4 Limit sensor b+ operates and generates a command to disconnect group I and select group II.

5 Group II gives the command B– and cylinder B retracts. Note that because group I has been switched off, there is no opposing signal from a+.

6 Limit sensor b– operates and gives the command A–. Cylinder A retracts.

7 Limit sensor a– operates and gives the command to start the sequence again.

If at any time the start/stop valve is switched to stop, the current cycle will be completed before the operation comes to a halt.

For more complex pneumatic systems, while the cascade form of circuit can be used, the trend is to use solenoid valves with a programmable logic controller (PLC). This method is discussed in Chapter 9.

Problems

1 Explain the significance of the differences in the valves in Figure 7.31(a) and (b) on the control of the cylinder.

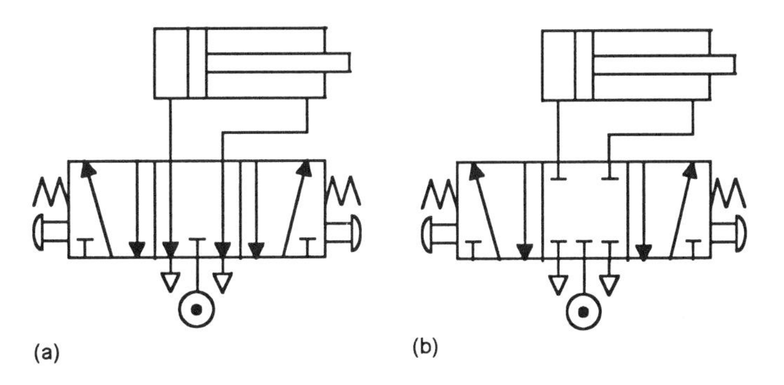

Figure 7.31 *Problem 1*

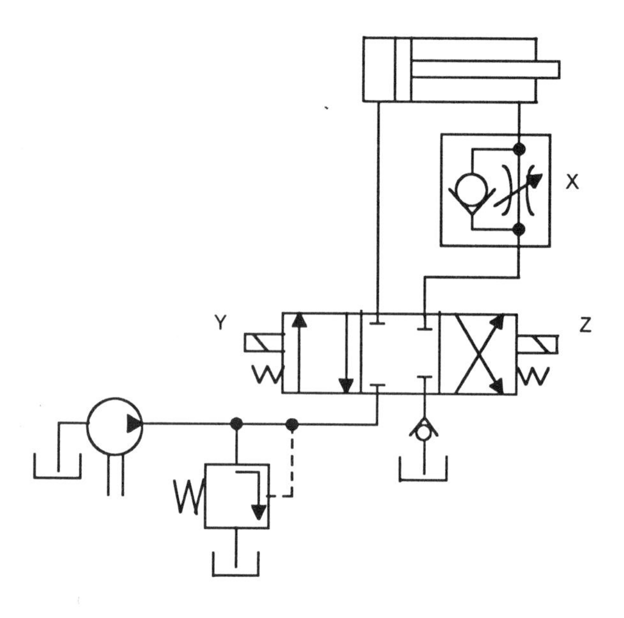

Figure 7.32 *Problem 2*

2 Identify the function of valve X in Figure 7.32 and explain what happens when (a) solenoid Y and (b) solenoid Z are energised.

3 Devise a system that can be used to give slow, smooth control of a double-acting cylinder for the extension but a faster, smooth-controlled return.

4 An air–oil intensifier has an intensification ratio of 10:1. If the maximum air supply gauge pressure is 5 bar, what will be the maximum oil gauge pressure produced?

5 What is the diameter of the hydraulic piston with an air–oil intensifier if it has an intensification ratio of 10:1, exerts a force of 5 kN, and is supplied with air at a gauge pressure of 5 bar?

6 Describe what happens when the push-button is pressed for the circuit shown in Figure 7.33.

7 Devise an emergency shutdown system which will park a cylinder in the fully retracted position when an emergency stop-button is pressed.

8 For the circuit shown in Figure 7.34 state which valves need to be operated to (a) extend the piston, (b) retract the piston.

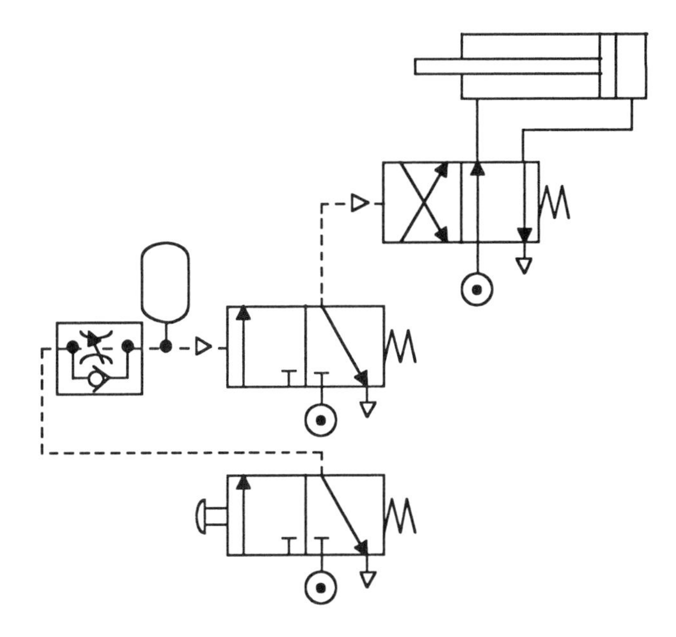

Figure 7.33 *Problem 6*

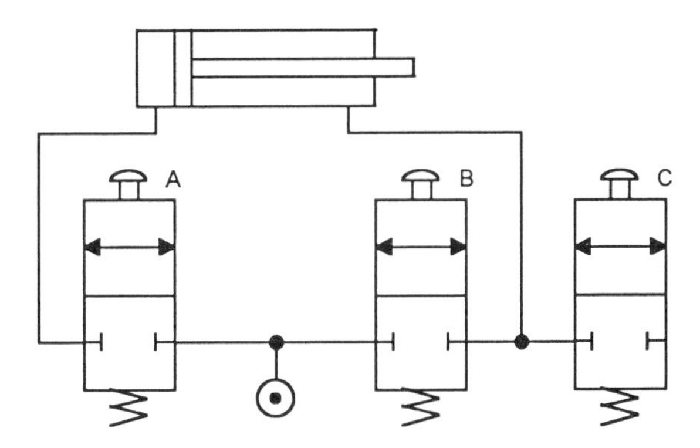

Figure 7.34 *Problem 8*

9 A pneumatically operated machine is required to have cylinder A push a component from a magazine into position, this action also ejecting the previous component, cylinder B clamps it, cylinder A retracts and then B unclamps so that the cycle can be repeated. Specify the actuator sequence required and suggest a circuit that could be used.

10 Determine the cylinder sequence for the circuit shown in Figure 7.35.

11 Design a circuit to give the sequence A+, B+ and then simultaneously A– and B–.

12 State for the circuit shown in Figure 7.36, the sequence in which the cylinders will operate.

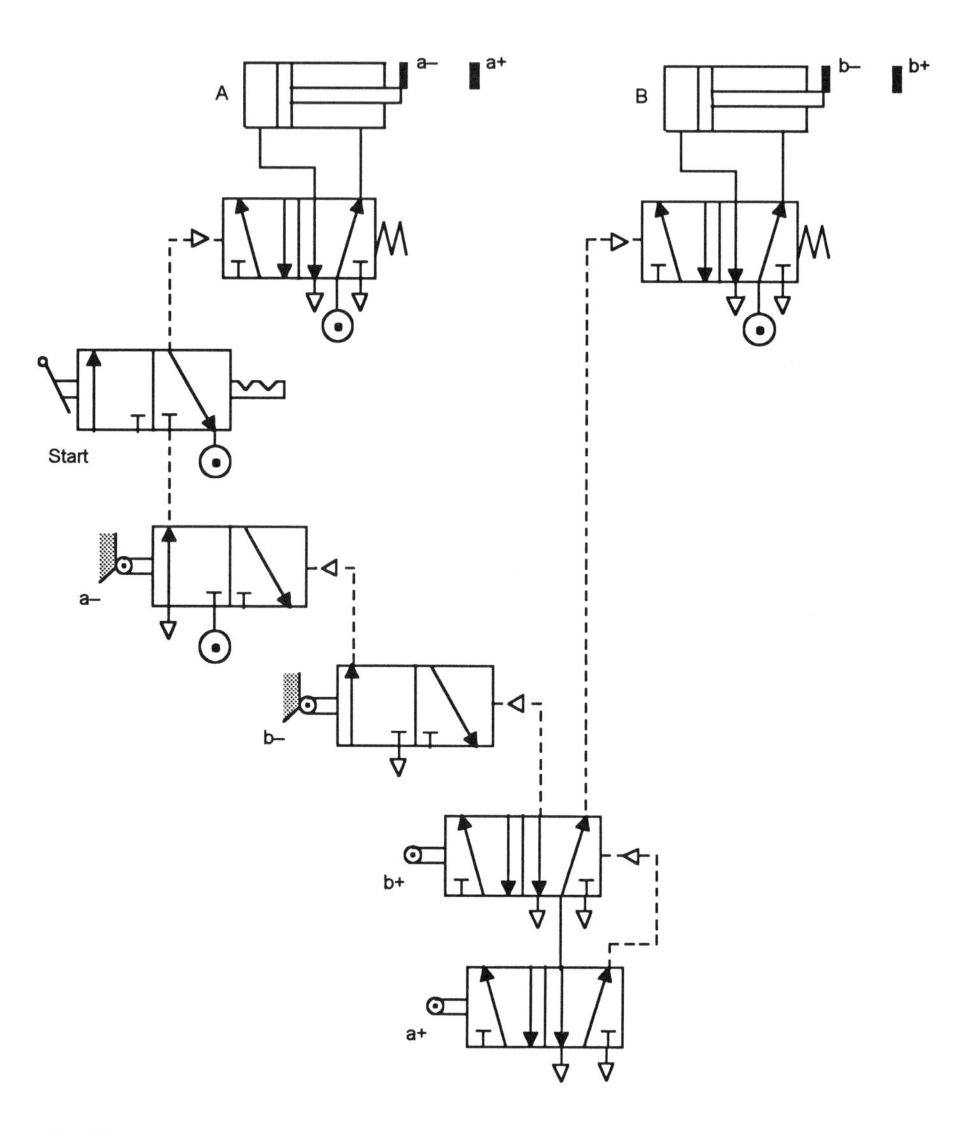

Figure 7.35 *Problem 10*

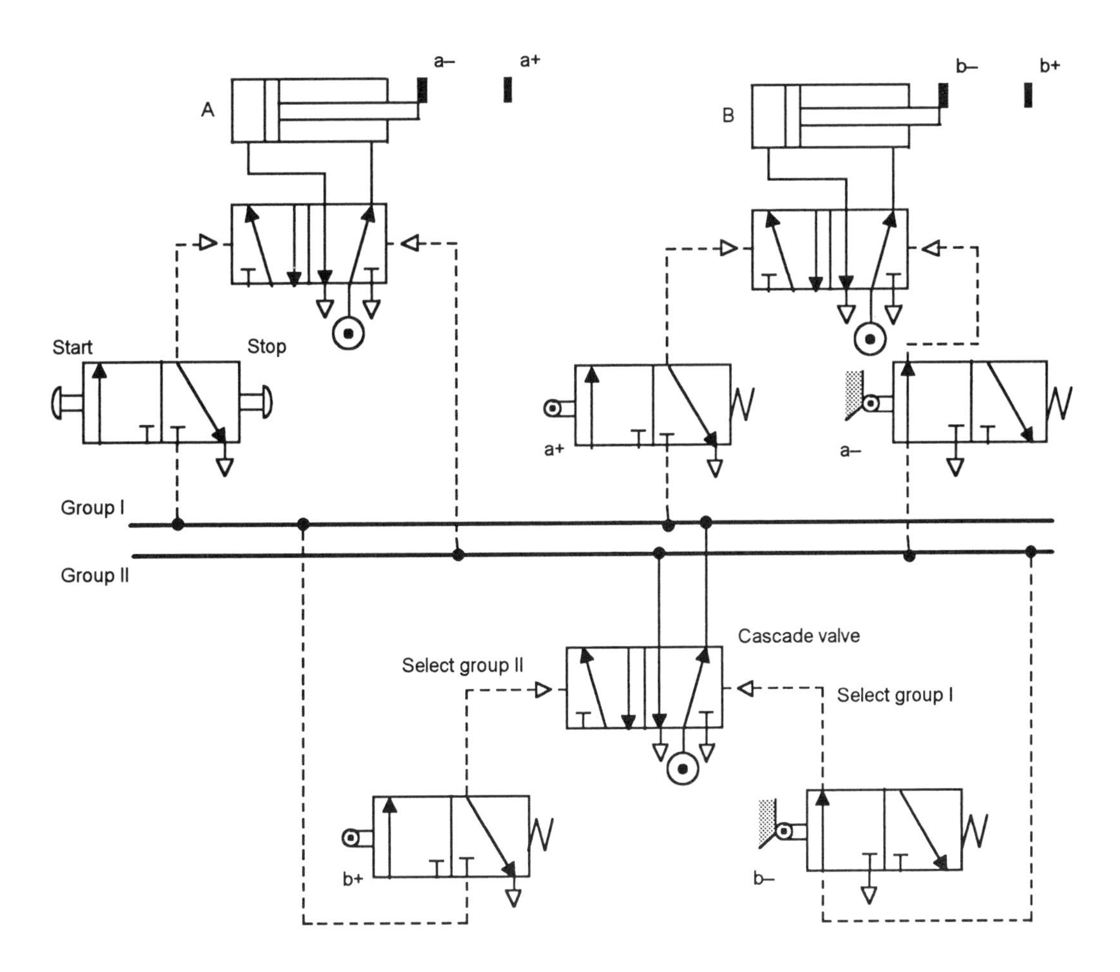

Figure 7.36 *Problem 12*

13 Draw a displacement-event diagram for the sequence A+, B–, C+, A–, B+, C–.

14 What is the smallest number of groups that can be used for the sequence A+, B–, C–, C+, B+, A– ? Hence draw a group circuit.

8 Logic systems

8.1 Introduction

A *logic system* is one in which events occur in a reasoned sequence. We can think of such a system as involving an array of on–off switches and the condition for an outcome depends on which of the switches are activated. Logic-switching circuits can be developed with electrical, electronic or pneumatic/hydraulic systems. In this chapter the basic mathematics necessary to analyse and synthesise such circuits is introduced. Chapter 9 takes the consideration further. The mathematics involved is named after George Boole (1815–64), who first developed the modern ideas of the mathematics concerned with the manipulation of logic statements, and is termed *Boolean algebra*.

8.2 Switching

Before considering pneumatic/hydraulic logic-switching elements, consider the switching produced with electrical switches. A simple on–off electrical switch (Figure 8.1) can be considered to have two possible states, either on or off, and these states can be denoted by the digits 1 or 0. Thus we can denote a closed contact by 1 and an open contact by 0. Note that the 0 and the 1 do not represent actual numbers but the state of the voltage or current in the circuit controlled by the switch. With a pneumatic switch the 0 might be assigned to no pressure output and the 1 to a pressure output.

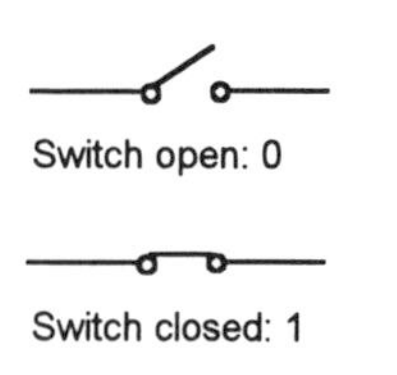

Figure 8.1 *The two states of a switch*

8.2.1 AND gate

Consider a combination of two switches A and B in series. Each switch has two possible states, 0 and 1. Figure 8.2 shows the various possibilities for switches. In (a) both switches are open, in (b) A is open and B is closed, in (c) A is closed and B is open and in (d) A and B are both closed. With (a) the effect of both switches being open is the same as would be obtained by a single open switch; (b) and (c) likewise are equivalent to a single open switch but (d) is equivalent to a single closed switch. Thus we can say that the two elements are equivalent to 0 for (a), (b) and (c) but 1 for (d).

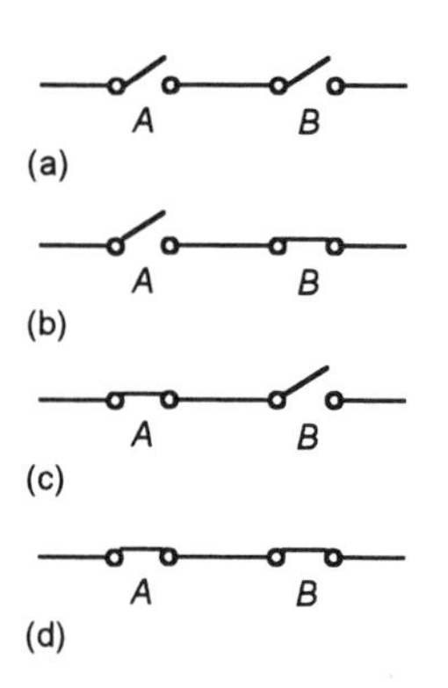

Figure 8.2 *Switches in series*

In tabular form we can represent the states of the circuit by Table 8.1; this stating that a 0 input for A and a 0 input for B give a 0 output, a 0 input for A and a 1 input for B give a 0 output, a 1 input for A and a 0 input for B give a 0 output and a 1 input for A and a 1 input for B give a 1 output. Such a table is known as a *truth table* and we can state that if A AND B are 1 then the result is 1, the combination of switches being termed an AND gate.

For series connections, i.e. an AND gate, A is considered to be *multiplied* by B and the symbol $\cdot$ used for the product. From the table we thus have the rules for the AND gate:

Table 14.1 *A* AND *B*

A	B	$A \cdot B$
0	0	0
0	1	0
1	0	0
1	1	1

$$0 \cdot 0 = 0, \quad 0 \cdot 1 = 0, \quad 1 \cdot 0 = 0, \quad 1 \cdot 1 = 1 \qquad [1]$$

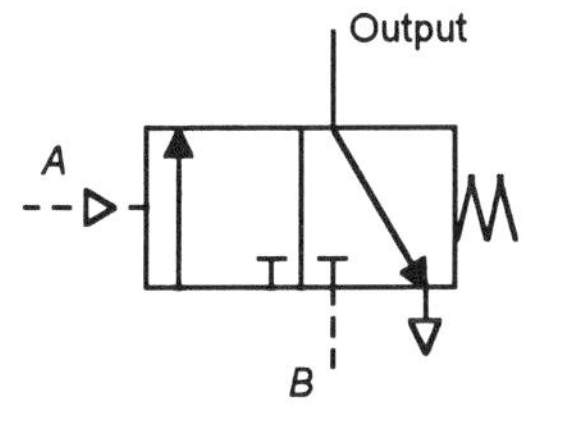

Figure 8.3 *AND gate*

Figure 8.3 shows how we can realise an AND gate with a 3/2 valve. To obtain an output of pressure we need input *A* and input *B* to be both at pressure. Special AND function valves are available, Figure 8.4 showing the form such a valve can take and the symbol used. For there to be an output from such a valve, input *A* and input *B* must both be at pressure.

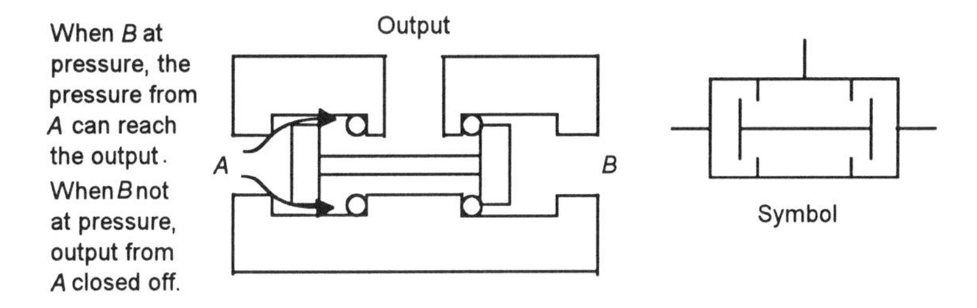

Figure 8.4 *AND function valve*

Different sets of standard circuit symbols for logic gates have been used with the main form being that originated in the United States, though an international standard form has now been developed. Figure 8.5(a) shows the symbol traditionally used for an AND gate with (b) showing the new standardised format. Because the traditional format is still widely used, in this book both forms of the symbols are used.

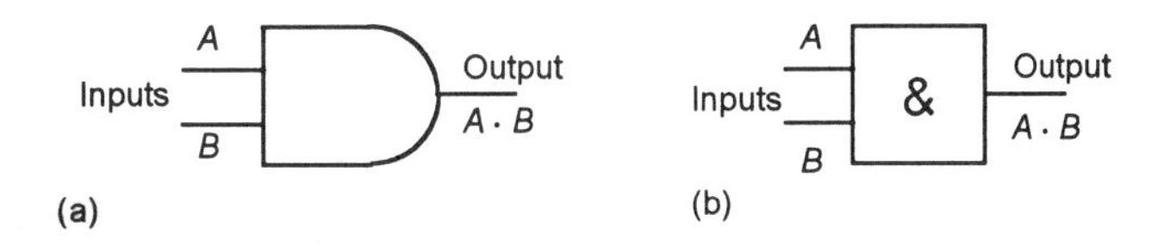

Figure 8.5 *Standard symbols for AND gates*

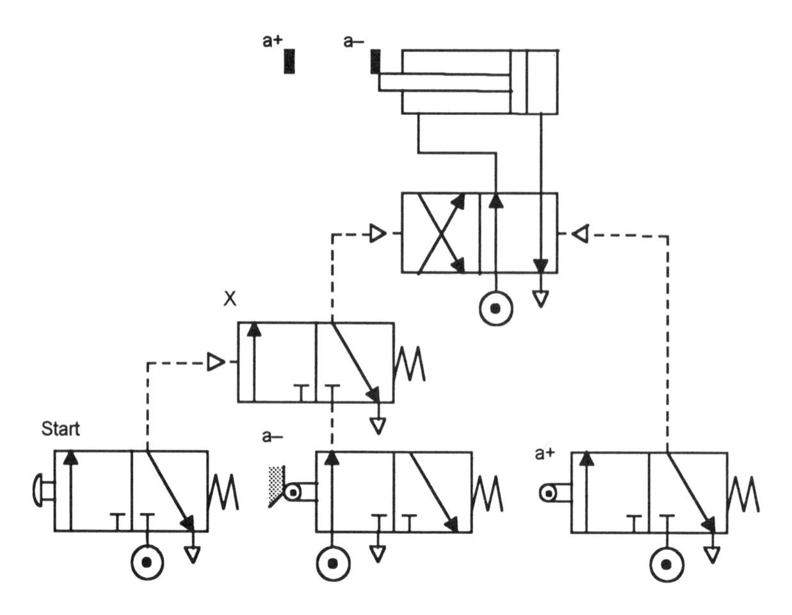

Figure 8.6 *Example*

Example

Explain the function of the valve X in the circuit of Figure 8.6.

Valve X has the AND function. When the start signal and the a_0 signal are both received by the valve there is an output to the cylinder directional control valve. There is no output if the start signal but not the a– signal, the a– signal but not the start signal, or neither signal are received.

Example

Explain the function of the valve X in the circuit of Figure 8.7.

Valve X is an AND function valve. For there to be an input to the cylinder directional control valve, start button 1 and start button 2 must both be pressed. If only one of the start buttons is pressed, or neither of them is pressed, there is no pilot signal supplied to the cylinder directional control valve; both buttons have to be pressed for this to happen.

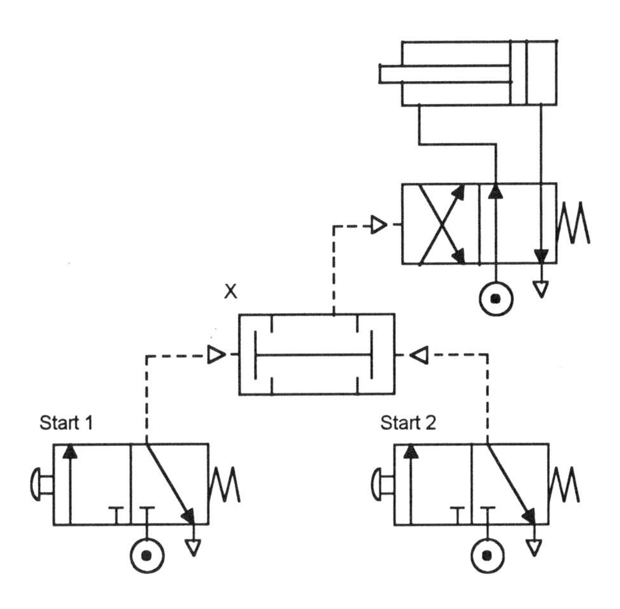

Figure 8.7 *Example*

8.2.2 OR gate

Consider two on–off switches in parallel. Figure 8.8 shows the various possibilities for the switches. In (a) both switches are open, in (b) A is open and B is closed, in (c) A is closed and B is open and in (d) A and B are both closed. With (a) the effect of both switches being open is the same as would be obtained by a single open switch; (b), (c) and (d) are equivalent to a single closed switch. Thus we can say that the two elements are equivalent to 0 for (a), and 1 for (b), (c) and (d). We can represent the state of the circuit by the truth table shown in Table 8.2.

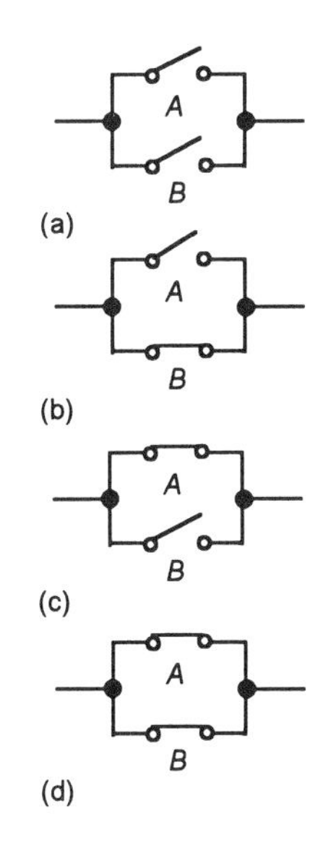

Figure 8.8 *Parallel switches*

Table 8.2 A OR B

A	B	$A+B$
0	0	0
0	1	1
1	0	1
1	1	1

We can state that if A OR B is 1 then the result is 1, the combination being termed an OR gate.

For an OR gate, i.e. parallel connections of switches, A is considered to be *added* to B. In this text the symbol + is used for such addition. From the above table for the OR gate we thus have the rules:

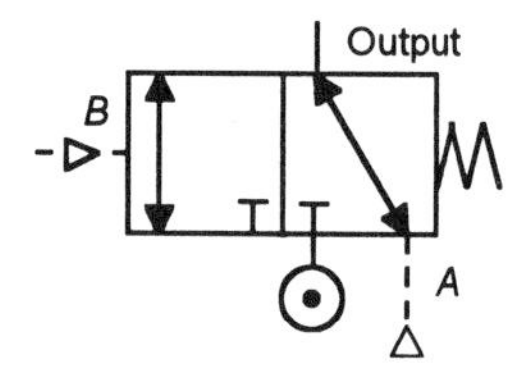

Figure 8.9 *OR gate*

$$0 + 0 = 0, \quad 0 + 1 = 1, \quad 1 + 0 = 1, \quad 1 + 1 = 1 \qquad [2]$$

Figure 8.9 shows how the OR function can be realised with a 3/2 valve. When input A or B is at pressure then there is a pressure output. Figure 8.10 shows a special OR function valve and its valve symbol. An input of either A or B will move the shuttle and give an output.

Figure 8.11 shows the traditional and the newer international standard symbols for the OR gate. The newer symbol was chosen to indicate that at least one active input is needed to activate the output.

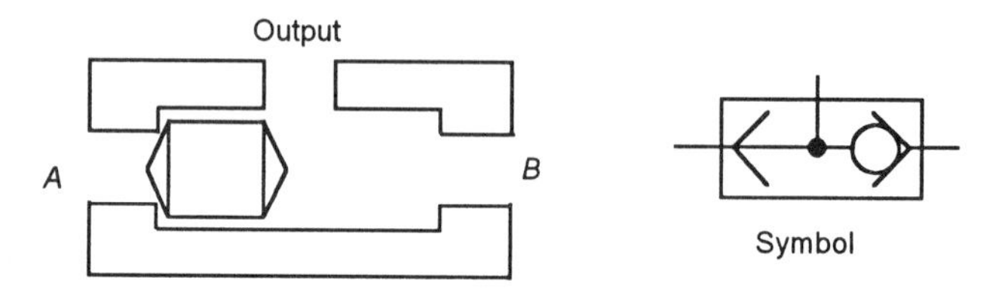

Figure 8.10 *OR function valve*

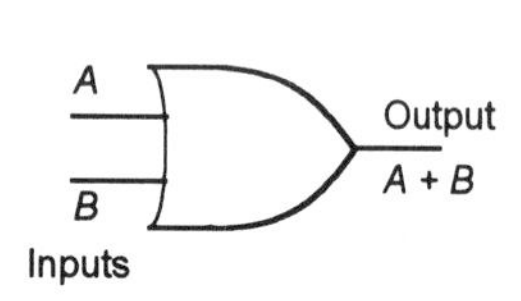

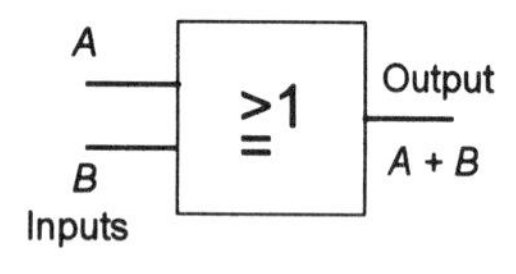

Figure 8.11 *Symbols for OR gate*

Example

Explain the function of valve X in Figure 8.12.

When either the start push-button 1 or the start push-button 2 are operated, a pilot pressure signal is applied to the cylinder direction control valve to cause the cylinder to extend. X thus provides the OR function.

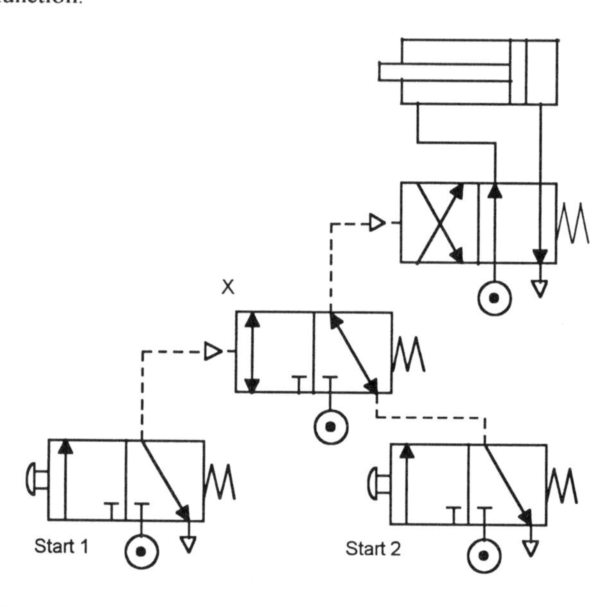

Figure 8.12 *Example*

Example

Describe how the circuit in Figure 8.13 will behave when the start switches are activated.

When either the push-button of start 1 or the push-button of start 2 are pressed, pilot pressure is supplied to the cylinder directional control valve and the cylinder extends.

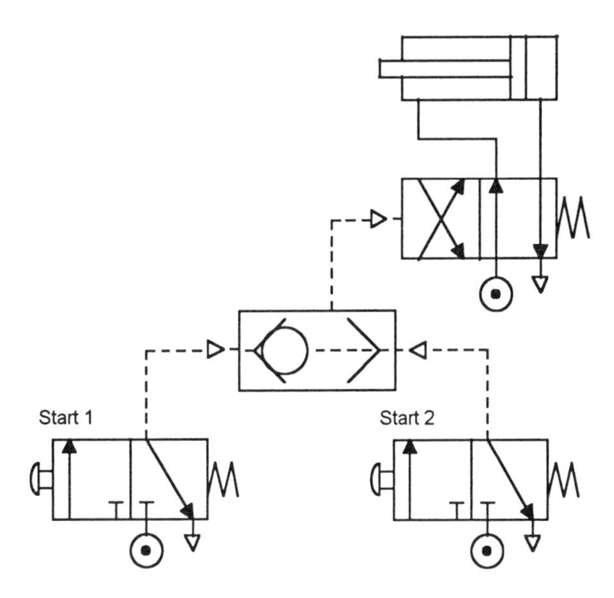

Figure 8.13 *Example*

8.2.3 NOT gate

Another possible form of switch circuit is where two switches are connected together so that the closing of one switch results in the opening of the other. Figure 8.14(a) illustrates the switch action with (b) showing the upper switch open when the lower switch is closed and (c) the upper switch closed when the lower switch is open. The lower switch is said to give the *complement* of the upper switch. If the upper switch is denoted by A then the lower switch is denoted by $\bar{A}$. A bar over a symbol indicates the complement. Table 8.3 is the truth table:

Table 8.3 NOT

A	$\bar{A}$
0	1
1	0

(a)

(b)

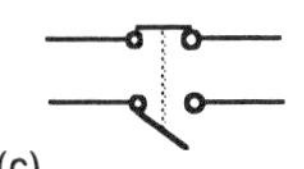

Figure 8.14 *Complement*

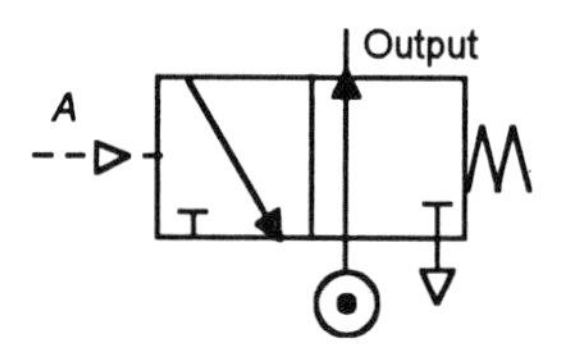

Figure 8.15 *NOT gate*

If one switch is 1 then the other switch is NOT 1, the arrangement being termed a NOT gate. From the above table for the NOT gate we thus have the rules:

$$\bar{0} = 1, \qquad \bar{1} = 0 \tag{3}$$

Figure 8.15 shows how a 3/2 valve can act as a NOT gate. When there is no input A there is an output. When there is an input A, the valve switches to give no output.

Figure 8.16 shows the traditional symbol (a) used and the newer international standard one (b).

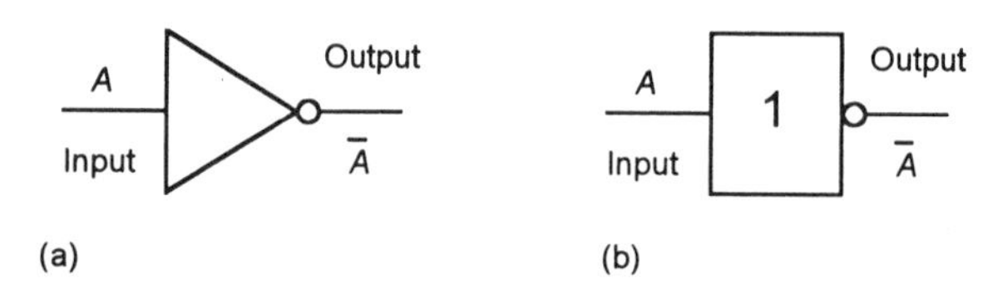

Figure 8.16 *Symbols for NOT gate*

8.2.4 NAND gate

This gate is logically equivalent to a NOT gate in series with an AND gate (Figure 8.17(a)), NAND standing for NotAND. The symbol for the gate (Figure 8.17(b)) is the AND symbol followed by a small circle, the small circle being used to indicate negation. The gate represents the Boolean function $\overline{A \cdot B}$ and has the truth table shown in Table 8.4. There is a 1 output when A and B are both not 1.

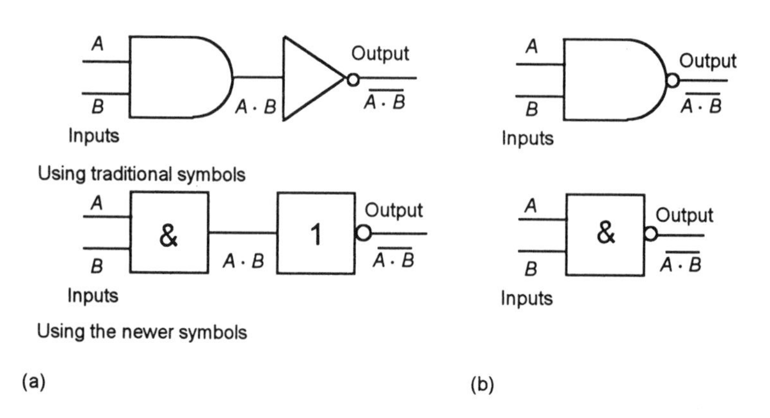

Figure 8.17 *NAND gate*

Table 8.4 NAND gate

A	B	$\overline{A \cdot B}$
0	0	1
0	1	1
1	0	1
1	1	0

Figure 8.18(a) shows how a NAND gate can be realised with two 3/2 valves or (b) an AND valve and a 3/2 valve. For the NAND gate, when there is no A pressure input and no B pressure input or just one of them gives an input, there is a pressure output. If there is pressure at both input A and input B there is no output.

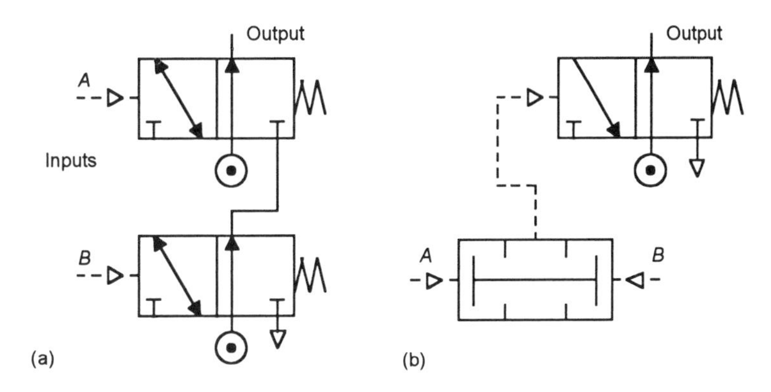

Figure 8.18 *NAND gate*

8.2.5 NOR gate

This gate is logically equivalent to a NOT gate in series with an OR gate (Figure 8.19(a)). It is represented by the OR gate symbol followed by a small circle to indicate negation (Figure 8.19(b)). The gate represents the Boolean function $\overline{A+B}$. Table 8.5 gives the truth table, there being a 1 output when neither A nor B is 1.

Table 8.5 NOR gate

A	B	$\overline{A+B}$
0	0	1
0	1	0
1	0	0
1	1	0

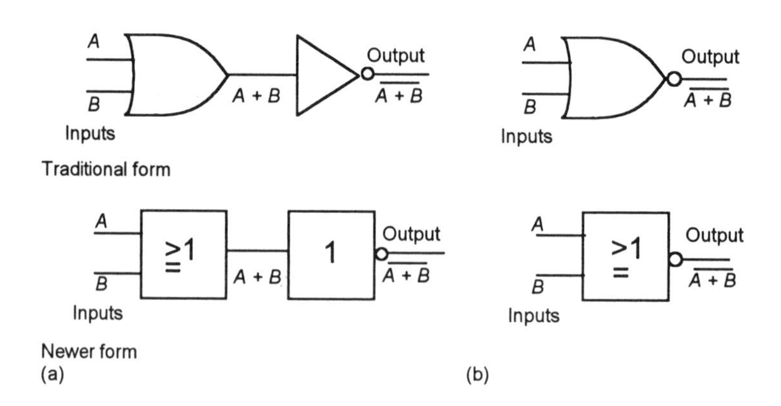

Figure 8.19 *NOR gate*

Figure 8.20(a) shows how a NOR gate can be realised with two 3/2 valves, (b) with an OR valve and a 3/2 valve. If there is a pressure input to either *A* or *B*, or both, there is no output. There is only an output when neither *A* nor *B* have an input.

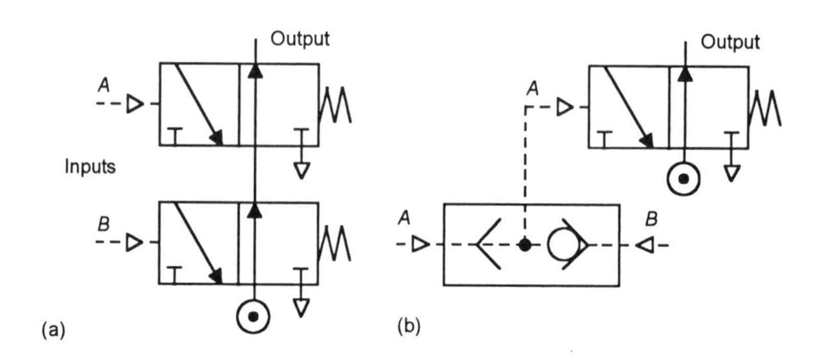

Figure 8.20 *NOR gate*

8.2.6 INHIBITION gate

Figure 8.21(a) shows an AND gate for which one of the inputs is supplied via a NOT gate. Such a gate is termed an INHIBITION gate, Figure 8.21(b) showing the symbol for such a gate. The gate has the truth table shown in Table 8.6.

Figure 8.22 shows how such a gate can be realised with a 3/2 valve. For there to be a pressure output, there must be a pressure input *A* and no pressure input *B*.

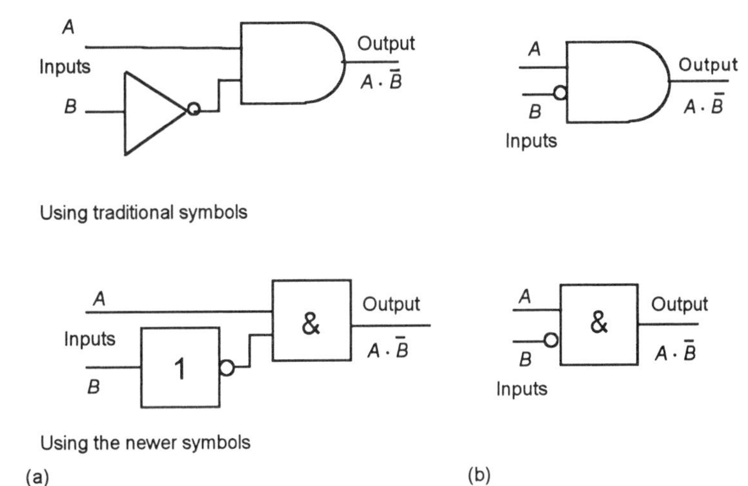

Figure 8.21 *INHIBITION gate*

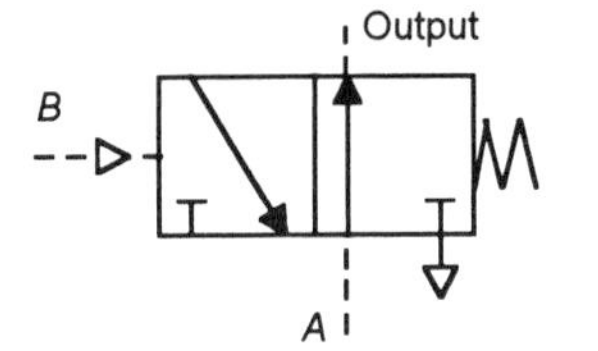

Figure 8.22 *INHIBITION gate*

Table 8.6 INHIBITION gate

A	B	$A \cdot \bar{B}$
0	0	0
0	1	0
1	0	1
1	1	0

8.2.7 Memory function

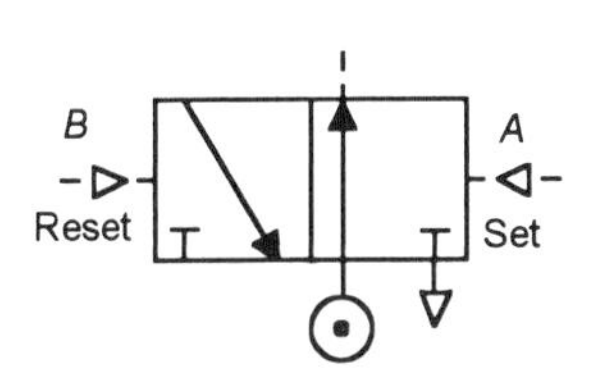

Figure 8.23 *Memory*

Consider the double pilot-operated 3/2 valve shown in Figure 8.23. One pilot line is used to set or enable the valve while the other pilot line resets or disables the valve. When the valve is set there is pressure output which remains as the output even when the set signal is removed. When it is reset the pressure output is cancelled. The valve remembers the last input.

Figure 8.24(a) shows a 5/2 valve. This has two inputs and can be set to one or reset to the other, Figure 8.24(b) showing the symbol used.

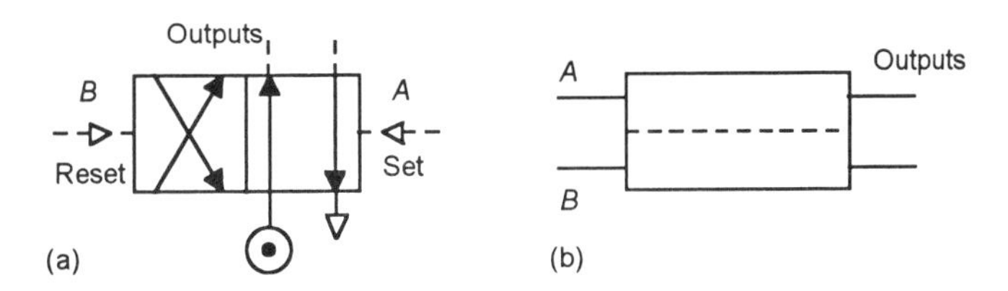

Figure 8.24 *Memory*

8.3 Laws of Boolean algebra

The binary digits 1 and 0 are the *Boolean variables* and, together with the operations ·, + and the complement, form what is known as *Boolean algebra*. The laws of Boolean algebra can be derived from a consideration of truth tables. The following derives the laws and considers their significance in terms of valve circuits.

As Table 8.7 indicates, for an OR function we have:

Table 8.7

A	A	$A+A$
0	0	0
1	1	1

$$A + A = A \qquad [4]$$

Anything ORed with itself is equal to itself. Thus if we have the same input to each of the inputs of an OR valve, the output will be the same as the input (Figure 8.25).

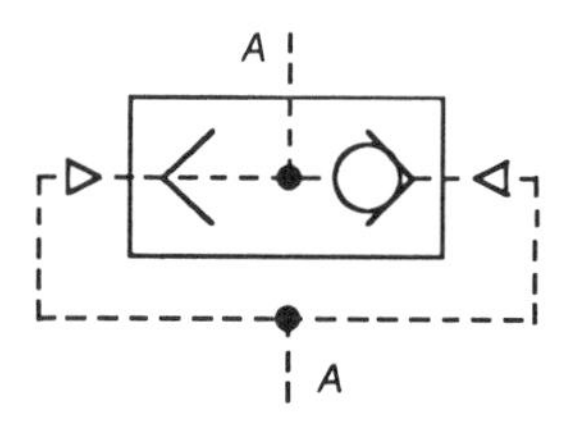

Figure 8.25 $A + A = A$

As Table 8.8 indicates, for an AND function we have:

Table 8.8

A	A	$A \cdot A$
0	0	0
1	1	1

$$A \cdot A = A \qquad [5]$$

Anything ANDed with itself is equal to itself. Thus if we have the same input A to each of the two inputs of an AND gate, the output is A. Figure 8.26 illustrates this with an AND valve.

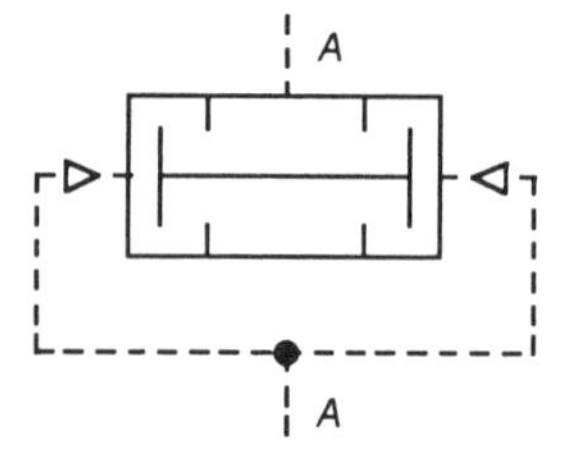

Figure 8.26 $A \cdot A = A$

Table 8.2 for the OR gate shows that:

$$A + B = B + A \qquad [6]$$

Table 8.1 for the AND gate shows that:

$$A \cdot B = B \cdot A \qquad [7]$$

Equations [6] and [7] indicate that it does not matter in which order we consider inputs for OR and AND gates.

As Table 8.9 indicates:

$$A + (B \cdot C) = (A + B) \cdot (A + C) \qquad [8]$$

Figure 8.27 shows the two equivalent valve circuits described by the above equation; for comparison, Figure 8.28 shows them in logic symbol form.

Table 8.9 $A + (B \cdot C) = (A + B) \cdot (A + C)$

A	B	C	$B \cdot C$	$A + B \cdot C$	$A + B$	$A + C$	$(A + B) \cdot (A + C)$
0	0	0	0	0	0	0	0
0	0	1	0	0	0	1	0
0	1	0	0	0	1	0	0
0	1	1	1	1	1	1	1
1	0	0	0	1	1	1	1
1	0	1	0	1	1	1	1
1	1	0	0	1	1	1	1
1	1	1	1	1	1	1	1

As Table 8.10 indicates:

$$A \cdot (B + C) = A \cdot B + A \cdot C \qquad [9]$$

Figure 8.29 shows the two equivalent valve circuits described by the above equation.

Equations [8] and [9] are very useful as a means of designing valve circuits with the minimum number of AND and OR gates.

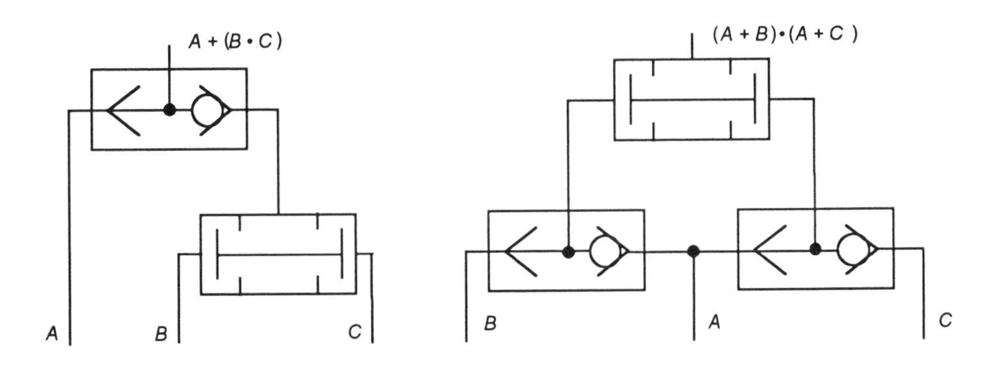

Figure 8.27 $A + (B \cdot C) = (A + B) \cdot (A + C)$

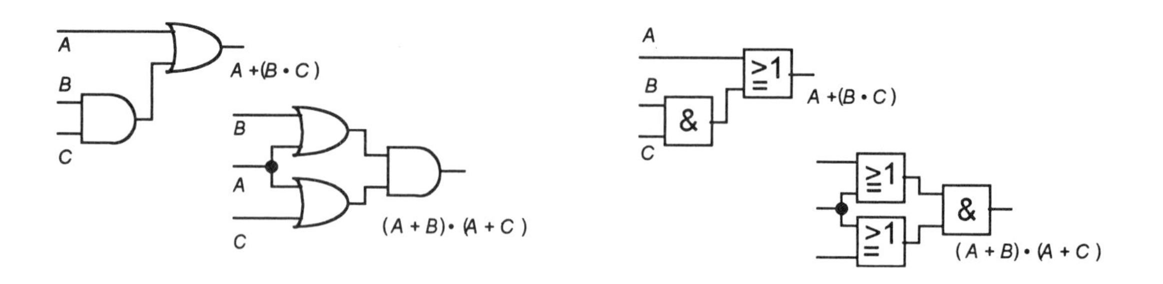

Figure 8.28 *Logic symbolic representation of Figure 8.27*

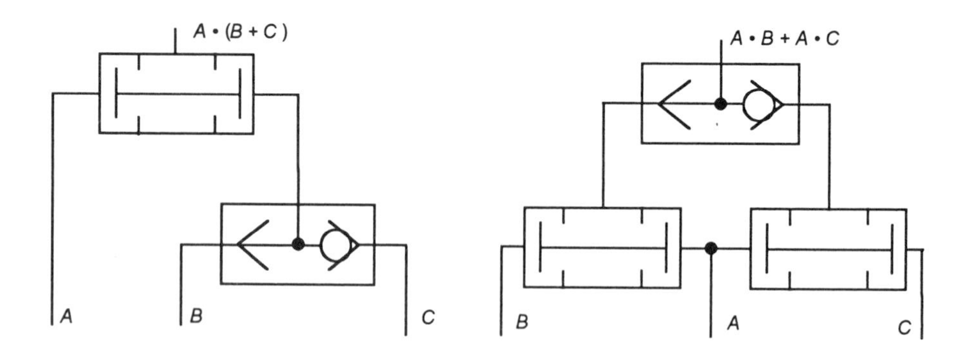

Figure 8.29 $A \cdot (B + C) = A \cdot B + A \cdot C$

Table 8.10 $A \cdot (B + C) = A \cdot B + A \cdot C$

A	B	C	$B + C$	$A \cdot (B + C)$	$A \cdot B$	$A \cdot C$	$A \cdot B + A \cdot C$
0	0	0	0	0	0	0	0
0	0	1	1	0	0	0	0
0	1	0	1	0	0	0	0
0	1	1	1	0	0	0	0
1	0	0	0	0	0	0	0
1	0	1	1	1	0	1	1
1	1	0	1	1	1	0	1
1	1	1	1	1	1	1	1

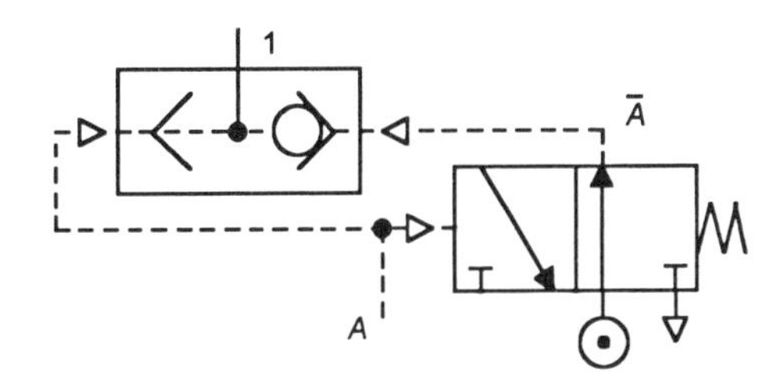

Figure 8.30 $A + \bar{A} = 1$

Table 8.11 $A + \bar{A} = 1$

A	$\bar{A}$	$A + \bar{A}$
0	1	1
1	0	1

As Table 8.11 indicates:

$$A + \bar{A} = 1 \quad [10]$$

Anything ORed with its own complement equals 1. Figure 8.30 illustrates this with valves. Whether the input A is 0 or 1, the output is 1.

As Table 8.12 indicates:

$$A \cdot \bar{A} = 0 \quad [11]$$

Table 8.12 $A \cdot \bar{A} = 0$

A	$\bar{A}$	$A \cdot \bar{A}$
0	1	0
1	0	0

Anything ANDed with its own complement equals 0. Figure 8.31 illustrates this with valves. When A is 0 we have 0 AND 1 giving 0. When A is 1 we have 1 AND 0 giving 0.

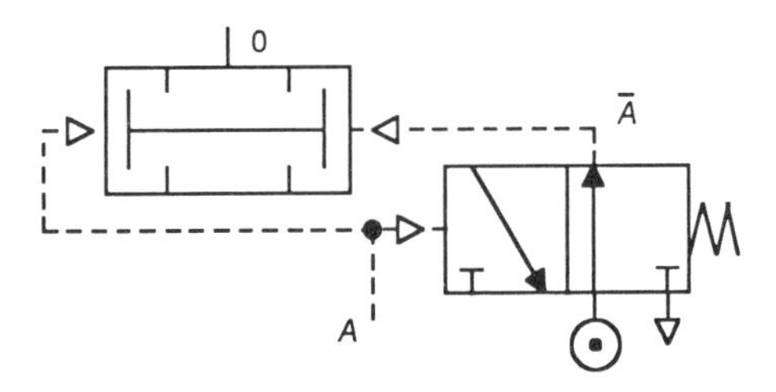

Figure 8.31 $A \cdot \bar{A} = 0$

Table 8.13 $A + 0$, $A + 1$

A	$A + 0$	$A + 1$
0	0	1
1	1	1

As Table 8.13 indicates, for the OR function:

$$A + 0 = A, \quad A + 1 = 1 \quad [12]$$

Figure 8.32 illustrates the above equations with valves. When we have an input of A to an OR valve, when the other input is 0 then the output will be A. Thus if A is 0 then the output is 0, if A is 1 the output is 1. Thus anything ORed with a 0 is equal to itself. With an input A to the OR valve, when the other input is 1 then the output must be 1, regardless of whether A is 0 or 1. Anything ORed with a 1 is equal to 1.

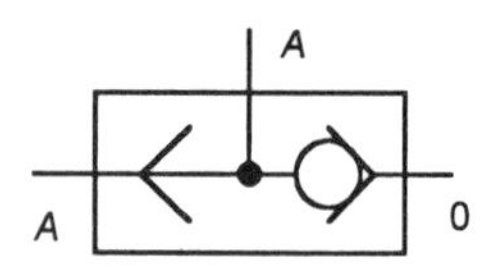

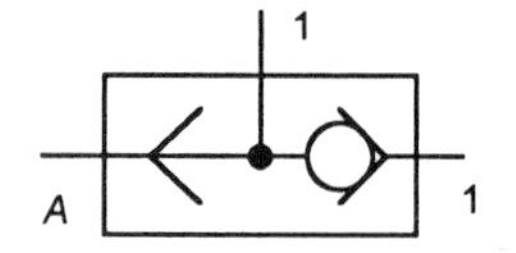

Figure 8.32 $A + 0 = A, \quad A + 1 = 1$

As Table 8.14 indicates, for the AND function:

Table 8.14 $A \cdot 1, A \cdot 0$

A	$A \cdot 1$	$A \cdot 0$
0	0	0
1	1	0

$$A \cdot 1 = A, \quad A \cdot 0 = 0 \qquad [13]$$

Figure 8.33 illustrates the above equations with valves. With one input to the AND valve of A and the other 1, then the output will be A and so 0 if A is 0 and 1 if A is 1. Anything ANDed with a 1 is equal to itself. With one input to the AND valve of A and the other 0, then the output will be 0 regardless of whether A is 0 or 1. Anything ANDed with a 0 is equal to 0.

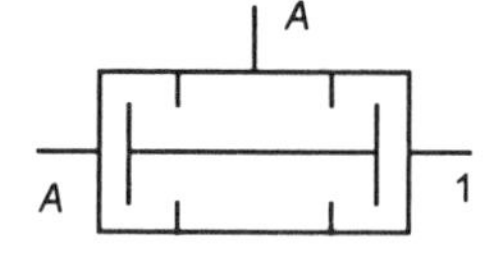

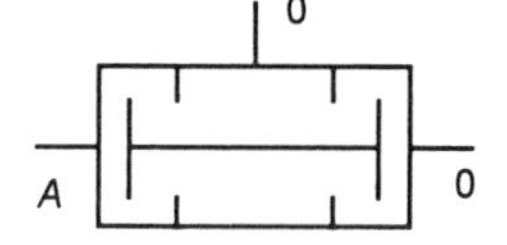

Figure 8.33 $A \cdot 1 = A, \quad A \cdot 0 = 0$

Note that as equation [9] in the above indicates, brackets can be used in the same way as in normal algebra. Thus, for example, we can use equation [9] to write:

$$A \cdot B + A \cdot \bar{B} = A \cdot (B + \bar{B})$$

Boolean algebra can be used to manipulate switching functions into many forms. However, the form to which most functions can be minimised involve two AND gates driving an OR gate (Figure 8.34(a)), this giving a Boolean expression of the form (termed the *sum of products* form):

$$A \cdot B + A \cdot C \qquad [14]$$

or two OR gates driving an AND gate (Figure 8.34(b)), giving a Boolean expression of the form (termed the *product of sums* form):

$$(A + B) \cdot (A + C) \qquad [15]$$

Thus in considering possible functions to fit truth tables, these forms are worth trying.

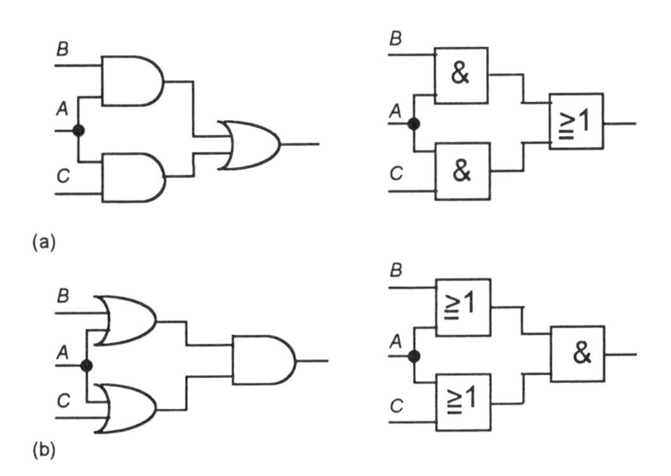

Figure 8.34 *(a) Sum of products, (b) product of sums*

Example

Simplify the following Boolean function $A \cdot C + (A + B) \cdot \bar{C}$.

Using equation [9] for $(A + B) \cdot \bar{C}$:

$$(A + B) \cdot \bar{C} = A \cdot \bar{C} + B \cdot \bar{C}$$

Hence we can write for the function:

$$A \cdot C + A \cdot \bar{C} + B \cdot \bar{C}$$

Using equation [9] for the first two terms gives:

$$A \cdot (C + \bar{C}) + B \cdot \bar{C}$$

Then using equation [7]:

$$A \cdot 1 + B \cdot \bar{C}$$

and so equation [10] then gives:

$$A + B \cdot \bar{C}$$

Example

Simplify the Boolean function $A + A \cdot B \cdot C + \bar{A} \cdot \bar{C}$.

Using equation [13] we can replace A by $A \cdot 1$. The function can then be written as:

$$A \cdot 1 + A \cdot (B \cdot C) + \bar{A} \cdot \bar{C}$$

Then using equation [9]:

$$A \cdot (1 + (B \cdot C)) + A \cdot \bar{C}$$

Using the second of the equations in [12] gives $1 + (B \cdot C) = 1$ and so the function becomes:

$$A \cdot 1 + \bar{A} \cdot \bar{C}$$

Since $A \cdot 1 = A$ (equation [10]), and applying equation [8]:

$$A + \bar{A} \cdot \bar{C} = (A + \bar{A}) \cdot (A + \bar{C})$$

But $A + \bar{A} = 1$ (equation [10]) and so, using equation [13], the function becomes:

$$A + \bar{C}$$

Example

Write, for the circuit shown in Figure 8.35, (a) the truth table and (b) the Boolean function to describe that truth table.

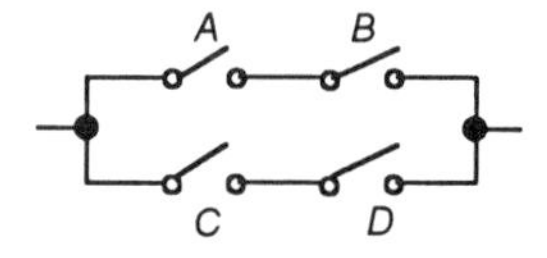

Figure 8.35 *Example*

(a) A and B are in series, and in parallel with the series arrangement of C and D. The result of using the switches is that only when either A and B are closed or C and D are closed will there be an output. Table 8.15 shows the truth table.

(b) The Boolean function for two switches in series is $A \cdot B$, the AND function, and thus, since the function for two items in parallel is OR, the function for the circuit as a whole is:

$$A \cdot B + C \cdot D$$

Table 8.15 *Example*

A	B	C	D	Outcome
0	0	0	0	0
0	0	0	1	0
0	0	1	0	0
0	0	1	1	1
0	1	0	0	0
0	1	0	1	0
0	1	1	0	0
0	1	1	1	0
1	0	0	0	0
1	0	0	1	0
1	0	1	0	0
1	0	1	1	1
1	1	0	0	1
1	1	0	1	1
1	1	1	0	1
1	1	1	1	1

Example

Derive the Boolean function for the switching circuit shown in Figure 8.36.

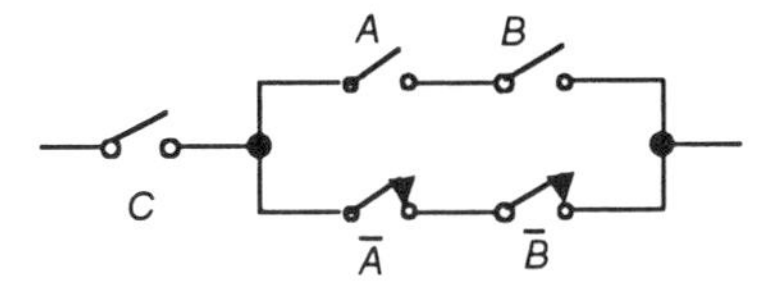

Figure 8.36 *Example*

In the upper parallel arm of the circuit, the switches A and B are in series and so have a Boolean expression of $A \cdot B$. In the lower arm, the complements of A and B are in series. Thus the Boolean expression for that part of the circuit is $\bar{A} \cdot \bar{B}$. Because the two arms are in parallel the expression for that part of the circuit is $A \cdot B + \bar{A} \cdot \bar{B}$. In series with this is switch C. Thus the Boolean function for the circuit is:

$$C \cdot (A \cdot B + \bar{A} \cdot \bar{B})$$

Example

A door is to be opened by a pneumatic cylinder when either of two push-button valves A or B is pressed and closed when either of them is pressed again. Thus we might press A to open the door and then later press A again to close it, or we might press A to open it and then B to close it, or B to open it and B to close it, or B to open it and then A to close it. Determine the truth table for the situation and the logic symbol diagram.

We will take it that the door is open when the output is 1 and closed when it is 0. Thus the above conditions give Table 8.16.

Table 8.16 *Example*

A	B	Output
0	0	0
1	0	1
1	1	0
0	1	1

Lines 2 and 4 of the truth table have a 1 output. For the output to be 1, line 2 indicates that we must have $A = 1$ and $B = 0$. For an AND gate with $A \cdot B$ we would need $A = 1$ and $B = 1$. Thus we need to have A ANDed with the complement of B. For line 4 we can generate the 1 output with the complement of A ANDed with B. Since there has to be a 1 output for either of these conditions, we can OR them to produce the required result and thus have the sum of products expression:

$$A \cdot \bar{B} + \bar{A} \cdot B$$

Table 8.17 shows the result. It fits the truth table of Table 8.16.

This approach of considering the lines in the truth table which result in a 1 output and finding the AND condition for such a line to be realised, then ORing all such lines, can be used as a general method for establishing a Boolean expression from a truth table.

Thus the logic gate circuit to fit the above Boolean expression is of the form shown in Figure 8.37.

Table 8.17 *Example*

A	B	$\bar{A}$	$\bar{B}$	$A \cdot \bar{B}$	$\bar{A} \cdot B$	$A \cdot \bar{B} + \bar{A} \cdot B$
0	0	1	1	0	0	0
1	0	0	1	1	0	1
1	1	0	0	0	0	0
0	1	1	0	0	1	1

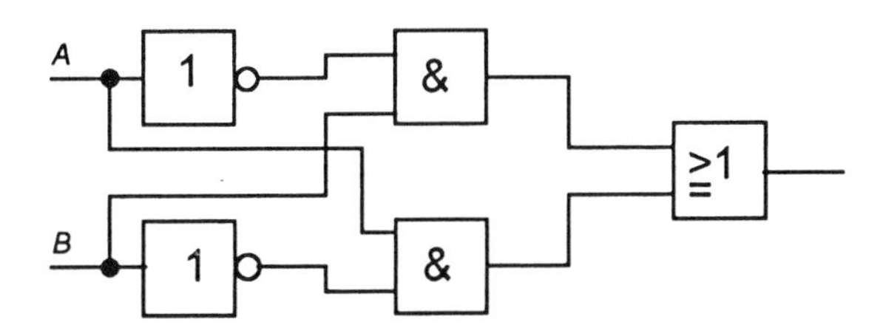

Figure 8.37 *Example*

8.3.1 De Morgan laws

In addition to the above ways of simplifying Boolean expression, we have what are known as the *De Morgan* laws:

1 The complement of the outcome of switches A and B in parallel, i.e. an OR situation, is the same as when the complements of A and B are separately combined in series, i.e. the AND situation. Table 8.18 shows the validity of this.

$$\overline{A+B}=\bar{A}\cdot\bar{B} \qquad [16]$$

Figure 8.38 shows the above equation with valve circuits.

Table 8.18 $\overline{A+B}=\bar{A}\cdot\bar{B}$

A	B	$A+B$	$\overline{A+B}$	$\bar{A}$	$\bar{B}$	$\bar{A}\cdot\bar{B}$
0	0	0	1	1	1	1
0	1	1	0	1	0	0
1	0	1	0	0	1	0
1	1	1	0	0	0	0

2 The complement of the outcome of switches A and B in series, i.e. the AND situation, is the same as when the complements of A and B are separately considered in parallel, i.e. the OR situation. Table 8.19 shows the validity of this.

$$\overline{A\cdot B}=\bar{A}+\bar{B} \qquad [17]$$

Figure 8.39 shows the above equation with valve circuits.

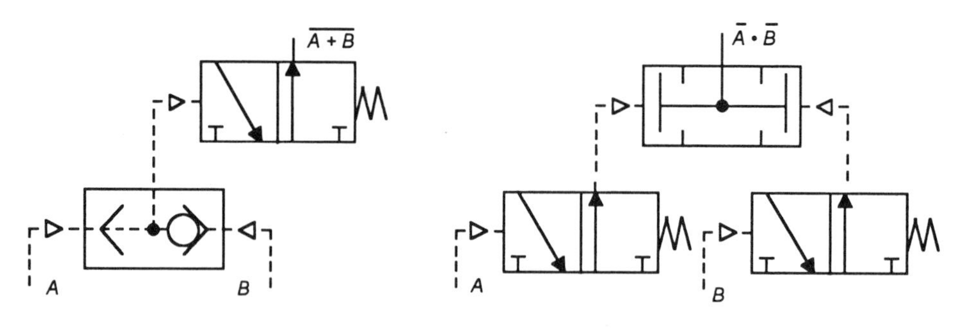

Figure 8.38 $\overline{A+B} = \bar{A} \cdot \bar{B}$

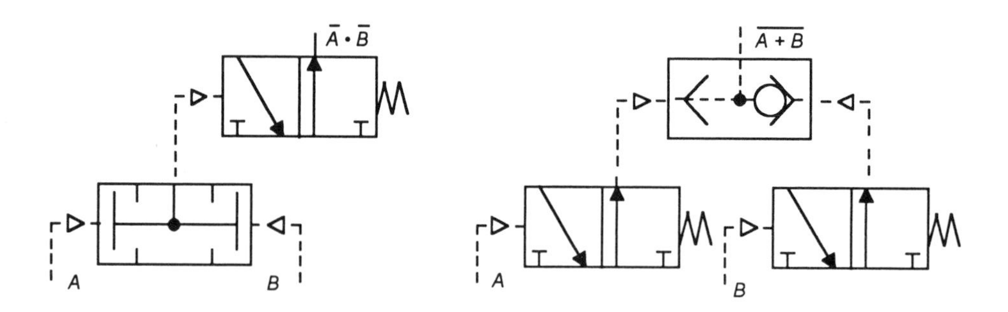

Figure 8.39 $\overline{A \cdot B} = \bar{A} + \bar{B}$

Table 8.19 $\overline{A \cdot B} = \bar{A} + \bar{B}$

A	B	$A \cdot B$	$\overline{A \cdot B}$	$\bar{A}$	$\bar{B}$	$\bar{A}+\bar{B}$
0	0	0	1	1	1	1
0	1	0	1	1	0	1
1	0	0	1	0	1	1
1	1	1	0	0	0	0

What the De Morgan rules state is that AND can be replaced by OR and vice versa, provided that each term is replaced by its complement, the result then being the complement of the original. Using these rules, complicated switching circuits can be reduced to simpler equivalent circuits.

Example

Determine the OR equivalent of $A \cdot B \cdot C$.

We can write $A \cdot B \cdot C$ as

$$\overline{\overline{A \cdot B \cdot C}}$$

i.e. the inverse complement of the inverse complement. Now we can use De Morgan equation [17] to give:

$$\overline{\overline{A \cdot B \cdot C}} = \overline{\bar{A} + \bar{B} + \bar{C}}$$

Example

Simplify the Boolean function $A + \bar{A} \cdot B$.

Using the De Morgan equation [16] which can be written as:

$$\overline{A + \bar{B}} = \bar{A} \cdot \bar{\bar{B}}$$

then:

$$A + \bar{A} \cdot B = A + \overline{(A + \bar{B})}$$

Using the De Morgan equation [17] gives:

$$\overline{\bar{A} \cdot (A + \bar{B})}$$

This can be multiplied out to give:

$$\overline{\bar{A} \cdot A + \bar{A} \cdot \bar{B}} = \overline{0 + \bar{A} \cdot \bar{B}}$$

Using the De Morgan equation [16]:

$$\overline{\bar{A} \cdot \bar{B}} = \overline{\overline{A + B}} = A + B$$

8.4 Karnaugh maps

The Karnaugh map is a graphical method of simplifying Boolean expressions which have been obtained as sums of products from truth tables. The truth table has a line for the value of the output for each combination of input values, the number of combinations depending on the number of variables. With two variables there are four lines in the truth table, with three variables six lines and with four variables 16 lines. The Karnaugh map gives the same information but in a different format. Each output value is placed in a separate cell of the Karnaugh map. The number of cells depends on the number of variables: with two variables there are four cells, with three variables six cells and with four variables 16 cells.

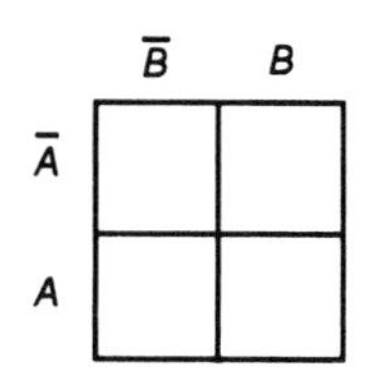

Figure 8.40 *Two-variable Karnaugh map*

Figure 8.40 shows the map when there are two variables. The cells have values which are:

upper left cell corresponds to $\bar{A} \cdot \bar{B}$,

the lower left cell corresponds to $A \cdot \bar{B}$,

the upper right cell corresponds to $\bar{A} \cdot B$,

the lower right cell corresponds to $A \cdot B$

The map squares are so labelled that horizontally adjacent squares differ only in one variable and, likewise, vertically adjacent squares differ in only one variable. Thus horizontally with our two-variable map, the variables differ in only the variable A, vertically they differ only in B.

Suppose for our two variables we have the truth table given in Table 8.20. We consider what conditions must be realised for the inputs for each row to be ANDed and give the indicated output and we will consider any 0 input to be a complement of a 1 input. The output for the first row is what we consider we would obtain with the Boolean product of the complements of each of the variables A and B. The output for the second row is what we consider we would obtain with the Boolean product of the complement of A and uncomplemented B. The output for the third row is what we consider we would obtain with the Boolean product of A and the complement of B. The fourth row is what we consider we would obtain with the Boolean product of the two uncomplemented variables.

If we now put the values given for these products in the Karnaugh map, only indicating where a cell has a 1 value and leaving blank those with a 0 value, then the map shown in Figure 8.41 is obtained. Because the only 1 entry is in the lower right square, the truth table can be represented by the Boolean expression:

Figure 8.41 *Two-variable Karnaugh map*

$$\text{output} = A \cdot \bar{B}$$

Hence Figure 8.42 shows the logic gate system that would generate such a truth table.

Table 8.20 *Truth table*

A	B	Output	Boolean expression
0	0	0	$\bar{A} \cdot \bar{B}$
0	1	0	$\bar{A} \cdot B$
1	0	1	$A \cdot \bar{B}$
1	1	0	$A \cdot B$

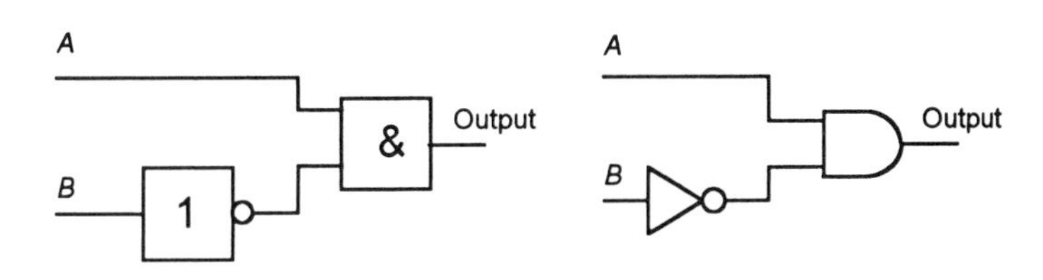

Figure 8.42 $A \cdot \bar{B}$

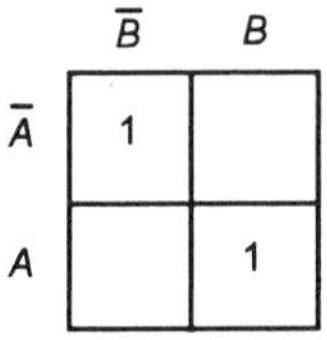

	$\bar{B}$	B
$\bar{A}$	1	
A		1

Figure 8.43 *Two-variable Karnaugh map*

With the above two-variable Karnaugh map, only one cell contained a 1. Now consider the truth table given in Table 8.21. The resulting Karnaugh map is shown in Figure 8.43. As a consequence the Boolean expression is:

$$\text{output} = \bar{A} \cdot \bar{B} + A \cdot B$$

Table 8.21 *Truth table*

A	B	Output	Boolean expression
0	0	1	$\bar{A} \cdot \bar{B}$
0	1	0	$\bar{A} \cdot B$
1	0	0	$A \cdot \bar{B}$
1	1	1	$A \cdot B$

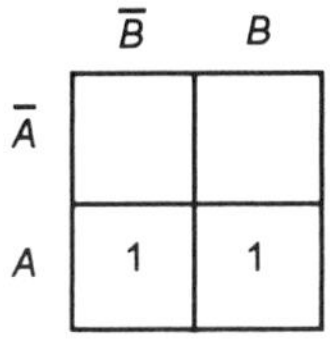

	$\bar{B}$	B
$\bar{A}$		
A	1	1

Figure 8.44 *Two-variable Karnaugh map*

As a further example, consider the truth table in Table 8.22. It gives the Karnaugh map shown in Figure 8.44. This then gives the Boolean expression:

$$\text{output} = A \cdot \bar{B} + A \cdot B$$

But, the two cells containing a 1 have a common vertical edge. The only variable common to both these cells is A. When this happens, the rule is that we can simplify the Boolean expression to just the common variable A. This is because:

$$A \cdot \bar{B} + A \cdot B = A \cdot (\bar{B} + B) = A$$

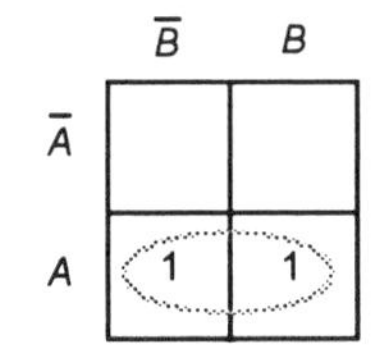

Figure 8.45 *Two-variable Karnaugh map*

We can indicate which cell entries in a Karnaugh map can be simplified by drawing loops round them. Thus for Figure 8.44 we have the result shown in Figure 8.45.

Table 8.22 *Truth table*

A	B	Output	Boolean expression
0	0	0	$\bar{A}\cdot\bar{B}$
0	1	0	$\bar{A}\cdot B$
1	0	1	$A\cdot\bar{B}$
1	1	1	$A\cdot B$

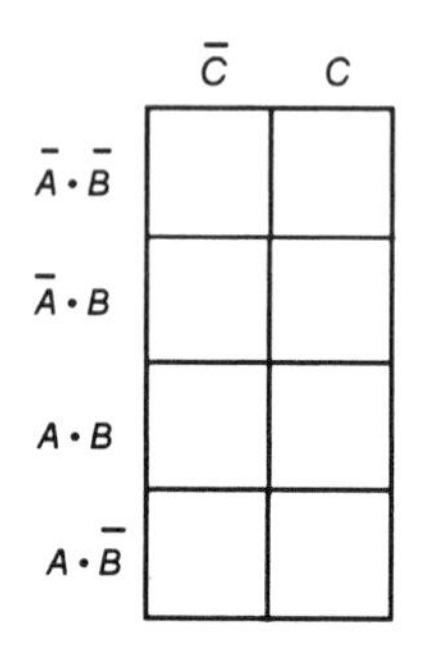

Figure 8.46 *Three-variable Karnaugh map*

Looping a pair of adjacent 1s in a map eliminates the variable that appears in complemented and uncomplemented form.

Now consider the drawing of a three-variable Karnaugh map. The basic map is shown in Figure 8.46. The cells are drawn so that a horizontal movement from a cell to its neighbour results in the change of just one variable. Likewise, a vertical movement from a cell to its neighbour results in the change of just one variable.

Consider the truth table shown in Table 8.23. As before, we consider each row of the truth table and the conditions to be realised for obtaining the outcome of 1 by ANDing the inputs, with 0s being considered to be complements of 1s. When entered on the Karnaugh map, the result is as shown in Figure 8.47. Ignoring for the moment the loops, the Boolean expression indicated by the map is:

$$\text{output} = \bar{A}\cdot\bar{B}\cdot\bar{C}+\bar{A}\cdot B\cdot\bar{C}$$

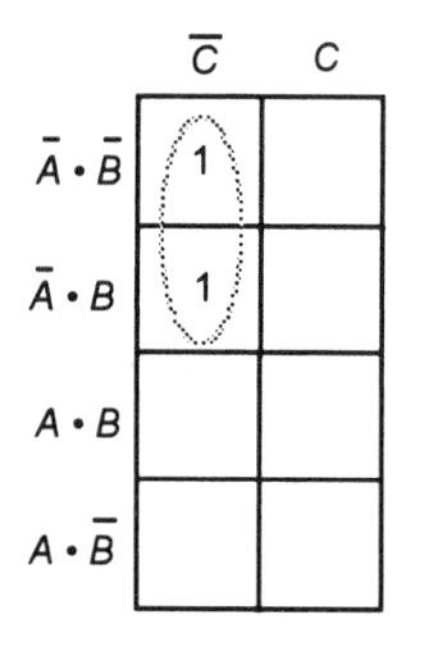

Figure 8.46 *Three-variable Karnaugh map*

For the two vertically looped cells we can eliminate B. Thus the expression can be written as:

$$\text{output} = \bar{A}\cdot\bar{C}+\bar{A}\cdot\bar{C}$$

Table 8.23 *Three-variable truth table*

A	B	C	Output	Boolean expression
0	0	0	1	$\bar{A}\cdot\bar{B}\cdot\bar{C}$
0	0	1	0	$\bar{A}\cdot\bar{B}\cdot C$
0	1	0	1	$\bar{A}\cdot B\cdot\bar{C}$
0	1	1	0	$\bar{A}\cdot B\cdot C$
1	0	0	0	$A\cdot\bar{B}\cdot\bar{C}$
1	0	1	0	$A\cdot\bar{B}\cdot C$
1	1	0	0	$A\cdot B\cdot\bar{C}$
1	1	1	0	$A\cdot B\cdot C$

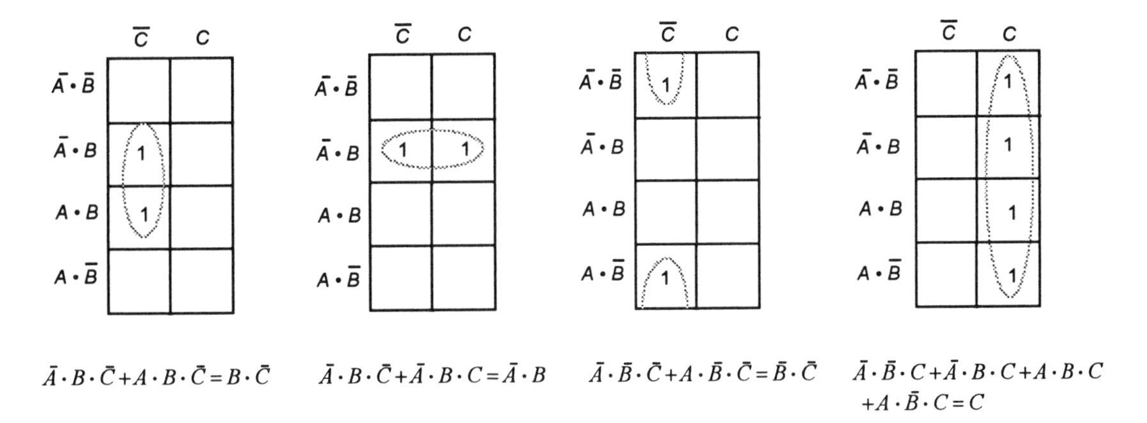

Figure 8.47 *Three-variable maps*

As further examples of looping, consider the Karnaugh maps shown in Figure 8.47 and the outcomes indicated, both before and after looping. Note that adjacent cells can be considered to be in the top and bottom rows or the left- and right-hand columns. Think of opposite edges of the map being joined together.

Looping a pair of adjacent 1s in a map eliminates the variable that appears in complemented and uncomplemented form.

Looping a quad of adjacent 1s in a map eliminates the two variables that appear in both complemented and uncomplemented form.

Now consider the drawing of a four-variable Karnaugh map. Figure 8.48 shows the basic map.

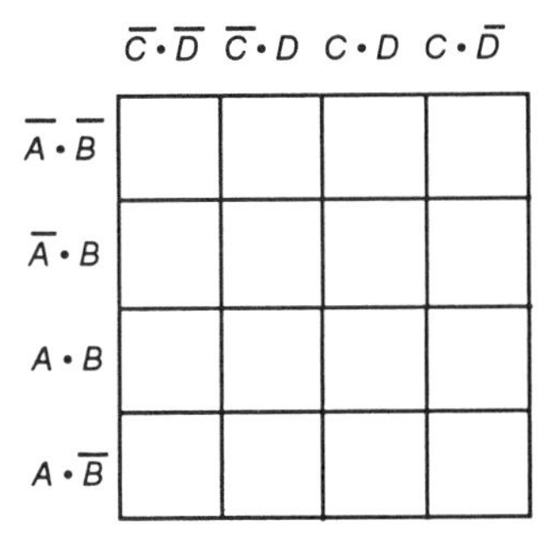

Figure 8.48 *Four-variable Karnaugh map*

Table 8.24 *Four-variable truth table*

A	B	C	D	Output	Boolean expression
0	0	0	0	0	
0	0	0	1	1	$\bar{A}\cdot\bar{B}\cdot\bar{C}\cdot D$
0	0	1	0	0	
0	0	1	1	0	
0	1	0	0	0	
0	1	0	1	1	$\bar{A}\cdot B\cdot\bar{C}\cdot D$
0	1	1	0	0	
0	1	1	1	0	
1	0	0	0	0	
1	0	0	1	0	
1	0	1	0	0	
1	0	1	1	0	
1	1	0	0	0	
1	1	0	1	0	
1	1	1	0	1	$A\cdot B\cdot C\cdot\bar{D}$
1	1	1	1	1	$A\cdot B\cdot C\cdot D$

Table 8.24 shows the truth table for four variables, only the Boolean expressions for the 1 outputs being shown. The Karnaugh map for the truth table is shown in Figure 8.49. The resulting Boolean expression is:

$$\bar{A}\cdot\bar{B}\cdot\bar{C}\cdot D+\bar{A}\cdot B\cdot\bar{C}\cdot D+A\cdot B\cdot C\cdot D+A\cdot B\cdot C\cdot\bar{D}$$

Looping simplifies this to:

$$\bar{A}\cdot\bar{C}\cdot D+A\cdot B\cdot C$$

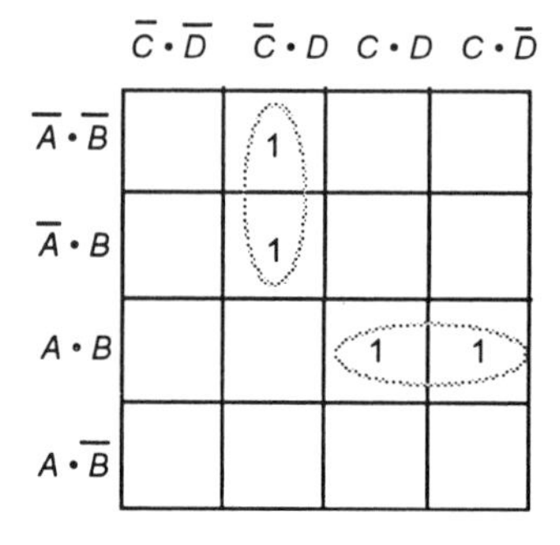

Figure 8.49 *Karnaugh map for four variables*

As further examples of looping, consider the Karnaugh maps shown in Figure 8.50 and the outcomes indicated, both before and after looping. Note that adjacent cells can be considered to be in the top and bottom rows or the left- and right-hand columns. Think of opposite edges of the map being joined together.

Looping a pair of adjacent 1s in a map eliminates the variable that appears in complemented and uncomplemented form.

Looping a quad of adjacent 1s in a map eliminates the two variables that appear in both complemented and uncomplemented form.

Looping an octet of adjacent 1s eliminates the three variables that appear in both complemented and uncomplemented form.

The procedure for using the Karnaugh map for simplification can be summarised as:

1 Draw the Karnaugh map, placing 1s in those cells which correspond to the 1s in the truth table.

2 Loop any octet of adjacent 1s.

3 Loop any quad containing 1s that have not already been looped.

4 Loop all pairs of adjacent 1s which are not entirely in any octet or quad or only partly within any octet or quad.

5 Identify any isolated 1s.

6 Form the OR sum of all the terms generated.

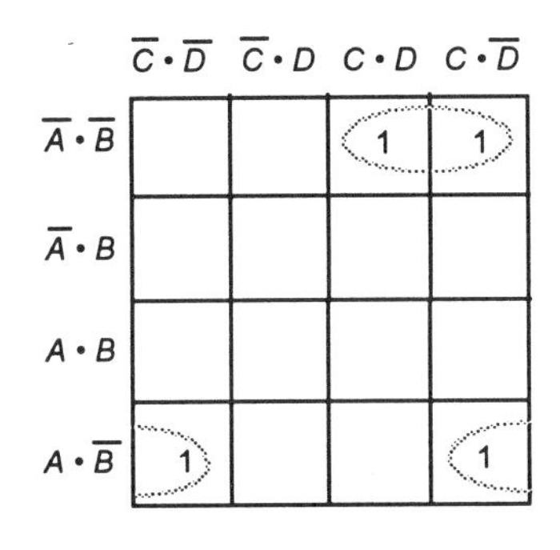

	$\bar{C}\cdot\bar{D}$	$\bar{C}\cdot D$	$C\cdot D$	$C\cdot\bar{D}$
$\bar{A}\cdot\bar{B}$				
$\bar{A}\cdot B$		1	1	
$A\cdot B$		1	1	
$A\cdot\bar{B}$				

$\bar{A}\cdot\bar{B}\cdot C\cdot D+\bar{A}\cdot\bar{B}\cdot C\cdot\bar{D}+A\cdot\bar{B}\cdot\bar{C}\cdot\bar{D}+A\cdot\bar{B}\cdot C\cdot\bar{D}$
$=\bar{A}\cdot\bar{B}\cdot C+A\cdot\bar{B}\cdot\bar{D}$

$\bar{A}\cdot B\cdot\bar{C}\cdot D+\bar{A}\cdot B\cdot C\cdot D+A\cdot B\cdot\bar{C}\cdot D+A\cdot B\cdot C\cdot D$
$=B\cdot D$

Figure 8.50 *Four-variable Karnaugh maps (continued on following page)*

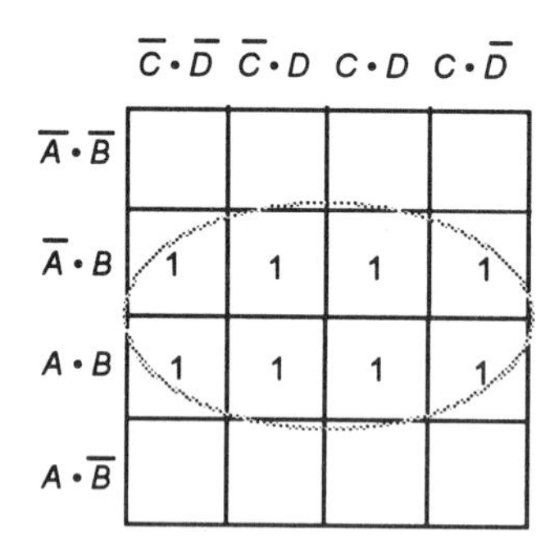

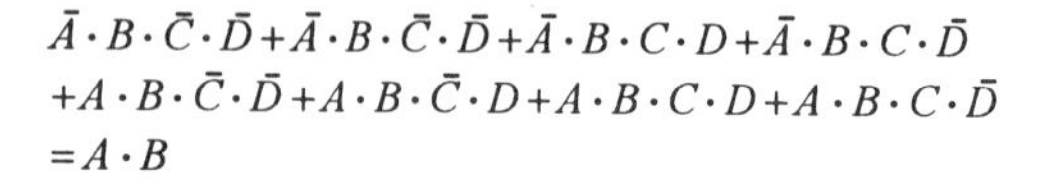

$\bar{A}\cdot B\cdot\bar{C}\cdot\bar{D}+\bar{A}\cdot B\cdot\bar{C}\cdot\bar{D}+\bar{A}\cdot B\cdot C\cdot D+\bar{A}\cdot B\cdot C\cdot\bar{D}$
$+A\cdot B\cdot\bar{C}\cdot\bar{D}+A\cdot B\cdot\bar{C}\cdot D+A\cdot B\cdot C\cdot D+A\cdot B\cdot C\cdot\bar{D}$
$=A\cdot B$

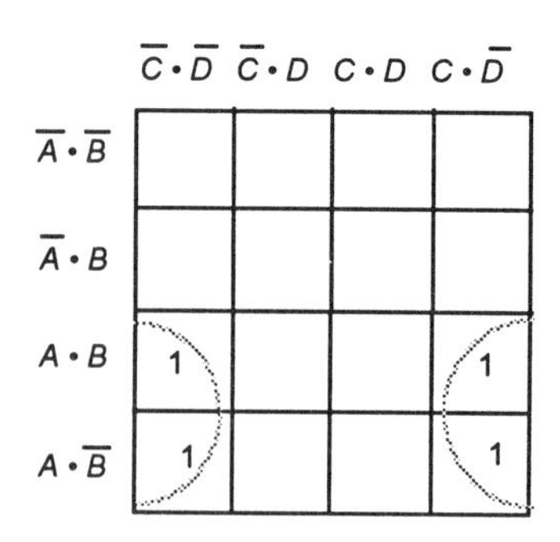

$A\cdot B\cdot\bar{C}\cdot\bar{D}+A\cdot\bar{B}\cdot\bar{C}\cdot\bar{D}+A\cdot B\cdot C\cdot\bar{D}+A\cdot\bar{B}\cdot C\cdot\bar{D}$
$=A\cdot\bar{D}$

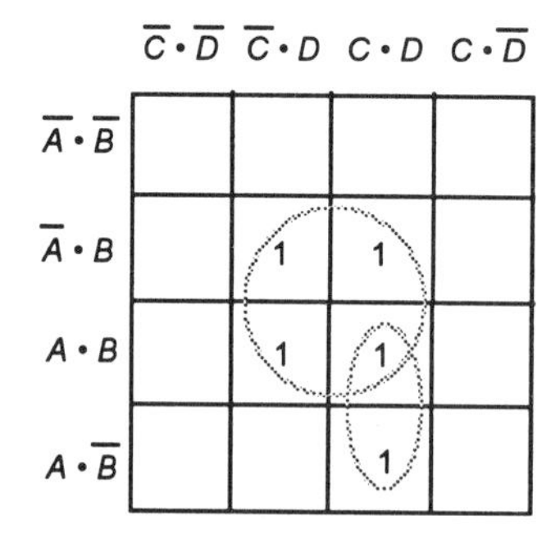

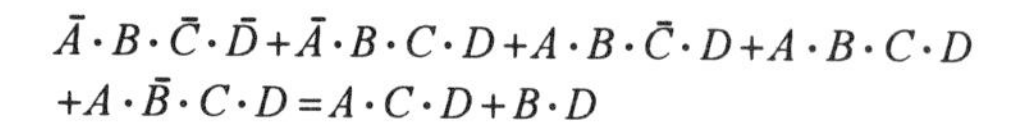

$\bar{A}\cdot B\cdot\bar{C}\cdot\bar{D}+\bar{A}\cdot B\cdot C\cdot D+A\cdot B\cdot\bar{C}\cdot D+A\cdot B\cdot C\cdot D$
$+A\cdot\bar{B}\cdot C\cdot D=A\cdot C\cdot D+B\cdot D$

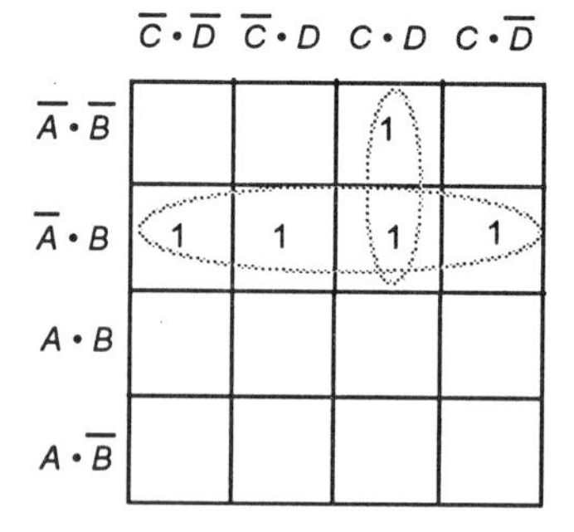

$\bar{A}\cdot\bar{B}\cdot C\cdot D+\bar{A}\cdot B\cdot\bar{C}\cdot\bar{D}+\bar{A}\cdot B\cdot\bar{C}\cdot D+\bar{A}\cdot B\cdot C\cdot D$
$+\bar{A}\cdot B\cdot C\cdot\bar{D}=A\cdot\bar{B}+\bar{A}\cdot C\cdot D$

Figure 8.50 *Four-variable Karnaugh maps (continued from previous page)*

Example

A system is required that will only start a cylinder to initiate some action when at least two of three sensors, A, B, C, give signals and will not start when less than two sensors give signals. Design a logic system to implement this.

For an output to start the cylinder, the truth table required is given by Table 8.25. A 1 is used to indicate a signal from a sensor. The resulting Karnaugh map is shown in Figure 8.51. The Boolean expression which fits the map is $A \cdot B + B \cdot C + A \cdot C$ which can be written as $A \cdot (B + C) + B \cdot C$. Figure 8.52 shows the logic system.

Table 8.25 *Example*

A	B	C	Output	Boolean expression
0	0	0	0	
0	0	1	0	
0	1	0	0	
0	1	1	1	$\bar{A}\cdot B\cdot C$
1	0	0	0	
1	0	1	1	$A\cdot\bar{B}\cdot C$
1	1	0	1	$A\cdot B\cdot\bar{C}$
1	1	1	1	$A\cdot B\cdot C$

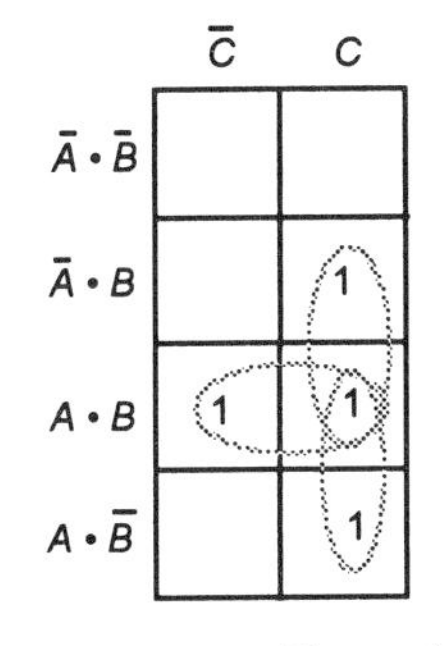

Figure 8.51 *Example*

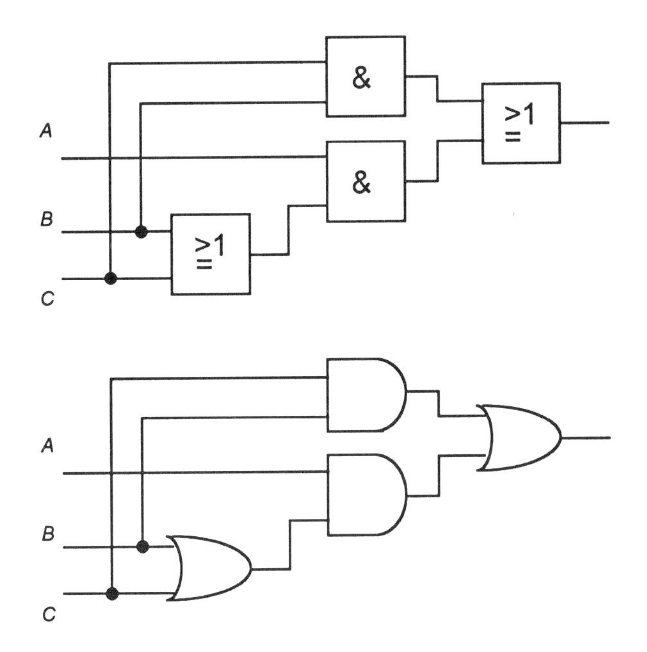

Figure 8.52 *Example*

Example

Determine the Boolean function describing the relation between the output from the logic circuit shown in Figure 8.53. Hence, consider how the circuit could be simplified. This might be a circuit used with a car warning buzzer so that it sounds when the key is in the ignition (A) and a car door is opened (B) or the headlights are on (C) and the car door open.

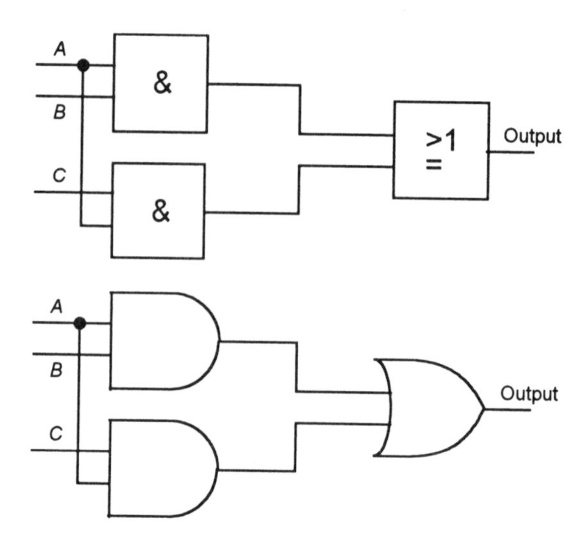

Figure 8.53 *Example*

We might argue it as follows. We have two AND gates and an OR gate. The output from the top AND gate is $A \cdot B$, and from the lower AND gate $C \cdot A$. These outputs are the inputs to the OR gate and thus the output is $A \cdot B + C \cdot A$. The circuit can be simplified by considering the Boolean algebra. Using equation [9] the Boolean function can be written as:

$$A \cdot B + C \cdot A = A \cdot (B + C)$$

Alternatively we might draw up the truth table (Table 8.26) and the corresponding Karnaugh map (Figure 8.54). The same result is obtained.

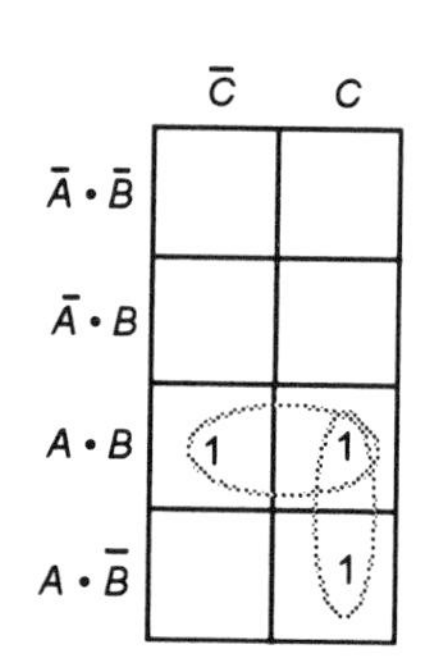

Figure 8.54 *Example*

Table 8.26 *Example*

A	*B*	*C*	Output	Boolean expression
0	0	0	0	
0	0	1	0	
0	1	0	0	
0	1	1	0	
1	0	0	0	
1	0	1	1	$A \cdot \bar{B} \cdot C$
1	1	0	1	$A \cdot B \cdot \bar{C}$
1	1	1	1	$A \cdot B \cdot C$

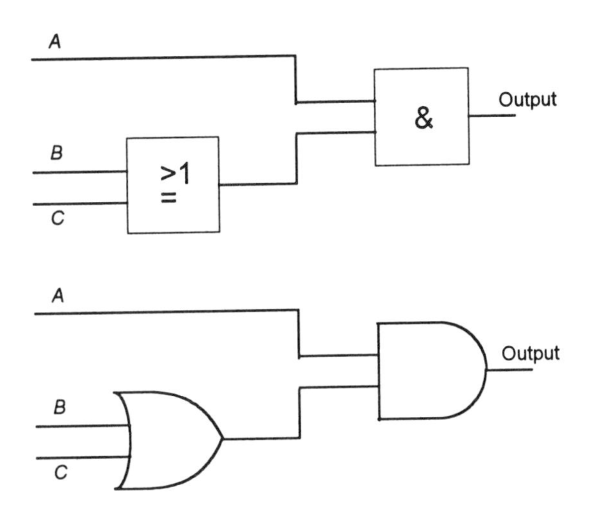

Figure 8.55 *Example*

We thus have A and B or C. This function now describes a logic circuit with just two gates, an OR gate and an AND gate. Figure 8.55 shows the circuit.

Problems

1 What conditions need to be realised for the cylinder in Figure 8.56 to retract?

2 What conditions need to be realised for the cylinder to retract in the circuit shown in Figure 8.57?

3 Design a system that will allow a cold-room door to be opened, as a result of the extension of a piston in a cylinder from either side, when push-buttons are pressed.

4 Simplify the following Boolean functions:

(a) $a \cdot b \cdot d + a \cdot b \cdot c \cdot d$, (b) $\overline{\bar{a}b\bar{c}}$, (c) $\bar{a} \cdot b \cdot \bar{c} + a \cdot b \cdot \bar{c} + b \cdot \bar{c} \cdot d$,

(d) $a + \bar{a} \cdot \bar{b} \cdot c$, (e) $\overline{(\overline{\bar{a}+b})(\overline{a \cdot b + \bar{c}})}$, (f) $(\overline{a+c}) + \overline{a \cdot b} \cdot (b+c)$

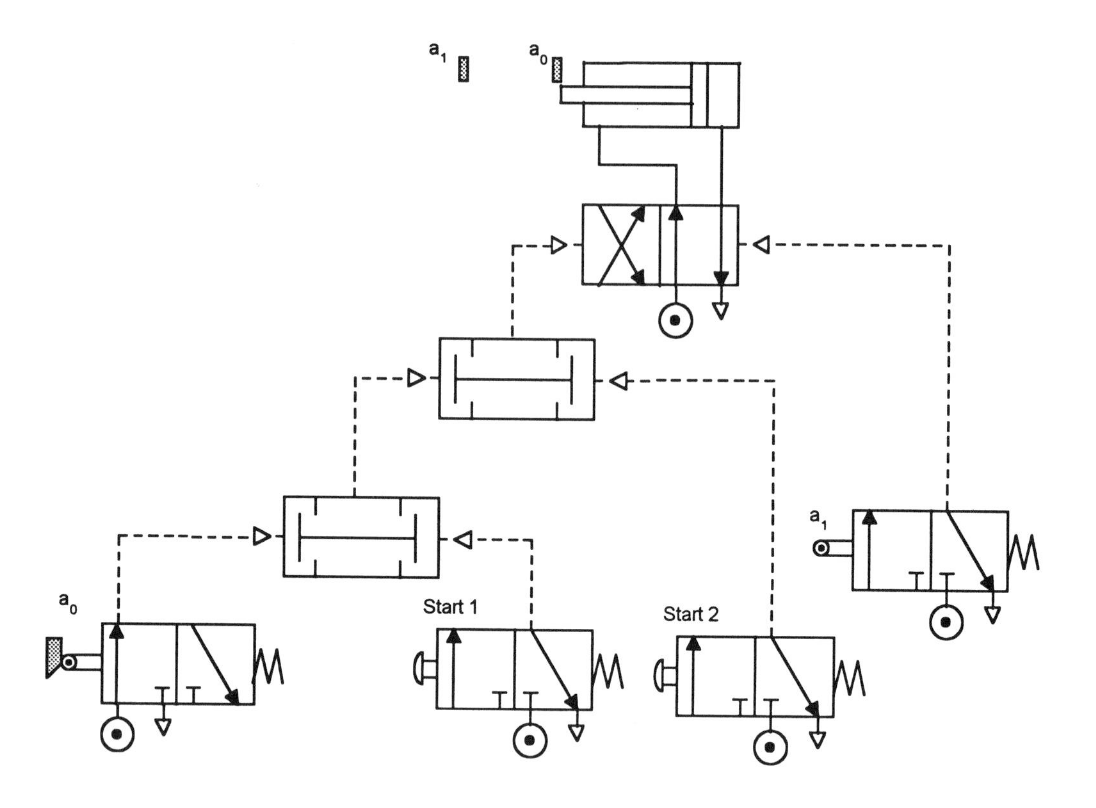

Figure 8.56 *Problem 1*

5 State a Boolean function that can be used to represent each of the switching circuits shown in Figure 8.58.

6 Give the truth tables for the switching circuits represented by the Boolean functions:

(a) $(a+\bar{b})+(a+\bar{c})$, (b) $\bar{a}\cdot(a\cdot b+\bar{b})\cdot\bar{b}$

7 Draw switching circuits to represent the Boolean functions:

(a) $a\cdot b$, (b) $a\cdot b + b$, (c) $c\cdot(a\cdot b+a\cdot\bar{b})$,

(d) $a\cdot(a\cdot b\cdot\bar{c}+a\cdot(\bar{b}+c))$

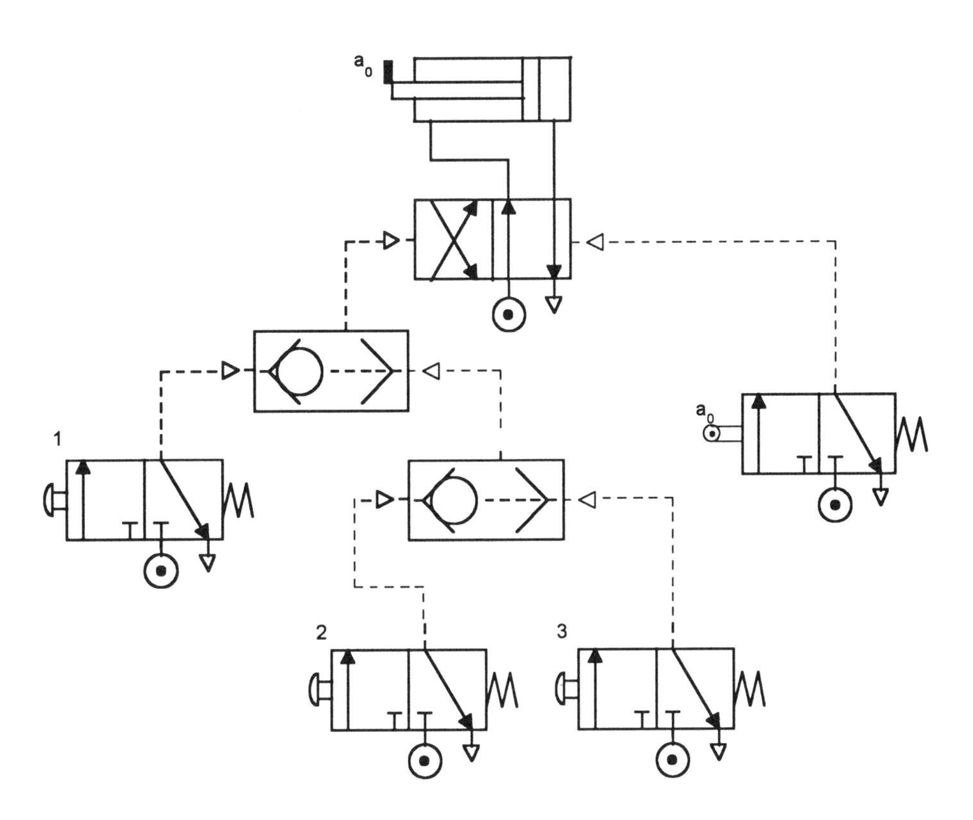

Figure 8.57 *Problem 2*

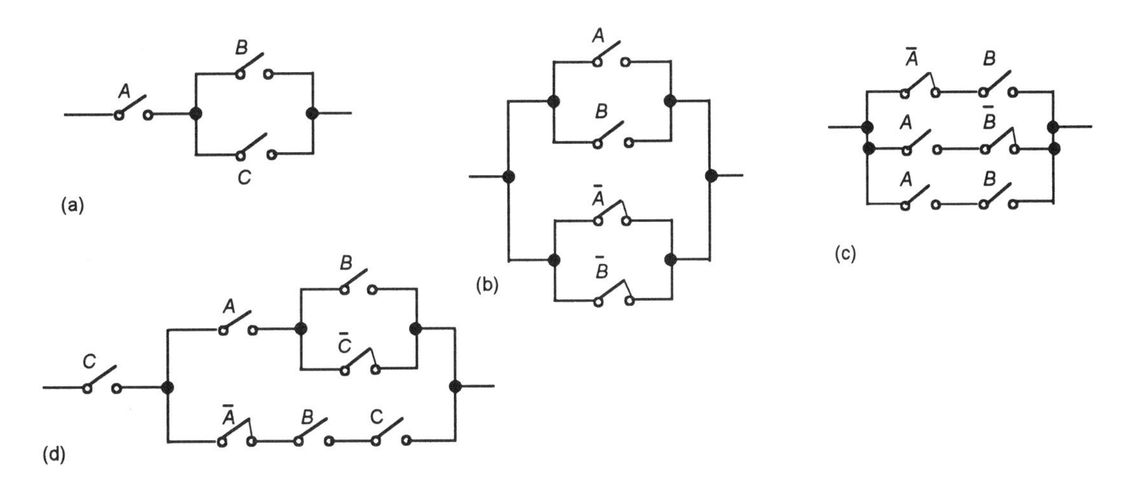

Figure 8.58 *Problem 5*

8 Derive the Boolean functions for the truth tables in Table 8.27(a) and (b).

Table 8.27(a)

A	*B*	*C*	Function
0	0	0	0
0	0	1	1
0	1	0	0
0	1	1	0
1	0	0	0
1	0	1	0
1	1	0	1
1	1	1	0

Table 8.27(b)

A	*B*	*C*	Function
0	0	0	0
0	0	1	0
0	1	0	0
0	1	1	1
1	0	0	0
1	0	1	0
1	1	0	0
1	1	1	1

9 State the Boolean functions for the logic circuits shown in Figure 8.59.

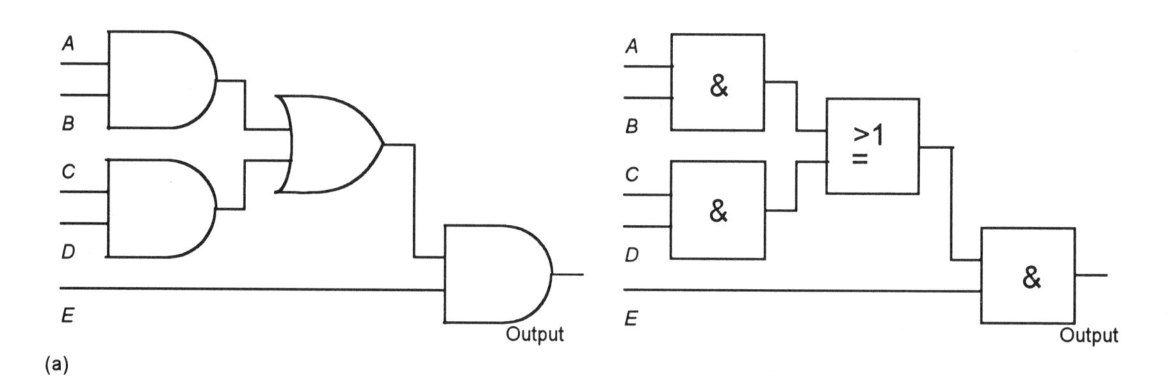

Figure 8.59 *Problem 9* *(continued on next page)*

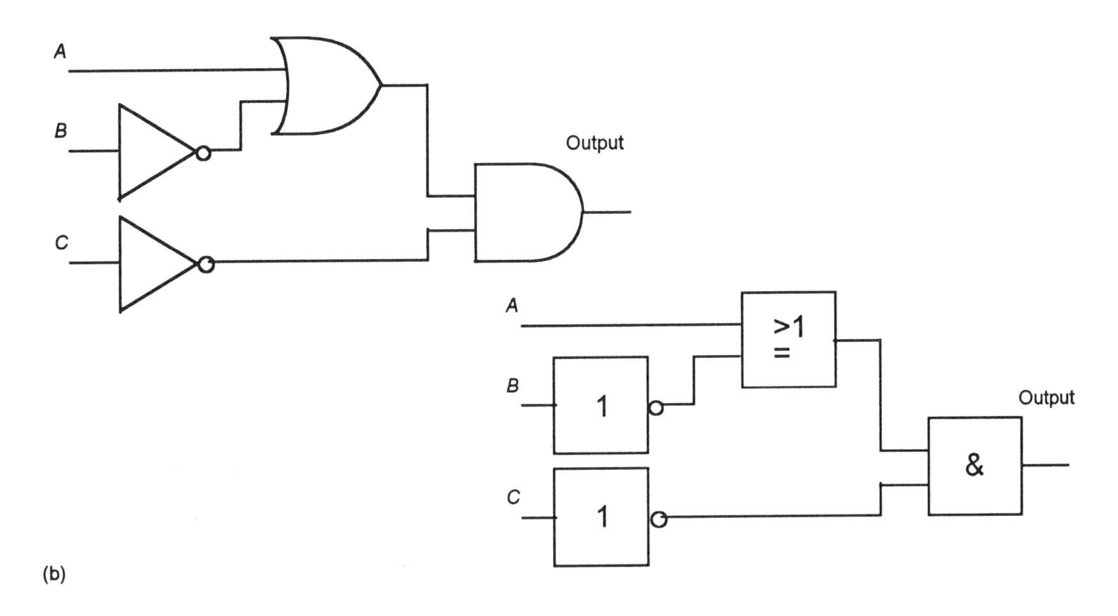

Figure 8.59 *Problem 9 (continued from previous page)*

10 Devise logic gate systems to give the following Boolean functions:

(a) $\bar{A}+B+C$, (b) $A \cdot B+C$, (c) $A \cdot \bar{B}+B \cdot C$, (d) $\bar{D} \cdot (\bar{A}+\bar{B} \cdot \bar{C})$

11 Determine the Boolean function for a logic circuit which is to give a high output only when any two of the three inputs A, B and C are low.

12 Determine the Boolean function for a logic circuit which is to give a high output when all three inputs A, B and C are low, a high output when all three inputs are high, a high output when just A or just B is high and a high output when A and B are high, otherwise the output is low for all the other conditions.

13 Determine the Boolean equations describing the logic circuits in Figure 8.60, then simplify the equations and hence obtain simplified logic circuits

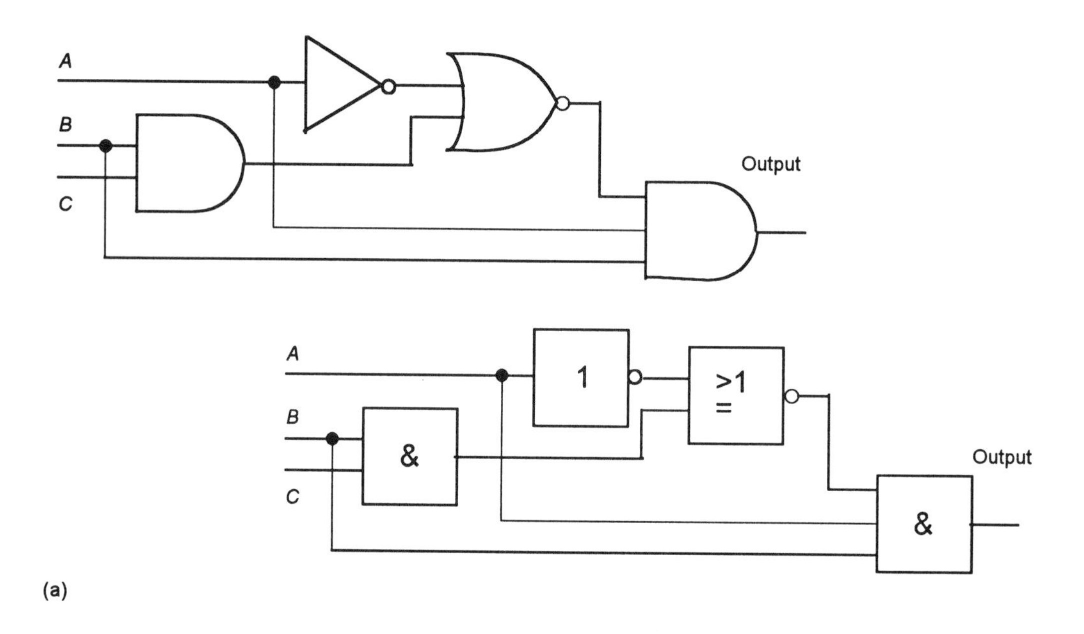

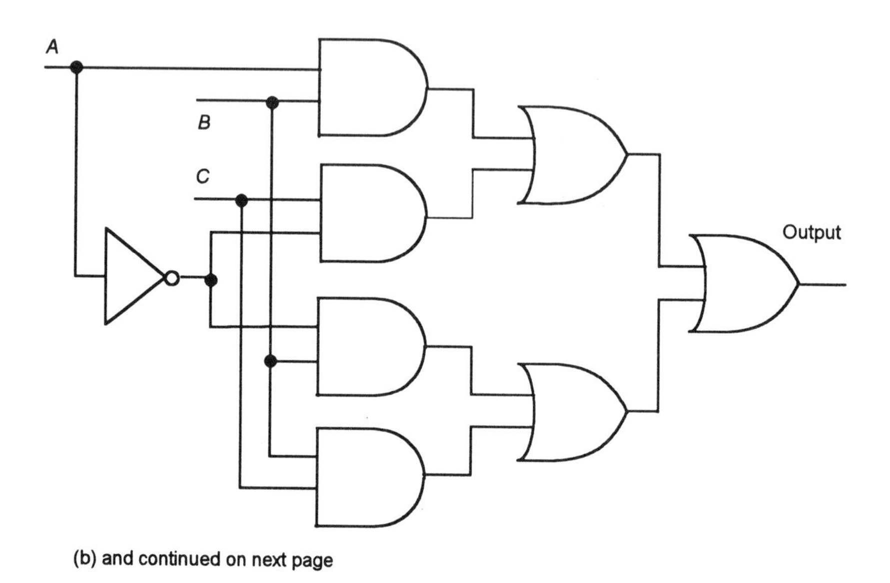

(b) and continued on next page

Figure 8.60 *Problem 13 (continued on next page)*

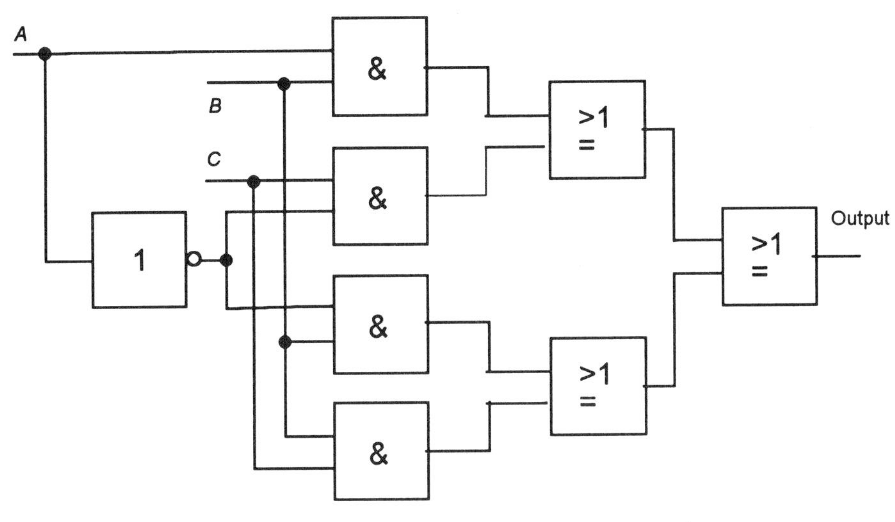

Figure 8.60 *Problem 13 (continued from previous page)*

14 Draw the Karnaugh maps for the truth tables in Table 8.28 and hence determine the simplified Boolean expression for the tables.

Table 8.28(a)

A	*B*	Output
0	0	1
0	1	1
1	0	1
1	1	1

Table 8.28(b)

A	*B*	*C*	Output
0	0	0	0
0	0	1	1
0	1	0	1
0	1	1	1
1	0	0	0
1	0	1	1
1	1	0	0
1	1	1	1

Table 8.28(c)

A	*B*	*C*	*D*	Output
0	0	0	0	1
0	0	0	1	0
0	0	1	0	0
0	0	1	1	0
0	1	0	0	1
0	1	0	1	0
0	1	1	0	0
0	1	1	1	0
1	0	0	0	0
1	0	0	1	0
1	0	1	0	1
1	0	1	1	1
1	1	0	0	0
1	1	0	1	0
1	1	1	0	1
1	1	1	1	1

15 Write simplified Boolean expressions for the Karnaugh maps in Figure 8.61.

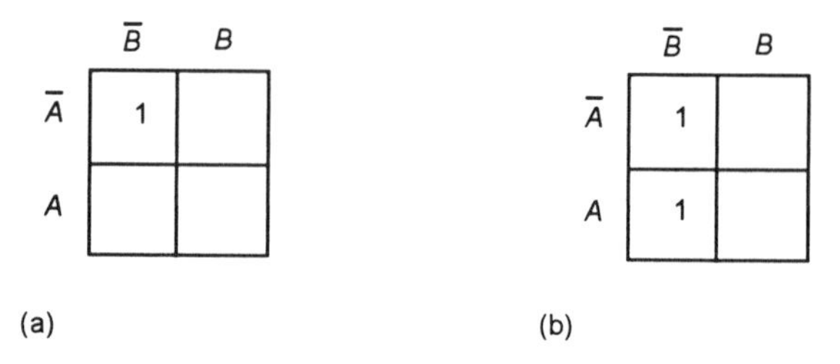

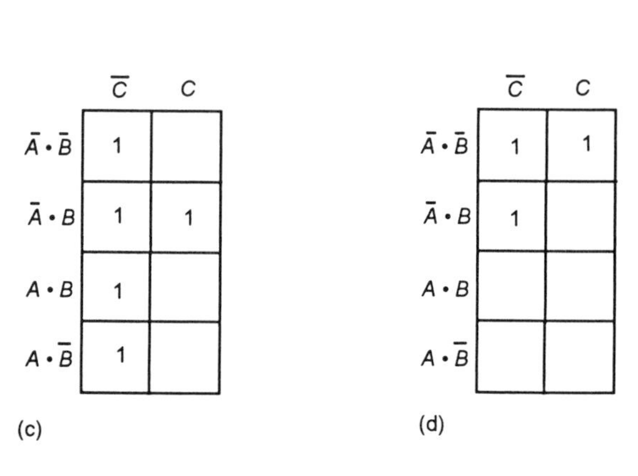

Figure 8.61 *Problem 15 (continued on next page)*

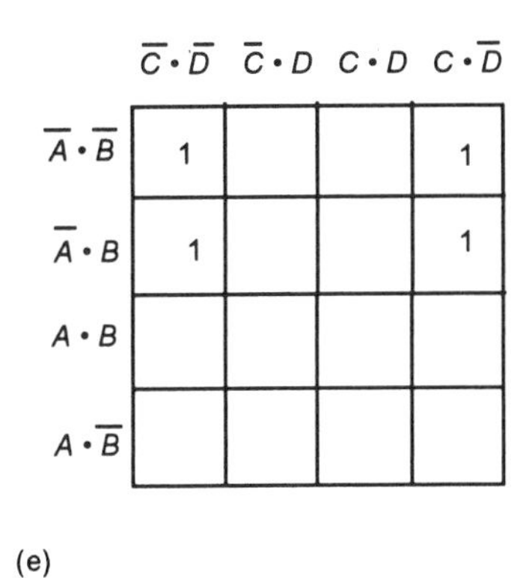

(e)

	$\overline{C}\cdot\overline{D}$	$\overline{C}\cdot D$	$C\cdot D$	$C\cdot\overline{D}$
$\overline{A}\cdot\overline{B}$			1	1
$\overline{A}\cdot B$				
$A\cdot B$				
$A\cdot\overline{B}$	1			1

(f)

Figure 8.61 *Problem 15 (continued from previous page)*

16 Devise a logic system to implement the truth table given in Table 8.29, minimising the number of gates used.

Table 8.29 *Problem 16*

A	*B*	*C*	Output
0	0	0	0
0	0	1	0
0	1	0	0
0	1	1	0
1	0	0	0
1	0	1	1
1	1	0	1
1	1	1	1

17 Devise a logic circuit with inputs *A*, *B*, *C* that will give a 1 output whenever *A* is 0 or whenever *B* and *C* are both 1.

18 A plant has an alarm system to warn of critical conditions occurring. It consists of four sensors *A*, *B*, *C* and *D* which give outputs of 0 or 1 depending on the conditions. *A* monitors liquid level in a tank, *B* monitors the flow rate, *C* monitors the temperature and *D* monitors the pressure. Design a system, giving the answer as the Boolean expression with the minimum number of gates, which will activate an alarm when any of the following conditions occur: high fluid level with high temperature and high pressure; low fluid level with high temperature and high flow rate; low fluid level with low temperature and high pressure; low fluid level with low flow rate and high temperature.

19 A system used for controlling a machine tool has two indicators, one of which is illuminated when the plant is in a fit state to be operated and the other is illuminated when it is not ready to be operated. The indicators are controlled by four sensors A, B, C and D which give either a 0 or 1 output. The condition for the plant to be not in a fit state to operate is when A and B are both 0, with C being either 0 or 1; when A and C are zero and B is 1. All other conditions allow the plant to be operated. Devise a logic system to give outputs to illuminate both indicators.

9 Programmable logic controllers

9.1 Introduction

Solenoid directional control valves are controlled by switching the current to them on or off. This might be done by the use of relays, the circuit being wired up to give the required sequence of signals to the valves, or by the use of a programmable logic controller. *Programmable logic controllers* (PLCs) are microprocessor-based controllers that use a programmable memory to store instructions and implement functions such as logic, sequencing, timing, counting and arithmetic in order to control machines and processes. Thus with inputs to the PLC of signals from sensors such as limit switches and start/stop switches, output signals are sent to the solenoids of the valves, the sequence of such signals being determined by the programme of instructions inputted to the PLC. This chapter is about such devices and how they are used with pneumatic and hydraulic systems.

To illustrate the above, Figure 9.1 shows a control system using solenoid-controlled directional valves to determine the movement of pistons in a cylinder with the sequence being determined by the use of relays sending signals to the solenoids. The arrangement shown gives the sequence:

A+ and B+ concurrently, C+, A– and B– concurrently, C–

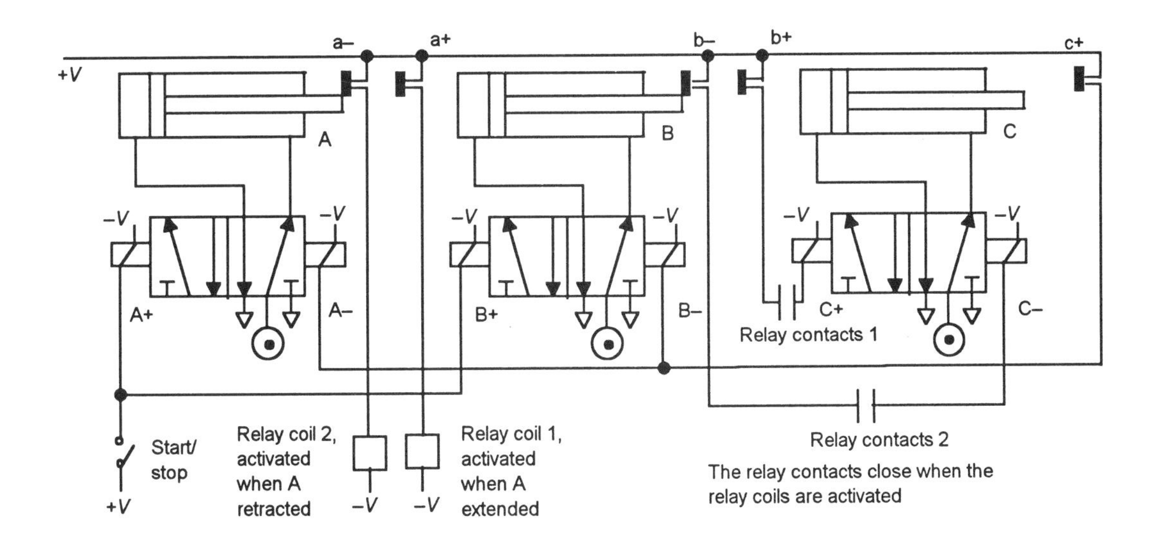

Figure 9.1 *A relay control system*

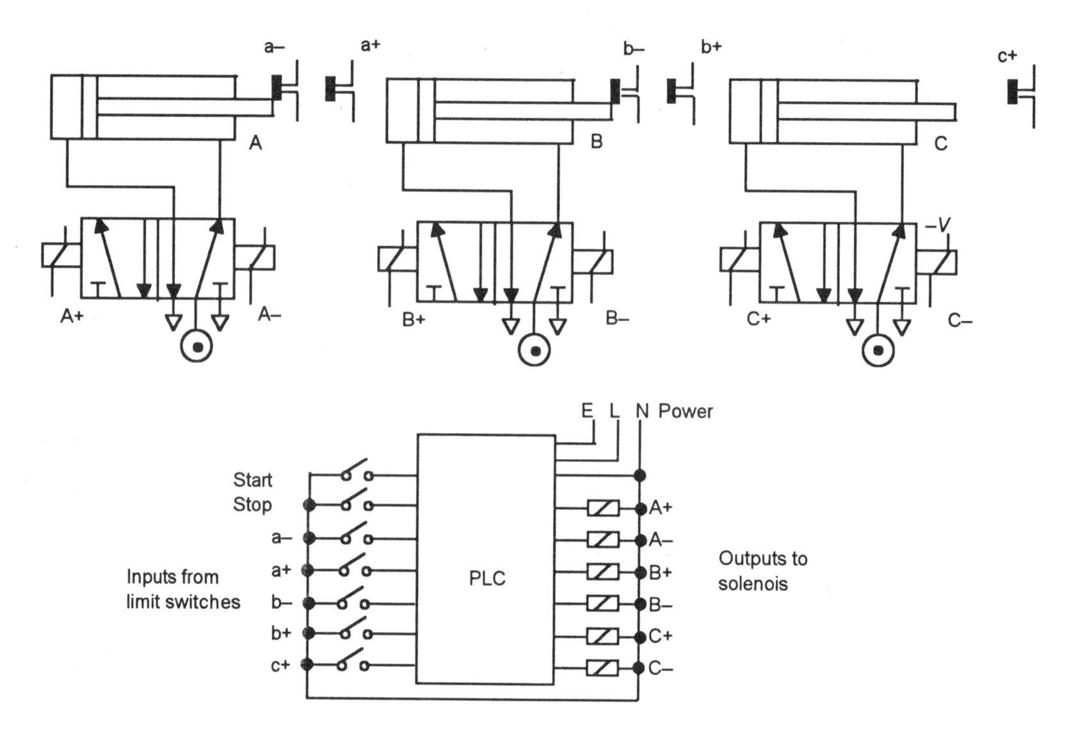

Figure 9.2 *A PLC system*

When the start switch is closed, current is supplied to the A+ solenoid and the B+ solenoid with the result that cylinders A and B extend. The extension closes limit switches a+ and b+. The closing of a+ feeds current to relay coil 1 which then becomes energised and closes its contacts. As a result, current is now fed to the C+ solenoid and results in C extending. When C is extended, c+ is closed and feeds current to the A– and B– solenoids, causing A and B to retract. When A is retracted, a– closes and energises relay coil 2. This closes the relay 2 contacts and so feeds current to the C– solenoid and results in C retracting.

If we wish to change the above sequence we have to rewire the circuit. This type of control is termed *hardwired* since the wiring is made and fixed for a particular form of control. Alternatively, instead of hardwired control, we can use a PLC which is programmed by inputting instructions as to how the control should be exercised. Figure 9.2 shows the arrangement. The PLC has the start and stop switches and the sensors connected as inputs and the solenoid coils as outputs. When the start switch is closed, the PLC starts its program. It monitors the inputs from the limit switches and, on the results of that monitoring, gives outputs to energise the relevant solenoid coils. Compared with the relay form of control, the PLC reduces the complexity of the wiring required and changes can be more easily implemented.

9.2 The PLC system

Typically a PLC system has five basic components. These are the processor unit, memory, the power supply unit, input/output interface section and the programming device (Figure 9.3).

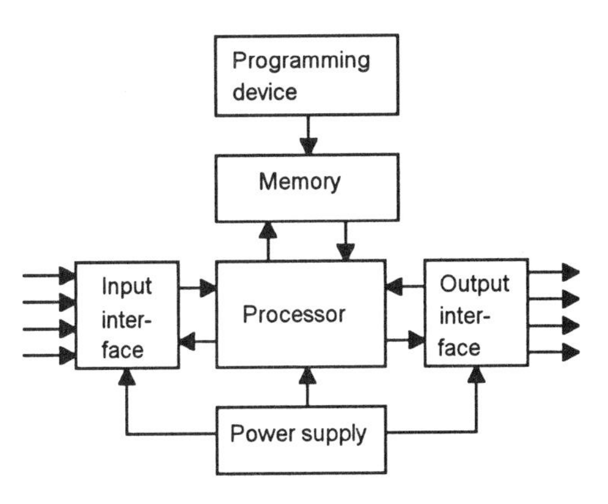

Figure 9.3 *The PLC system*

The *processor unit* interprets the input signals and carries out the control actions, according to the program stored in its memory, communicating the decisions as action signals to the outputs. The *power supply unit* is needed to convert the mains a.c. voltage to the low d.c. voltage (5 V) necessary for the processor and the circuits in the input and output interface modules. The *programming device* is used to enter the required program into the memory of the processor. The *memory unit* is where the rules are stored that are to be used for the control actions by the microprocessor. The *input and output sections* are where the processor receives information from external devices and communicates information to external devices.

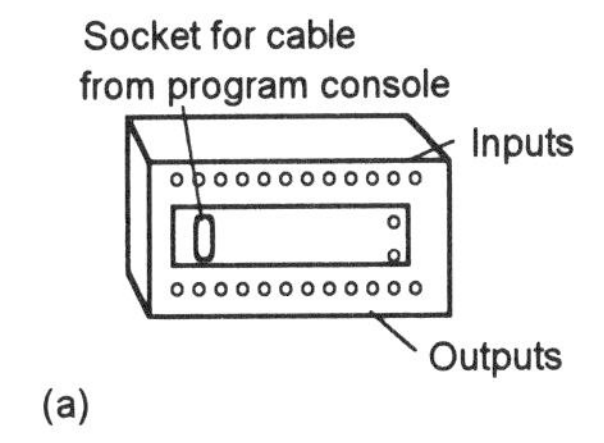

9.2.1 Mechanical design of PLC systems

There are two common types of mechanical design: a *single box*, and the *modular* and *rack types*. The single-box type is commonly used for small programmable controllers and is supplied as an integral compact package complete with power supply, processor, memory, and input/output units (Figure 9.4(a)). Typically such a PLC might have 40 input/output points and a memory which can store some 300 to 1000 instructions. The modular type consists of separate modules for power supply, processor, etc. which are often mounted on rails within a metal cabinet. The rack type can be used for all sizes of programmable controllers and has the various functional units packaged in individual modules which can be plugged into sockets in a base rack (Figure 9.4(b)). The mix of modules required for a particular purpose is decided by the user and the appropriate ones then plugged into the rack. Thus it is comparatively easy to expand the number

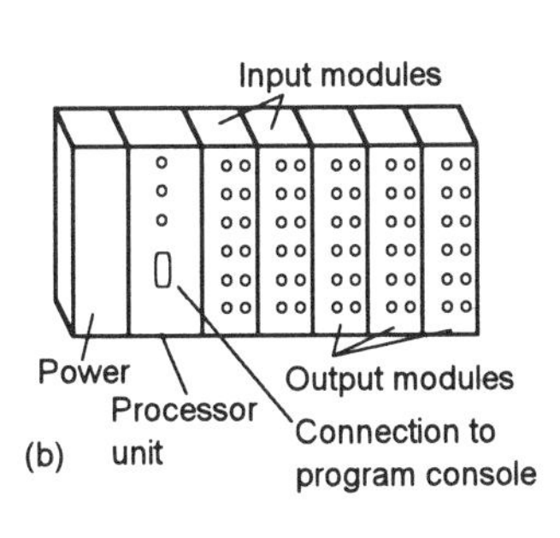

Figure 9.4 *(a) Single box, (b) modular/rack forms*

of input/output connections by just adding more input/output modules or to expand the memory by adding more memory units.

Programs are entered into a PLC's memory using a program device which is usually detachable and can be moved from one controller to the next without disturbing operations. It is not necessary for the programming device to be connected to the PLC. Programming devices can be a hand-held device, a desktop console or a computer. Hand-held systems incorporate a small keyboard and liquid crystal display, Figure 9.5 showing a typical form. Hand-held programming consoles will normally contain enough memory to allow the unit to retain programs while being carried from one place to another. Only when the program is ready is it transferred to the PLC. Desktop devices are likely to have a visual display unit with a full keyboard and screen display. Personal computers are widely used as program development workstations. Some PLCs only require the computer to have appropriate software, others special communication cards. The software may load to give a series of menu headings, which when selected give drop-down lists from which functions can be selected. A major advantage of using a computer is that the program can be stored on the hard disk or a floppy disk and copies easily made.

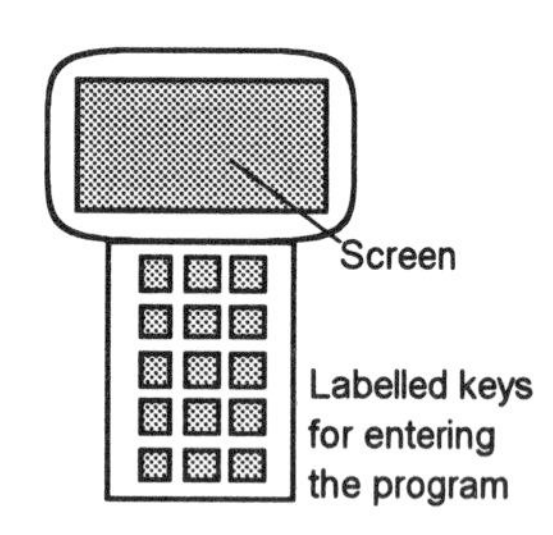

Figure 9.5 *Hand-held programmer*

9.2.2 Input/output unit

The input/output unit provides the interface between the system and the outside world, allowing for connections to be made through input/output channels to input devices such as sensors and output devices such as motors and solenoids. It is also through the input/output unit that programs are entered from a program panel. Every input/output point has a unique address which can be used by the processor.

The input/output channels provide signal conditioning and isolation functions so that sensors and actuators can often be directly connected to them without the need for other circuitry. Electrical isolation from the external world is usually by means of *optoisolators* (the term *optocoupler* is also often used). Figure 9.6 shows the principle of an optoisolator. When a digital pulse passes through the light-emitting diode, a pulse of infrared radiation is produced. This pulse is detected by the phototransistor and gives rise to a voltage in that circuit. The gap between the light-emitting diode and the phototransistor gives electrical isolation but the arrangement still allows for a digital pulse in one circuit to give rise to a digital pulse in another circuit.

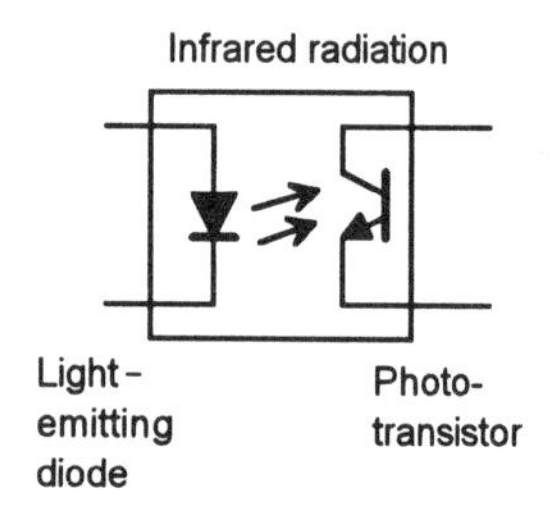

Figure 9.6 *Optoisolator*

The digital signal that is generally compatible with the microprocessor in the PLC is 5 V d.c. However, signal conditioning in the input channel, with isolation, enables a wide range of input signals to be supplied to it. A range of inputs might be available with a larger PLC, e.g. 5 V, 24 V, 110 V and 240 V digital/discrete, i.e. on–off, signals. A small PLC is likely to have just one form of input, e.g. 24 V. Figure 9.7 shows the basic form a d.c. input channel might take.

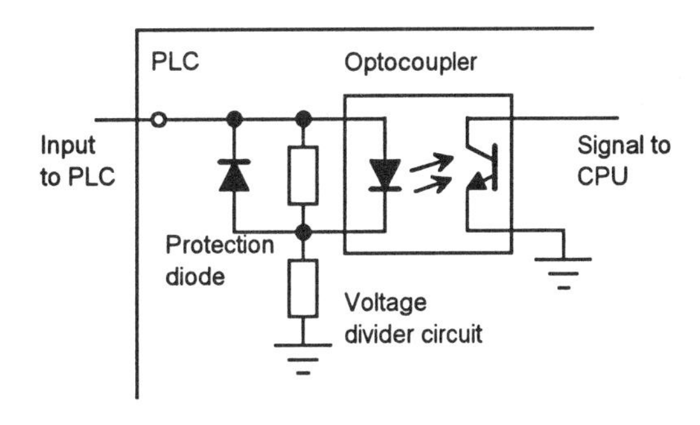

Figure 9.7 *Basic d.c. input circuit*

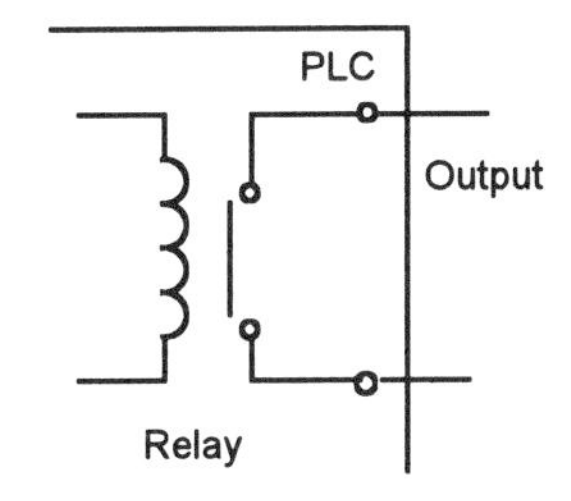

Figure 9.8 *Relay output*

Outputs are often specified as being of relay type, transistor type or triac type. With the *relay type*, the signal from the PLC output is used to operate a relay and so is able to switch currents of the order of a few amperes in an external circuit. The relay not only allows small currents to switch much larger currents but also isolates the PLC from the external circuit. Relays are, however, relatively slow to operate. Relay outputs are suitable for a.c. and d.c. switching. They can withstand high surge currents and voltage transients. Figure 9.8 shows the basic feature of a relay output.

The *transistor type* of output uses a transistor to switch current through the external circuit. This gives a considerably faster switching action. It is, however, strictly for d.c. switching and is destroyed by overcurrent and high reverse voltage. As a protection, either a fuse or built-in electronic protection are used. Optoisolators are used to provide isolation. Figure 9.9 shows the basic form of such a transistor output channel.

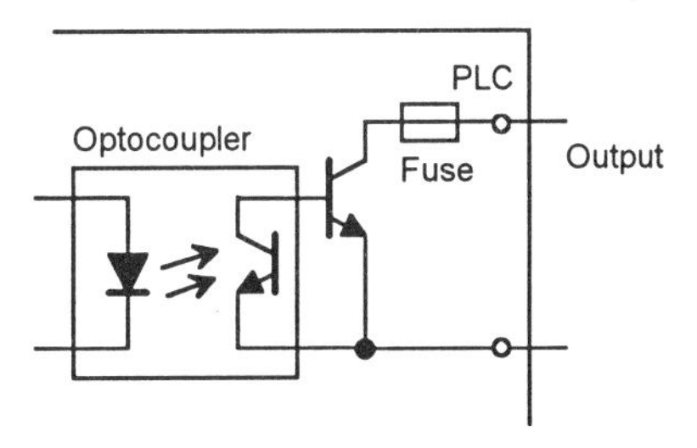

Figure 9.9 *Basic form of transistor output*

Triac outputs, with optoisolators for isolation, can be used to control external loads which are connected to the a.c. power supply. It is strictly for a.c. operation and is very easily destroyed by overcurrent. Fuses are virtually always included to protect such outputs.

The output from the input/output unit will be digital with a level of 5 V. However, after signal conditioning with relays, transistors or triacs, the output from the output channel might be a 24 V, 100 mA switching signal, a d.c. voltage of 110 V, 1 A or perhaps 240 V, 1 A a.c. or 240 V, 2 A a.c. from a triac output channel. With a small PLC, all the outputs might be of one type, e.g. 240 V a.c., 1 A. With modular PLCs, however, a range of outputs can be accommodated by selection of the modules to be used.

The following illustrates the types of inputs and outputs available with a small PLC, one of the Mitsubishi F2 series:

Number of inputs 12
Number of outputs 8
Input specification:
  voltage 24 V ± 4 V, d.c.
  operation current off → on, 4 mA d.c. max.,
  on → on, 1.5 mA d.c. max.
Output specification:
  Type: Relay
    Relay isolation
    2 A per point resistive load, 35 VA inductive load, 100 W lamp load
  Type: Transistor
    Optocoupler isolation
    1 A per point resistive load, 24 W inductive load, 100 W lamp load
  Type: Triac
    Optocoupler isolation
    1 A per point, 50 VA 110/120 V a.c., 100 VA 220/240 V a.c. inductive load, 100 W lamp load

9.2.3 Inputs and outputs

A PLC is continually running through its program and updating its outputs as a result of the current state of the input signals. Each such loop is termed a cycle (Figure 9.10). A typical cycle time is of the order of 10 to 50 ms.

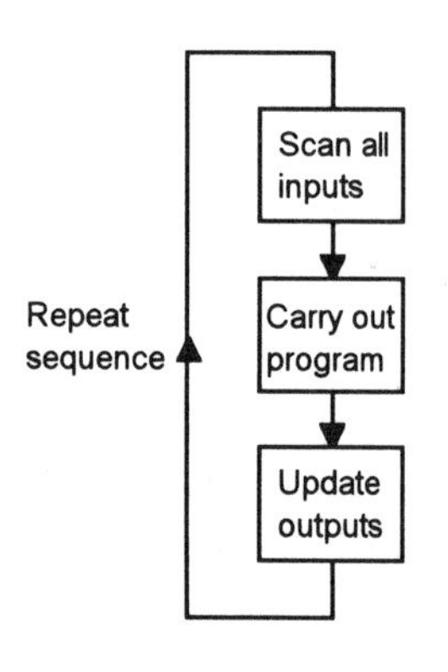

Figure 9.10 *PLC operation*

The PLC has to be able to identify each particular input and output. It does this by allocating unique addresses to each input and output. With a small PLC this is likely to be just a number, prefixed by a letter to indicate whether it is an input or an output. Thus for the Mitsubishi PLC we might have inputs with addresses X400, X401, X402, etc. and outputs with addresses Y430, Y431, Y432, etc. Toshiba also uses an X and Y, with inputs such as X000 and X001 and outputs Y000 and Y001.

With larger PLCs having several racks of input and output channels, the racks are numbered. With the Allen Bradley PLC-5, the rack containing the processor is given the number 0 and the addresses of the other racks are numbered 1, 2, 3, etc. according to how set-up switches are set. Each rack can have a number of modules and each one deals with a number of inputs and/or outputs. Thus addresses can be of the form shown in Figure 9.11.

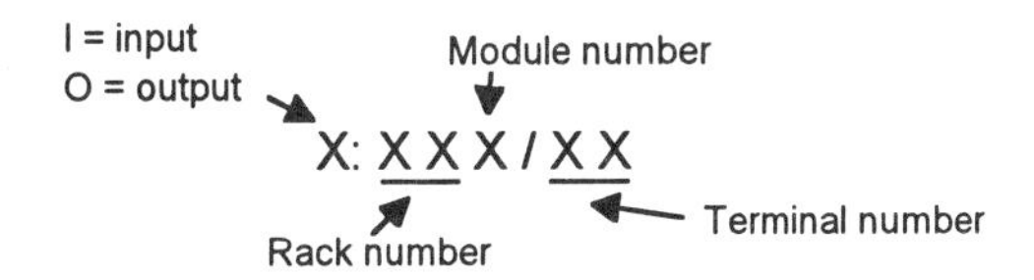

Figure 9.11 *Allen Bradley PLC-5 addressing*

For example, we might have an input with address I:012/03. This would indicate an input, rack 01, module 2 and terminal 03.

With the Siemens SIMATIC S5, the inputs and outputs are arranged in groups of 8. Each 8 group is termed a byte and each input or output with an 8 is termed a bit. The inputs and outputs thus have their addresses in terms of the byte and bit numbers, effectively giving a module number followed by a terminal number, a full stop (.) separating the two numbers. Figure 9.12 shows the system. Thus I0.1 is an input at bit 1 in byte 0, Q2.0 is an output at bit 0 in byte 2.

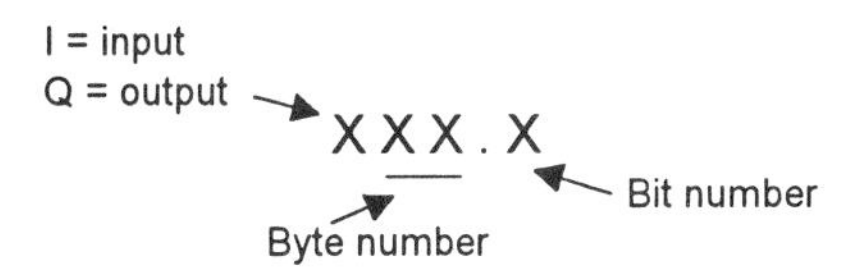

Figure 9.12 *Siemens SIMATIC S5 addressing*

9.3 Programming

A very common method of writing programs for PLCs is based on the use of *ladder diagrams*. As an introduction to ladder diagrams, consider the simple wiring diagram for an electrical circuit in Figure 9.13(a). The diagram shows the circuit for switching on or off an electric motor. We can redraw this diagram in a different way, using two vertical lines to represent the input power rails and stringing the rest of the circuit between them. Figure 9.13(b) shows the result. Both circuits have the switch in series with the motor and supplied with electrical power when the switch is closed. The circuit shown in Figure 9.13(b) can be termed a *ladder diagram*.

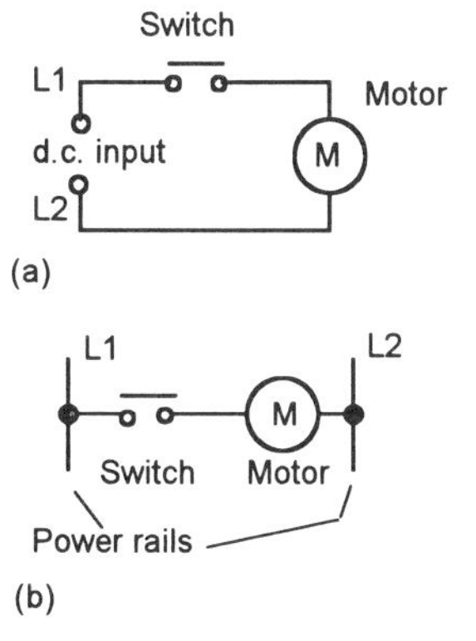

Figure 9.13 *Ways of drawing a circuit*

With such a diagram the power supply for the circuits is always shown as two vertical lines with the rest of the circuit as horizontal lines. The power lines, or rails as they are often termed, are like the vertical sides of a ladder with the horizontal circuit lines like the rungs of the ladder. The horizontal rungs show only the control portion of the circuit, in the case of Figure 9.13 it is just the switch in series with the motor. Circuit diagrams often show the relative physical location of the circuit components and how they are actually wired. With ladder diagrams no attempt is made to show the actual physical locations and the emphasis is on clearly showing how the control is exercised.

Writing a ladder program is equivalent to drawing a switching circuit. The ladder diagram consists of two vertical lines representing the power rails.

Circuits are connected as horizontal lines, i.e. the rungs of the ladder, between these two verticals.

In drawing a ladder diagram, certain conventions are adopted:

1 The vertical lines of the diagram represent the power rails between which circuits are connected.

2 Each rung on the ladder defines one operation in the control process.

3 A ladder diagram is read from left to right and from top to bottom, Figure 9.14 showing the scanning motion employed by the PLC. The top rung is read from left to right. Then the second rung down is read from left to right and so on. When the PLC is in its run mode, it goes through the entire ladder program to the end, the end rung of the program being clearly denoted, and then promptly resumes at the start.

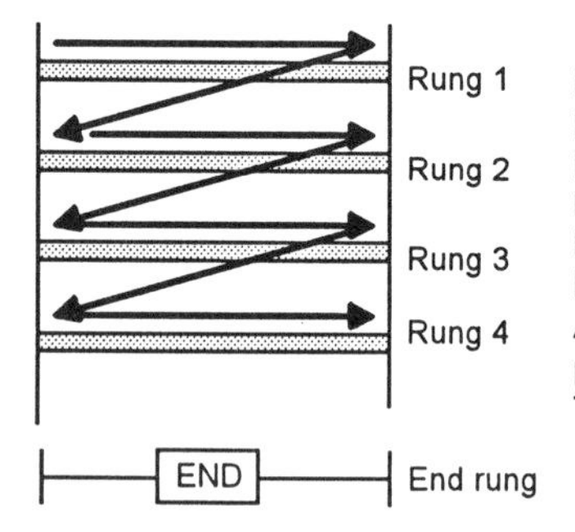

Figure 9.14 *Scanning the ladder program*

4 Each rung must start with an input or inputs and must end with at least one output. The term input is used for an action, such as closing the contacts of a switch, that is used as an input to the PLC. The term output is used for a device connected to the output of a PLC, e.g. a solenoid of a valve.

5 Electrical devices are shown in their normal condition. Thus a switch which is normally open until some object closes it, is shown as open on the ladder diagram. A switch that is normally closed is shown closed.

6 A particular device can appear in more than one rung of a ladder. For example, we might have a relay which switches on one or more devices. The same letters and/or numbers are used to label the device in each situation.

7 The inputs and outputs are all identified by their addresses, the notation used depending on the PLC manufacturer. This is the address of the input or output in the memory of the PLC.

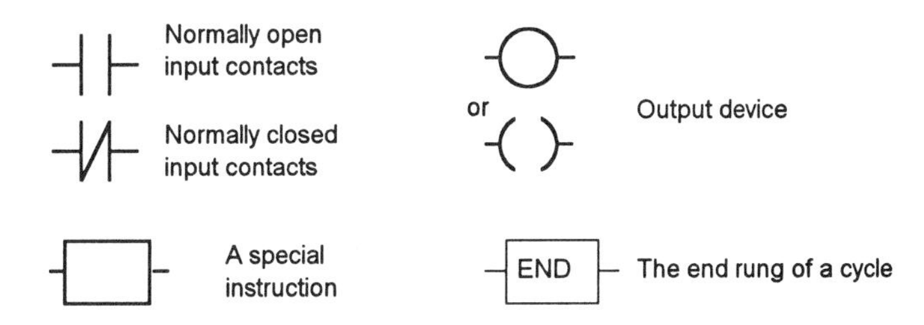

Figure 9.15 *Basic standard symbols*

Figure 9.15 shows basic standard symbols that are used for input and output devices. Further symbols will be introduced later.

Input Output

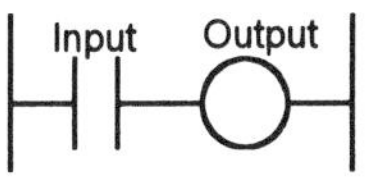

Figure 9.16 *A ladder rung*

To illustrate the drawing of the rung of a ladder diagram, consider a situation where the energising of an output device, e.g. a solenoid, depends on a normally open switch being activated by being closed. The input is thus the switch and the output the motor. Figure 9.16 shows the ladder diagram. Starting with the input, we have the normally open symbol | | for the input contacts. There are no other input devices and the line terminates with the output, denoted by the symbol O. When the switch is closed, i.e. there is an input, the output occurs to the solenoid.

Figure 9.17 shows the ladder diagram of Figure 9.16 using (a) Mitsubishi, (b) Siemens, (c) Allen Bradley notations for the addresses.

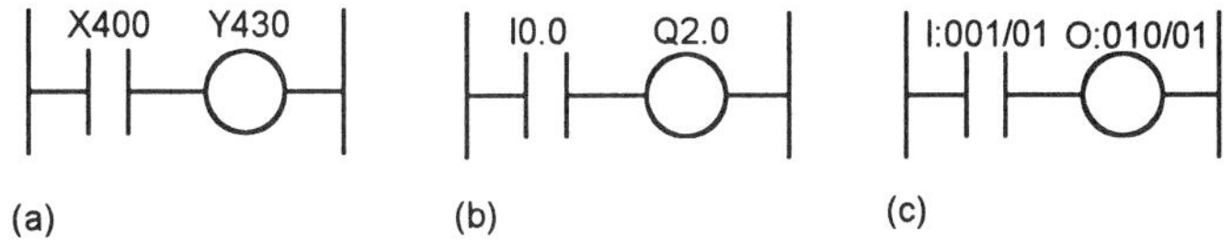

X400 Y430 (a) I0.0 Q2.0 (b) I:001/01 O:010/01 (c)

Figure 9.17 *Notation: (a) Mitsubishi, (b) Siemens, (c) Allen Bradley*

9.3.1 Logic functions

Figure 9.18 shows an AND gate system on a ladder diagram, Figure 9.19 the equivalent diagrams with Mitsubishi and Siemens notations. The ladder diagram starts with | |, a normally open set of contacts labelled input A and in series with it | |, another normally open set of contacts labelled input. The line then terminates with O to represent the output. For there to be an output, both input A and input B have to occur, i.e. input A and input B contacts have both to be closed. Thus if the output was a solenoid of a valve and A and B limit switches, for the solenoid to be energised we must have both the limit switches activated and closed.

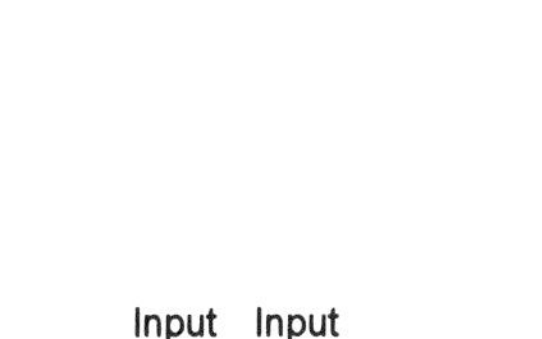

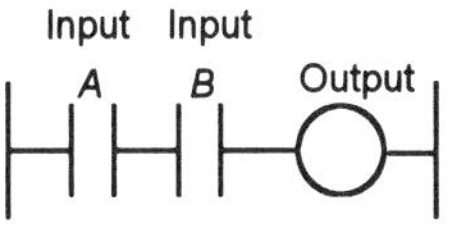

Figure 9.18 *AND gate*

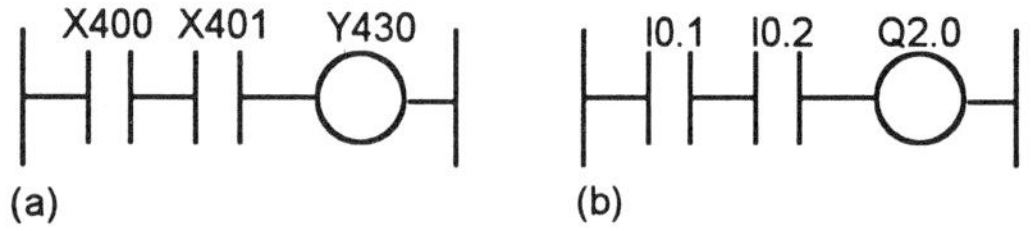

Figure 9.19 *AND gate: (a) Mitsubishi, (b) Siemens notations*

Figure 9.20(a) shows an OR logic gate system on a ladder diagram, Figure 9.20(b) showing an equivalent alternative way of drawing the same diagram. The ladder diagram starts with | |, normally open contacts labelled input *A*, and in parallel with it | |, normally open contacts labelled input B. Either input *A* <u>or</u> input *B* have to be closed for the output to be energised. The line then terminates with O to represent the output. Figure 9.21 shows how such a gate could appear with Mitsubishi and Siemens notations.

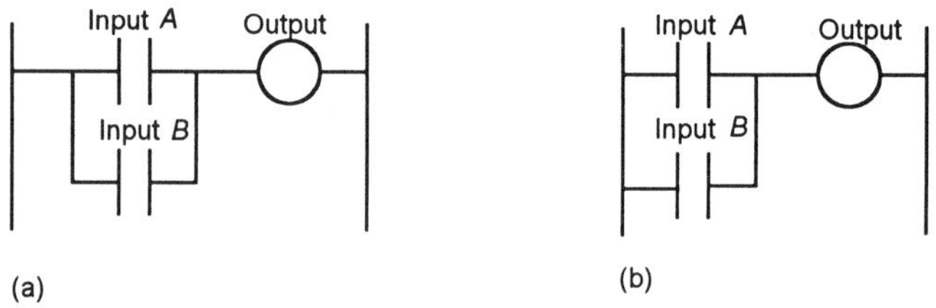

Figure 9.20 *OR gate*

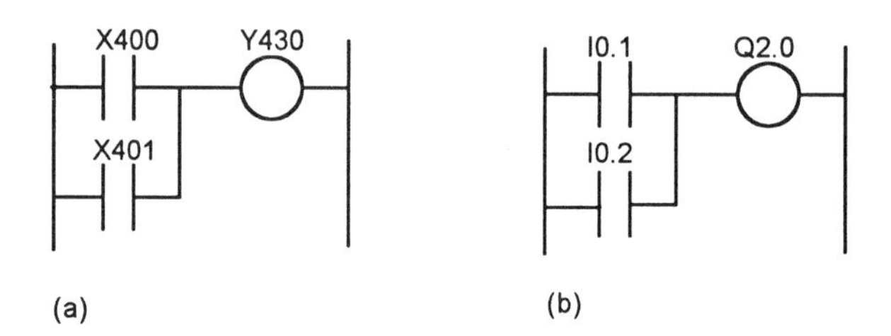

Figure 9.21 *OR gate: (a) Mitsubishi, (b) Siemens notations*

Figure 9.22 shows a NOT gate system on a ladder diagram. The input *A* contacts are shown as being normally closed. This is in series with the output O. With no input to input *A*, the contacts are closed and so there is an output. When there is an input to input *A*, it opens and there is then no output. Figure 9.23 shows Figure 9.22 in (a) Mitsubishi and (b) Siemens address notations.

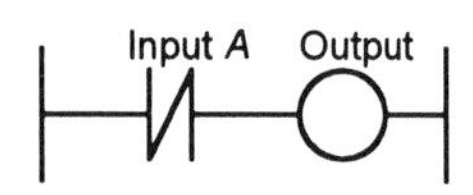

Figure 9.22 *NOT gate*

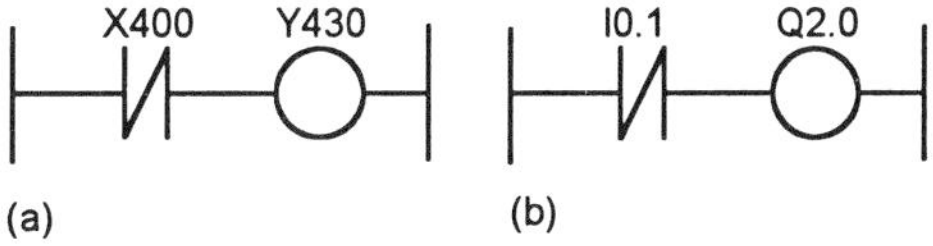

Figure 9.23 *NOT gate: (a) Mitsubishi, (b) Siemens notations*

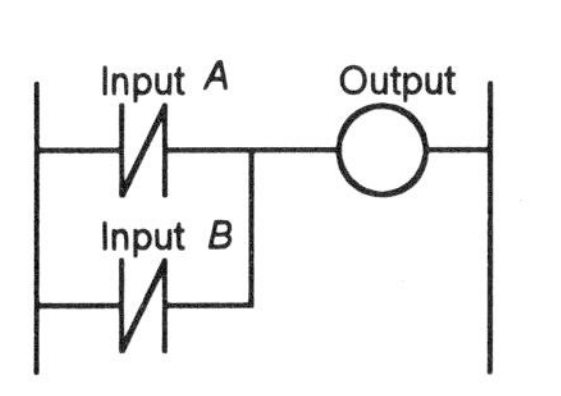

Figure 9.24 *A NAND gate*

Figure 9.24 shows a ladder diagram which gives a NAND gate. When the inputs to input *A* and input *B* are both 0 then the output is 1. When the inputs to input *A* and input *B* are both 1, or one is 0 and the other 1, then the output is 0. Figure 9.25 shows Figure 4.18 in (a) Mitsubishi and (b) Siemens notations.

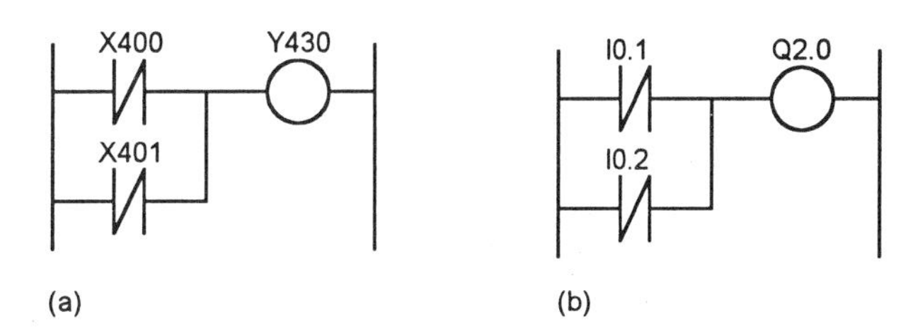

Figure 9.25 *NAND gate: (a) Mitsubishi, (b) Siemens notations*

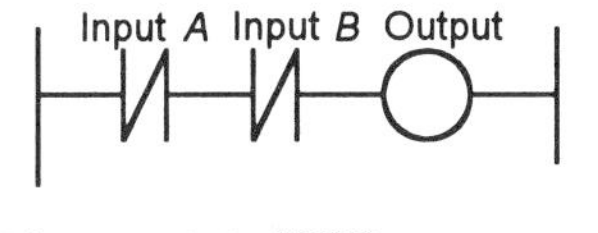

Figure 9.26 *NOR gate*

Figure 9.26 shows a ladder diagram of a NOR gate. When input *A* and input *B* are both not activated, there is a 1 output. When either *A* or *B* are activated there is a 0 output. Figure 9.27 shows the NOR gate system in (a) Mitsubishi, (b) Siemens notations.

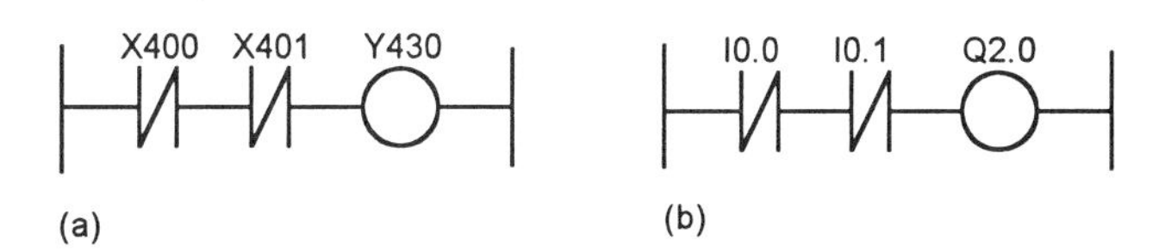

Figure 9.27 *NOR gate: (a) Mitsubishi, (b) Siemens notations*

Example

Draw a ladder diagram for a pneumatic system with double-solenoid controlled valves and involving two cylinders A and B, with limit switches a–, a+, b–, b+ detecting the limits of the piston rod movements (Figure 9.28), so that the sequence A+, B+, A–, B– is obtained. Include a start switch.

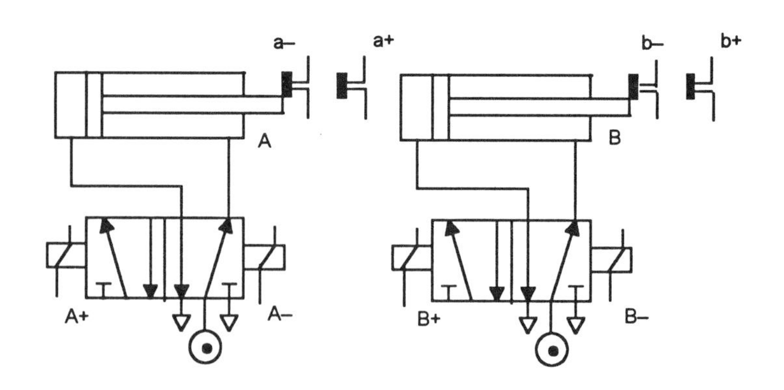

Figure 9.28 *Example*

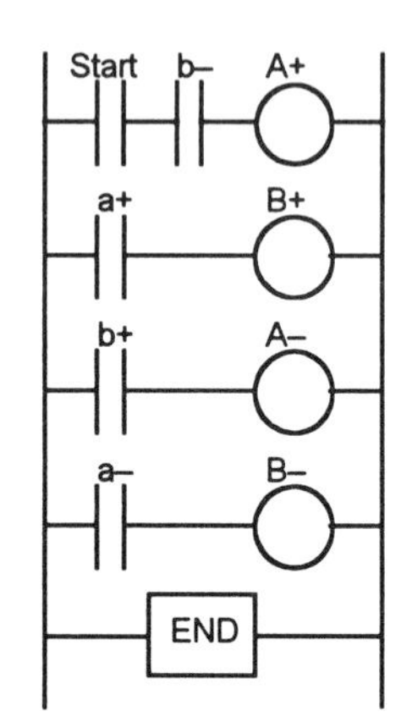

Figure 9.29 *Example*

The first line of the ladder diagram involves solenoid A+ being energised. This is to happen when the start switch and limit switch b– are activated. The next rung has solenoid B+ energised when limit switch a+ is activated. The next rung has solenoid A– energised when limit switch b+ is activated. The next rung has solenoid B– energised when limit switch a– is activated. Figure 9.29 shows the ladder diagram.

9.3.2 Latching

There are often situations where it is necessary to hold an output energised, even when the input ceases. A simple example of such a situation is a motor which is started by pressing a push-button switch. Though the switch contacts do not remain closed, the motor is required to continue running until a stop push-button switch is pressed. The term *latch circuit* is used for the circuit used to carry out such an operation. It is a self-maintaining circuit in that, after being energised, it maintains that state until another input is received.

An example of a latch circuit is shown in Figure 9.30(a), (b) showing the circuit in the Mitsubishi form of addresses. When the input *A* contacts close, there is an output. However, when there is an output, a set of contacts associated with the output closes. These contacts form an OR logic gate system with the input contacts. Thus, even if the input *A* opens, the circuit will still maintain the output energised. The only way to release the output is by operating the normally closed contact *B*.

As an illustration of the application of a latching circuit, consider a motor controlled by stop and start push-button switches and for which one signal light must be illuminated when the power is applied to the motor and another when it is not applied. Figure 9.31 shows the ladder diagram in Mitsubishi notation.

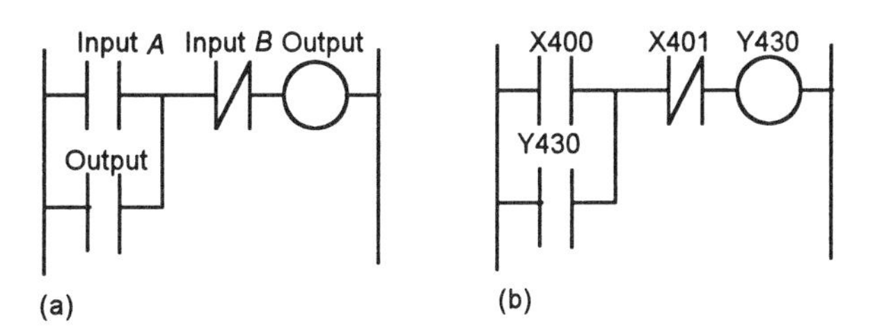

Figure 9.30 *Latched circuit*

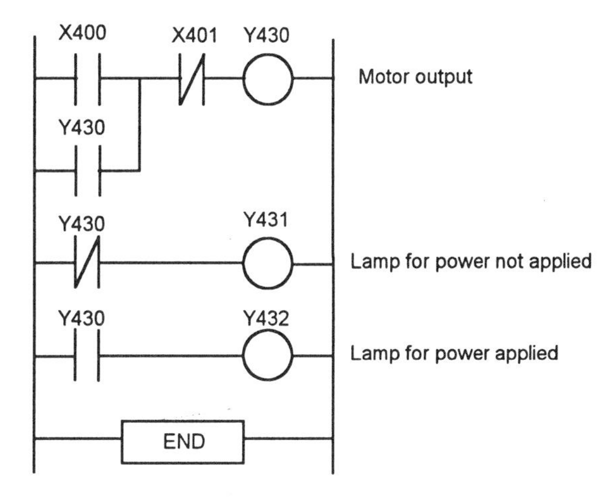

Figure 9.31 *Motor on–off, with signal lamps, ladder diagram*

When X400 is momentarily closed, Y430 is energised and its contacts close. This results in latching and also the switching off of Y431 and the switching on of Y432. To switch the motor off, X401 is pressed and opens. Y430 contacts open in the top rung and third rung, but close in the second rung. Thus Y431 comes on and Y432 off.

Example

Draw a ladder diagram for a pneumatic system with single-solenoid controlled valves and involving two cylinders A and B, with limit switches a–, a+, b–, b+ detecting the limits of the piston rod movements (Figure 9.32), so that the sequence A+, B+, A–, B– is obtained. Include a start switch.

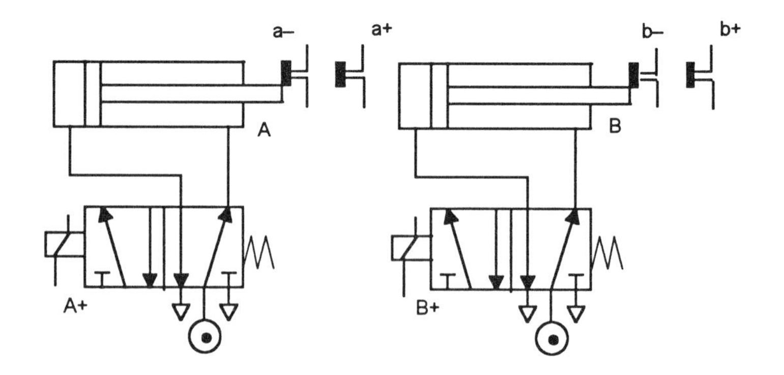

Figure 9.32 *Example*

Figure 9.33 shows the ladder diagram that can be used. The solenoid A+ is energised when the start switch is closed and limit switch b– closed. This provides latching to keep A+ energised as long as the normally closed contacts for limit switch b+ are not activated. When limit switch a+ is activated, solenoid B+ is energised. This provides latching which keeps B+ energised as long as the normally closed contacts for limit switch a– are not activated. When cylinder B extends, the limit switch b+ opens its normally closed contacts and unlatches the solenoid A+. Solenoid A thus retracts. When it has retracted and opened the normally closed contacts a–, solenoid B+ becomes unlatched and cylinder B retracts.

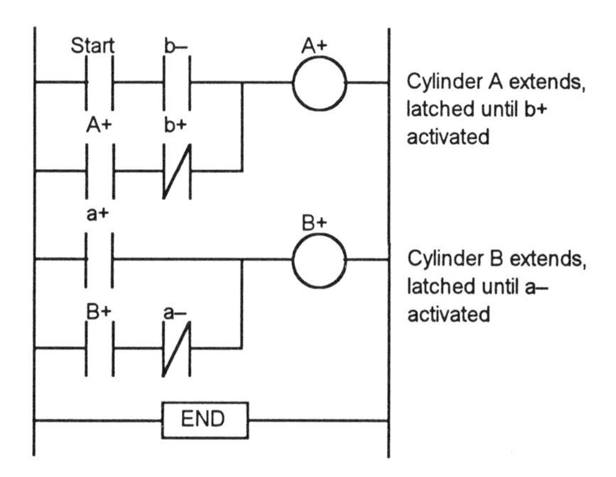

Figure 9.33 *Example*

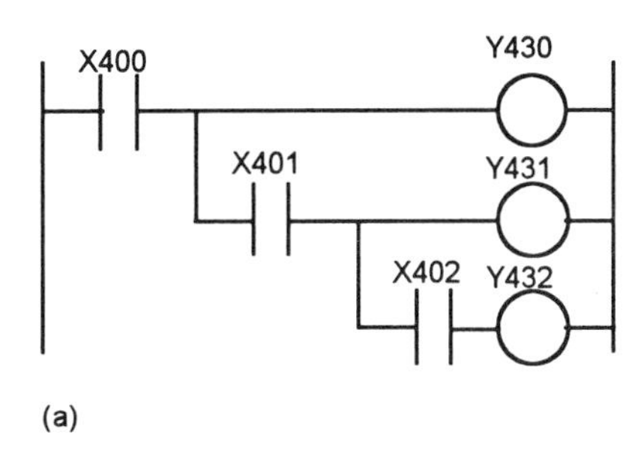

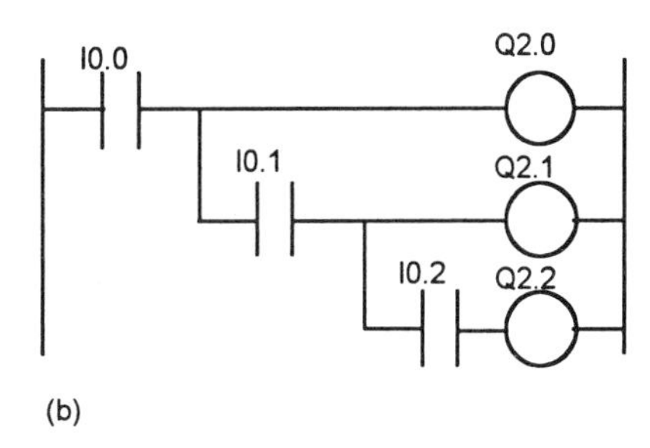

Figure 9.34 *Multiple outputs*

9.3.3 Multiple outputs

With ladder diagrams, there can be more than one output connected to a contact. Figure 9.34 illustrates this with the same ladder program in Mitsubishi and Siemens notations. Outputs Y430, Y431 and Y432 are switched on as the contacts in the sequence given by the contacts X400, X401 and X402 are being closed. Until X400 is closed, none of the other outputs can be switched on. When X400 is closed, Y430 is switched on. Then, when X401 is closed, Y431 is switched on. Finally, when X402 is closed, Y432 is switched on.

9.3.4 Instruction lists

Each horizontal rung on a ladder represents an instruction in the program to be used by the PLC, the entire ladder giving the complete program. A method which is very commonly used for loading ladder programs into PLCs is *instruction lists*. For this, mnemonic codes are used, each code corresponding to a ladder element. The codes used differ to some extent from manufacturer to manufacturer, though a standard (IEC 1131-3) has been proposed. Table 9.1 shows some of the codes used by manufacturers and the proposed standard (see later in this chapter for codes for other functions).

The following rules apply to representing ladder programs as instruction lists:

1 To identify the start of a rung on a ladder, a start rung code must be used. This might be LD, or perhaps A or L or STR, to indicate the rung is starting with open contacts, or LDI, or perhaps LDN or LD NOT or AN or LN or STR NOT, to indicate it is starting with closed contacts.

2 All rungs must end with the code for an output, the next item that follows will then be identified as a rung start instruction. This code for an output might be OUT or =.

Table 9.1 *Instruction code mnemonics*

IEC 1131-3	Mitsubishi	OMRON	Siemens	
LD	LD	LD	A	Start a rung with an open pair of contacts
LDN	LDI	LD NOT	AN	Start a rung with a closed pair of contacts
AND	AND	AND	A	A series element with an open pair of contacts
ANDN	ANI	AND NOT	AN	A series element with a closed pair of contacts
O	OR	OR	O.	A parallel element with an open pair of contacts
ORN	ORI	OR NOT	ON	A parallel element with a closed pair of contacts
ST	OUT	OUT	=	An output

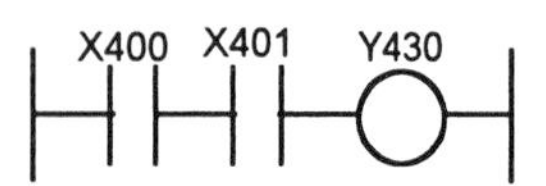

Figure 9.35 *AND gate*

The following shows how individual rungs on a ladder are entered using the Mitsubishi mnemonics. For the AND gate rung, shown in Figure 9.35:

0 Step 0 is the start of the rung with LD because it is starting with open contacts. Since the address of the input is X400, the instruction is LD X400.

1 This is followed by another open contacts input and so step 1 involves the instruction AND with the address of the element, thus the instruction is AND X401.

2 The rung terminates with an output and so the instruction OUT is used with the address of the output, i.e. OUT Y430.

The single rung of a ladder would thus be entered as:

Step	Instruction	
0	LD	X400
1	AND	X401
2	OUT	Y430

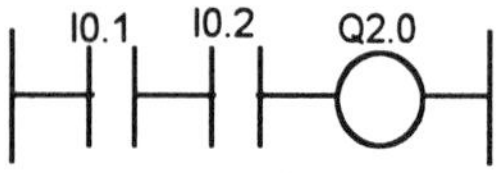

I0.1 I0.2 Q2.0

Figure 9.36 *AND gate*

For the same rung with Siemens notation (Figure 9.36) we have:

Step	Instruction	
0	A	I0.1
1	A	I0.2
2	=	Q2.0

Consider another example, an OR gate. Figure 9.37 shows the gate with Mitsubishi notation.

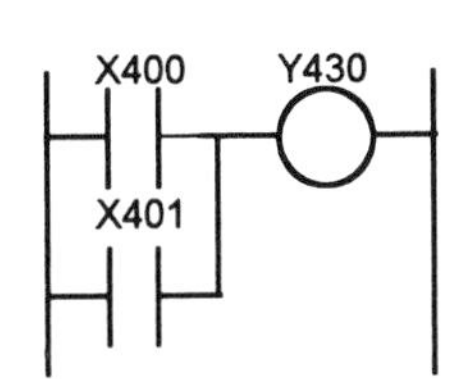

Figure 9.37 *OR gate*

0 The instructions for the rung start with an open contact is LD X400.

1 The next item is the parallel OR set of contacts X401. Thus the next instruction is OR X401.

2 The last step is the output, hence OUT Y430.

The instruction list would thus be:

Step	Instruction	
0	LD	X400
1	OR	X401
2	OUT	Y430

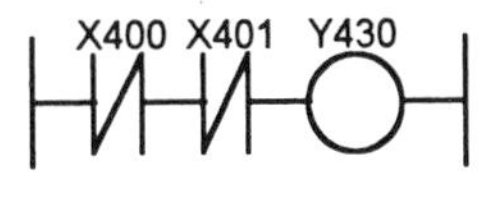

Figure 9.38 *NOR gate*

Figure 9.38 shows the ladder system for a NOR gate in Mitsubishi notation.

0 The rung starts with normally closed contacts and so the instruction is LDI. I when added to Mitsubishi instruction is used to indicate the inverse of the instruction.

1 The next step is a series normally closed contact and so ANI, again the I being used to make an AND instruction the inverse. I is also the instruction for a NOT gate.

2 The last step is the output, hence OUT Y430.

The instructions for the NOR gate rung of the ladder would thus be entered as:

Step	Instruction	
0	LDI	X400
1	ANI	X401
2	OUT	Y430

Consider the rung shown in Figure 9.39 in Mitsubishi notation, a NAND gate.

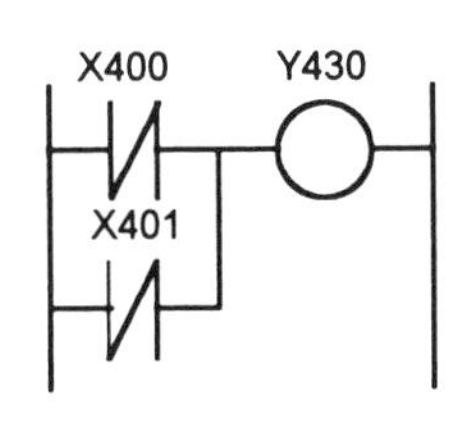

Figure 9.39 *NAND gate*

0 It starts with the normally closed contacts X400 and so starts with the instruction LDI X400.

1 The next instruction is for a parallel set of normally closed contacts, thus the instruction is ORI X401.

2 The last step is the output, hence OUT Y430.

The instruction list is thus:

Step	Instruction	
0	LDI	X400
1	ORI	X401
2	OUT	Y430

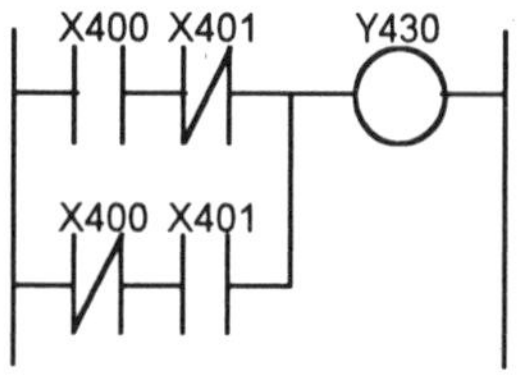

X400 X401 Y430

X400 X401

Figure 9.40 *Rung with parallel AND gates*

For the rung shown in Figure 9.40 in Mitsubishi notation, we have two parallel arms with an AND situation in each arm. In such a situation Mitsubishi uses an ORB instruction to indicate 'OR together parallel branches'.

0 The first instruction is for a normally open pair of contacts X400.

1 The next instruction is for a series set of normally closed contacts X401, hence ANI X401.

2 After reading the first two instructions, the third instruction starts a new line. It is recognised as a new line because it starts with LDI, all new lines starting with LD or LDI. But the first line has not been ended by an output. The PLC thus recognises that a parallel line is involved for the second line and reads together the listed elements until the ORB instruction is reached.

3 The AND instruction occurring before the ORB instruction is reached indicates that X401 is in series.

4 The mnemonic ORB (OR branches/blocks together) indicates to the PLC that it should OR the results of steps 0 and 1 with that of the new branch with steps 2 and 3.

5 The list concludes with the output OUT Y430.

The instruction list would thus be entered as:

Step	Instruction	
0	LD	X400
1	ANI	X401
2	LDI	X400
3	AND	X401
4	ORB	
5	OUT	Y430

Figure 9.41 shows the Siemens version of Figure 9.40. With Siemens, brackets are used to indicate that certain instructions are to be carried out

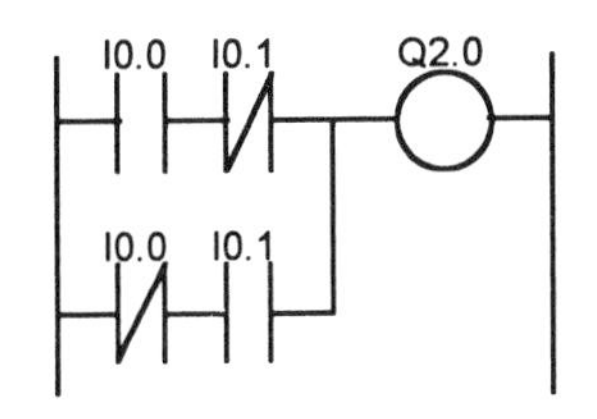

Figure 9.41 *Parallel AND gates*

as a block. They are used in the same way as brackets in any mathematical equation. For example, (2 + 3)/4 means that the 2 and 3 must be added before dividing by 4. Thus with the Siemens instruction list we have in step 0 the instruction A(. The brackets close in step 3. This means that the A in step 0 is applied only after the instructions in steps 1 and 2 have been applied.

Step	Instruction		
0	A(		Applies to the first set of brackets
1	A	I0.0	Steps 1 and 2 are in the first set of brackets
2	AN	I0.1	
3	)		
4	O(		Applies to the second set of brackets
5	AN	I0.0	Steps 5 and 6 are in the second set of brackets
6	A	I0.1	
7	)		
8	=	Q2.0	

Figure 9.42(a) shows a circuit, in Mitsubishi notation, which can be considered as two AND blocks, Figure 9.42(b) showing the same circuit in Siemens notation. The Mitsubishi instruction for this is ANB and thus the instruction list is:

Step	Instruction	
0	LD	X400
1	OR	X402
2	LD	X401
3	OR	X403
4	ANB	
5	OUT	Y430

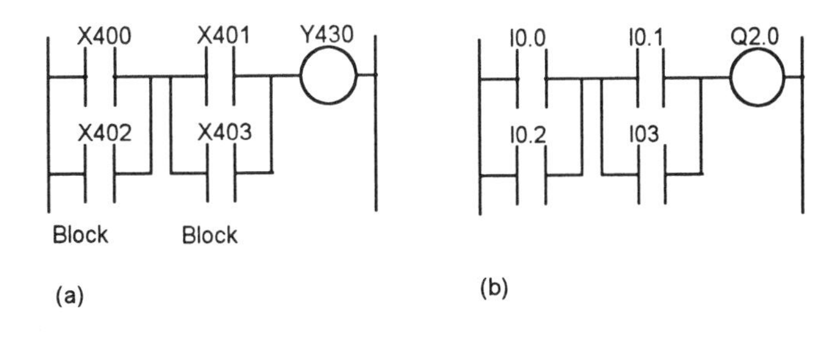

Figure 9.42 *(a) Mitsubishi, (b) Siemens*

In Siemens notation such a program is written as an instruction list using brackets. The A instruction in step 0 applies to the result of steps 1 and 2. The A instruction in step 4 applies to the result of steps 5 and 6. The program instruction list is:

Step	Instruction	
0	A(	
1	A	I0.0
2	O.	I0.1
3	)	
4	A(	
5	A	I0.2
6	O.	I0.3
7	)	
8	=	Q2.0

Example

Write the instruction list for the ladder program shown in Figure 9.43.

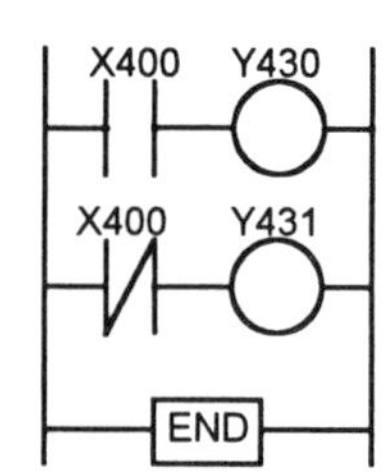

Figure 9.43 *Example*

Figure 9.43 shows a ladder, in Mitsubishi notation, with two rungs. In writing the instruction list we just write the instructions for each line in turn. The instruction LD or LDI indicates to the PLC that a new rung is starting. The instruction list is thus:

Step	Instruction	
0	LD	X400
1	OUT	Y430
2	LDI	X400
3	OUT	Y431
4	END	

The system is one where when X400 is not activated, there is an output from Y431 but not Y430. When X400 is activated, there is then an output from Y430 but not Y431.

9.4 Internal relays

In PLCs there are elements that behave like relays, being able to be switched on or off and switch other devices on or off. Hence the term *internal relay*. Such internal relays do not exist as real-world switching devices but are merely bits in the storage memory that behave in the same way as relays. For programming, they can be treated in the same way as an external relay output and input. Thus inputs to external switches on one rung of a ladder program can be used to give an output from an internal

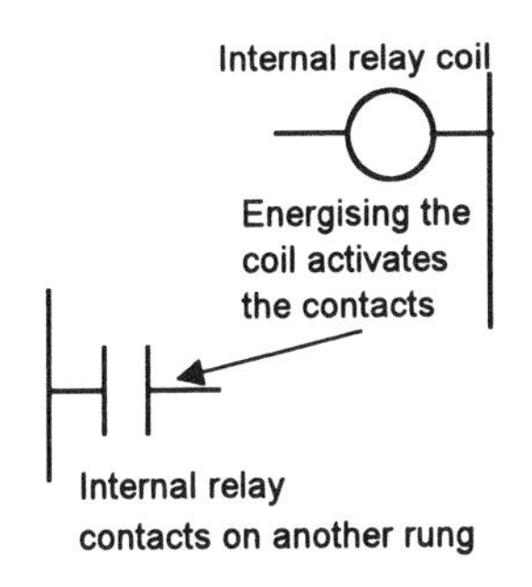

Figure 9.44 *Internal relay*

relay. This then results in the internal relay contacts being used on some other rung to give an output to some external device, e.g. a solenoid of a valve. Thus we might have (Figure 9.44):

On one rung of the program:
Inputs to external inputs activate the internal relay output.

On a later rung of the program:
As a consequence of the internal relay output:
internal relay contacts are activated and so control some output.

In using an internal relay, it has to be activated on one rung of a program and then its output used to operate switching contacts on another rung, or rungs, of the program. Internal relays can be programmed with as many sets of associated contacts as desired.

To distinguish internal relay outputs from external relay outputs, they are given different types of addresses. Different manufacturers tend to use different terms for internal relays and different ways of expressing their addresses. For example, Mitsubishi uses the term *auxiliary relay* or *marker* and the notation M100, M101, etc. Siemens uses the term *flag* and notation F0.0, F0.1, etc. Toshiba uses the term *internal relay* and notation R000, R001, etc. Allen Bradley uses the term *bit storage* and notation in the PLC-5 of the form B3/001, B3/002, etc.

9.4.1 Internal relays in programs

With ladder programs, an internal relay output is represented using the symbol for an output device, namely ( ) or O, with an address which indicates that it is an internal relay rather than an external relay, e.g. M100. The internal relay switching contacts are designated with the symbol for an input device, namely | |, and given the same address as the internal relay output, e.g. M100.

As an illustration of the use that can be made of internal relays, consider the following situation. A system is to be activated when two different sets of input conditions are realised. We might just program this as an AND logic gate system; however, if a number of inputs have to be checked in order that each of the input conditions can be realised, it may be simpler to use an internal relay. The first input conditions then are used to give an output to an internal relay. This has associated contacts which then become part of the input conditions with the second input. Figure 9.45 shows a ladder program for such a task. For the first rung: when input 1 or input 3 is closed and input 2 closed, then internal relay IR 1 is activated. This results in the contacts IR 1 closing. If input 4 is then activated, there is an output from output 1.

Figure 9.46(a) shows how Figure 9.45 would appear in Mitsubishi notation and Figure 9.46(b) in Siemens notation.

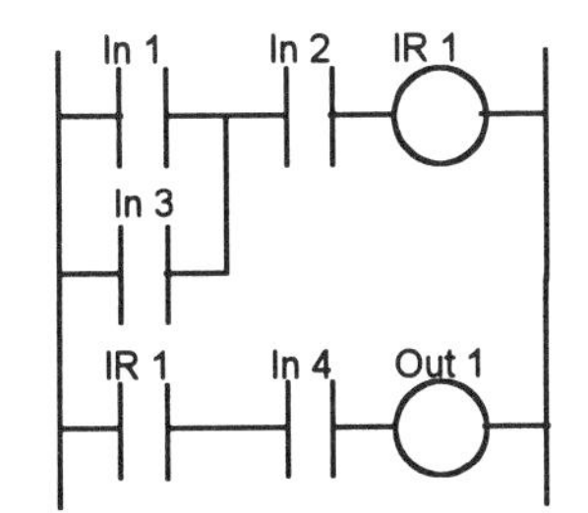

Figure 9.45 *Internal relay*

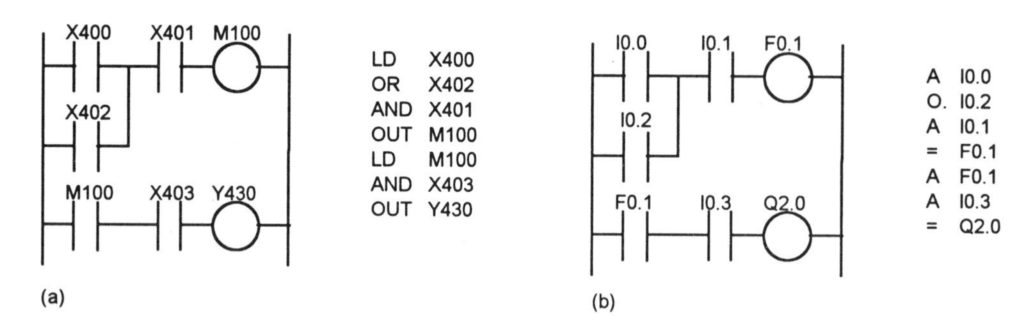

Figure 9.46 *(a) Mitsubishi notation, (b) Siemens notation*

Figure 9.47 is another example of this type of ladder program. The output 1 is controlled by two input arrangements. The first rung shows the internal relay IR 1 which is energised if the input In 1 or In 2 is activated and closed. The second rung shows the internal relay IR 2 which is energised if the inputs In 3 and In 4 are both energised. The third rung shows that the output Out 1 is energised if the internal relay IR 1 or IR 2 is activated. Thus there is an output from the system if either of two sets of input conditions is realised.

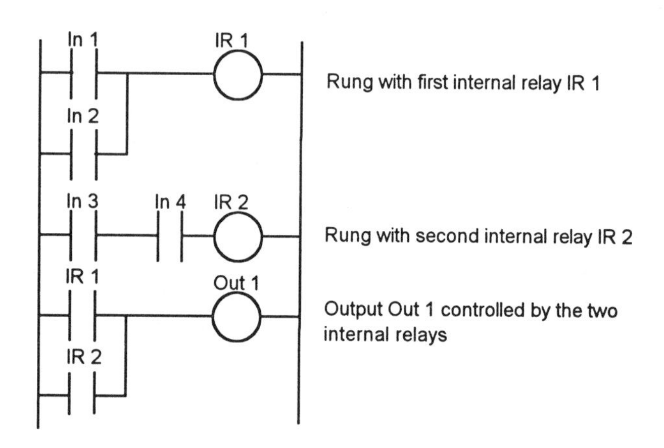

Figure 9.47 *Use of two internal relays*

9.4.2 Latching programs

Another use of internal relays is for resetting a latch circuit. Figure 9.47 shows an example of such a ladder program. When the input 1 contacts are momentarily closed, there is an output at Out 1. This closes the contacts for

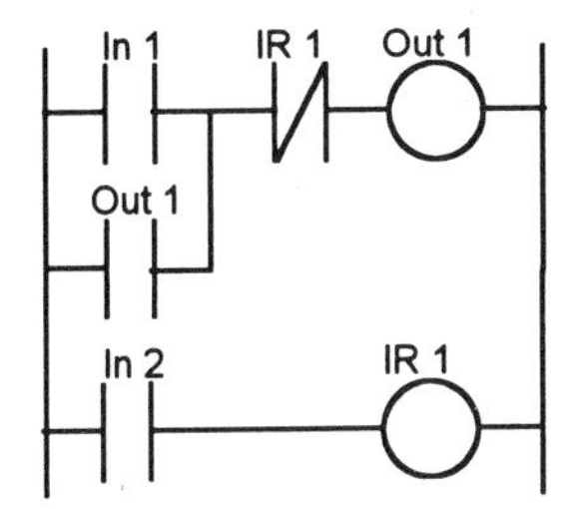

Figure 9.48 *Resetting latch*

Out 1 and so maintains the output, even when input 1 opens. When input 2 is closed, the internal relay IR 1 is energised and so opens the IR 1 contacts, which are normally closed. Thus the output Out 1 is switched off and so the output is unlatched.

Consider a situation requiring latch circuits where there is an automatic machine that can be started or stopped using push-button switches. A latch circuit can be used to start and stop the power being applied to the machine. The machine has several outputs which are to be turned on if the power has been turned on and off if the power is off. An internal relay can be used as the key feature of such a program. Figure 9.49 shows such a ladder diagram. The first rung has the latch for keeping the internal relay IR 1 on when the start switch gives a momentary input. The second rung will then use the contacts of the internal relay to switch the power on. The third rung will also switch on and give output Out 2 if input 2 contacts are closed. The third rung will also switch on and give output Out 3 if input 3 contacts are closed. Thus all the outputs can be switched on when the start push-button is activated. All the outputs will be switched off if the stop switch is opened. Thus all the outputs are latched by IR 1.

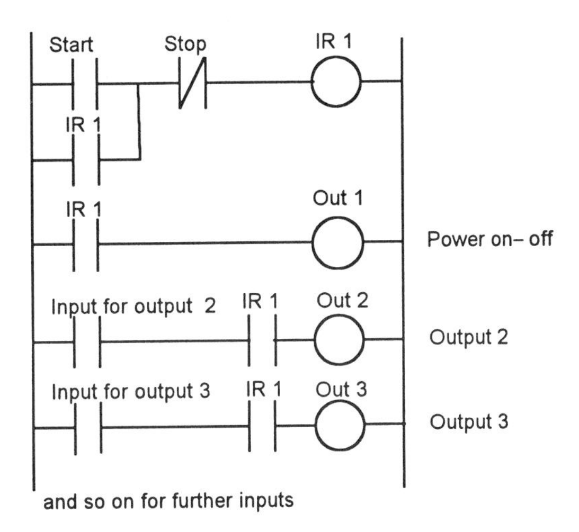

Figure 9.49 *Starting of multiple outputs*

Example

For the cylinders, valves and limit switches shown in Figure 9.50, devise a program to give the sequence A+, B+, B–, A–.

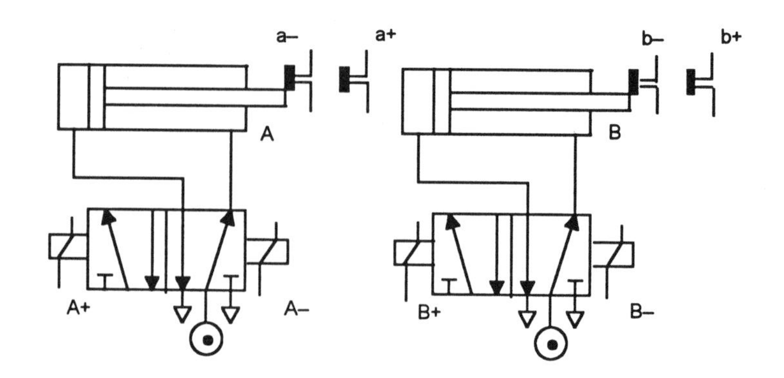

Figure 9.50 *Example*

We can consider the sequence as a cascade with group I as A+, B+ and group II as B–, A–. We can then use an internal relay to distinguish between the two groups. Figure 9.51 shows the ladder program that can be used. When the start switch is closed, solenoid A+ is energised, the internal relay contacts IR being closed. When limit switch a+ is activated, solenoid B+ is energised, the internal relay contacts IR being closed. When limit switch b+ is activated, the internal relay coil IR is energised and its contacts change their states. This switches off A+ and B+ solenoids and switches on B– and then, when limit switch b– is activated, solenoid A–.

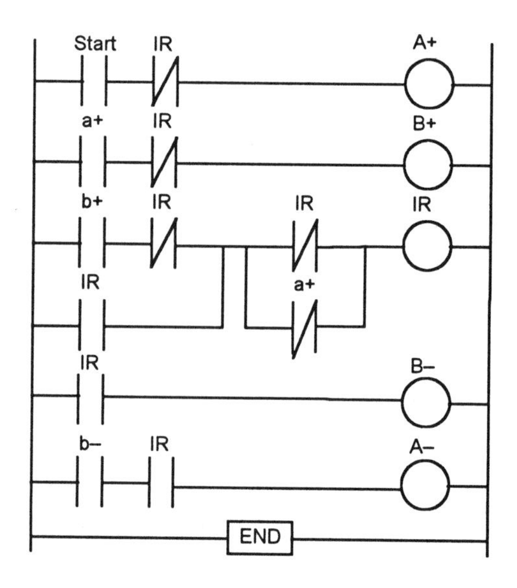

Figure 9.51 *Example*

9.5 Timers

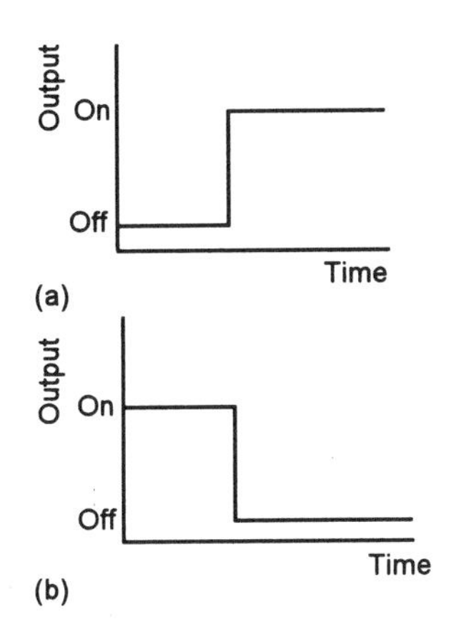

Figure 9.52 *Timers: (a) delay-on, (b) delay-off*

In many control tasks there is a need to control time of events, e.g. a cylinder might be needed to extend for a particular amount of time or perhaps remain extended for a particular time. There are a number of different forms of timers that can be found with PLCs. With small PLCs there is likely to be just one form, the *delay-on timers*. These are timers which come on after a particular time delay (Figure 9.52(a)). *Delay-off timers* are on for a fixed period of time before turning off (Figure 9.52(b)).

The time duration for which a timer has been set is termed the *preset* and is set in multiples of the time base used. Times bases are typically 10 ms, 100 ms, 1 s, 10 s and 100 s. Techniques for the entry of preset time values vary. Often it requires the entry of a constant K command followed by the time interval in multiples of the time base used. Thus, for example, an instruction of K5 would mean 5 multiples of the time base used. If this was the 1 s time base it would signify 5 s.

9.5.1 Programming timers

PLC manufacturers differ on how timers should be programmed. A common approach is to consider timers to behave like relays with coils which when energised result in the closure or opening of contacts after some preset time. The timer is thus treated as an output for a rung with control being exercised over pairs of contacts elsewhere. This is the predominant approach used in this book. Others treat a timer as a block which when inserted in a rung delays signals in that rung reaching the output.

Figure 9.53 shows these two different forms of ladder rung diagrams involving delay-on timers; all PLCs generally have delay-on timers with small PLCs often having only this type of timer. In (a) the timer is like a relay with a coil which is energised when the input In 1 occurs. It then opens, after some preset time delay, its contacts on rung 2. Thus the output from Out 1 occurs some preset time after the input In 1 occurs. In (b) the timer delays the passage of the signal through it from In 1 to Out 1, the results are the same as with (a).

Figure 9.53(a) is the form used by Mitsubishi. With their notation, In 1 having the address X400, timer T450, Out 1 Y430, the instruction list would be LD X400, OUT T450, K 5, LD T450, OUT Y430.

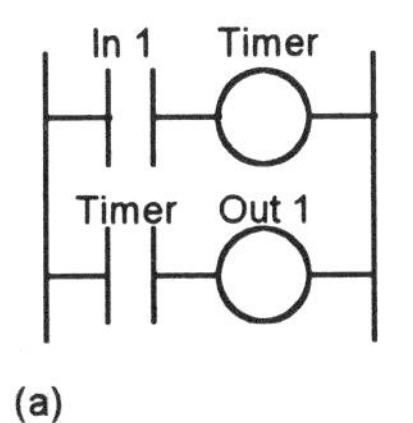

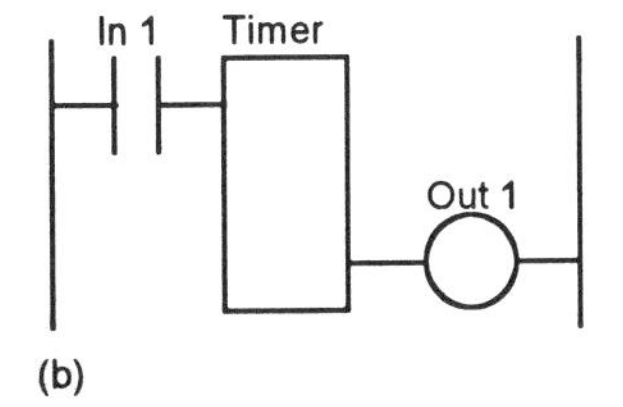

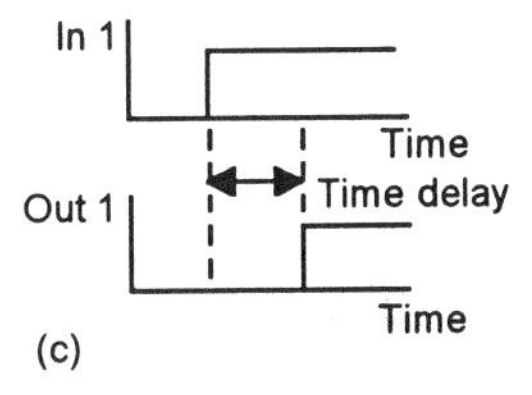

Figure 9.53 *Ladder diagram involving a delay-on timer*

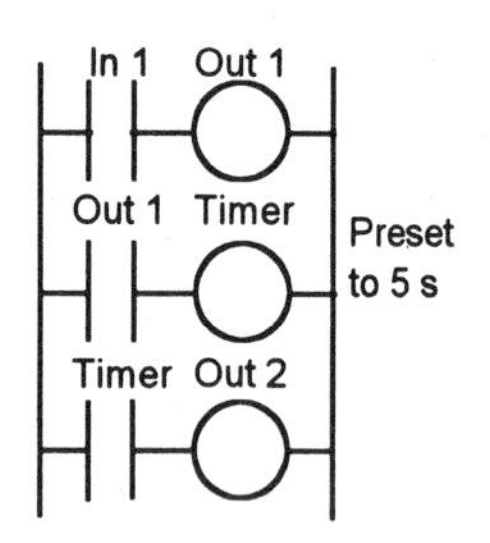

Figure 9.54 *Sequenced outputs*

As an illustration of the use of a timer, consider the *sequencing of outputs*. For the ladder diagram shown in Figure 9.54, when the input In 1 is switched on the output Out 1 is switched on. The contacts associated with this output then start the timer. The contacts of the timer will close after the preset time delay, in this case 5 s. When this happens, output Out 2 is switched on. Thus, following the input In 1, Out 1 is switched on and followed 5 s later by Out 2, thus giving a timed sequence of outputs.

Figure 9.55 shows how timers can be used to start three outputs, e.g. three motors, in sequence following a single start button being pressed. When the start push-button is pressed there is an output from internal relay IR 1. This latches the start input. It also starts both the timers, T1 and T2, and motor 1. When the preset time for timer 1 has elapsed then its contacts close and motor 2 starts. When the preset time for timer 2 has elapsed then its contacts close and motor 3 starts. The three motors are all stopped by pressing the stop push-button.

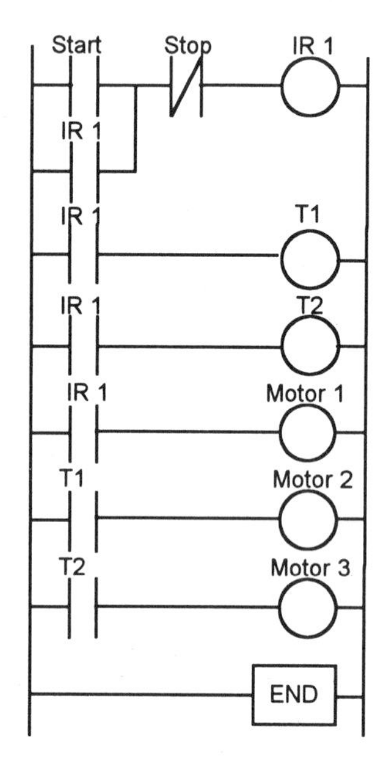

Figure 9.55 *Motor sequencing*

Figure 9.56 shows how on-delay timers can be used to produce an *on–off cycle timer*. The timer is designed to switch on an output for 5 s, then off for 5 s, then on for 5 s, then off for 5 s, and so on. When there is an input to In 1 and its contacts close, timer 1 starts. Timer 1 is set for a delay of 5 s. After 5 s, it switches on timer 2 and the output Out 1. Timer 2 has a delay of 5 s. After 5 s, the contacts for timer 2, which are normally closed, open. This results in timer 1, in the first rung, being switched off. This then causes its contacts in the second rung to open and switch off timer 2. This results in the timer 2 contacts resuming their normally closed state and so the input to In 1 causes the cycle to start all over again.

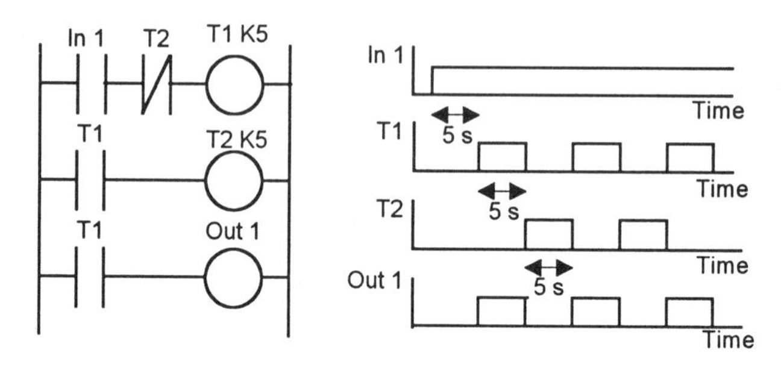

Figure 9.56 *On–off cycle timer*

Figure 9.57 shows how a delay-on timer can be used to produce a *delay-off timer*. With such an arrangement, when there is a momentary input to In 1, both the output Out 1 and the timer are switched on. Because the input is latched by the Out 1 contacts, the output remains on. After the preset timer time delay, the timer contacts, which are normally closed, open and switch off the output. Thus the output starts as on and remains on until the time delay has elapsed. Some PLCs have, as well as delay-on timers, built-in off-delay timers and thus there is no need to use a delay-on timer to produce a delay-off timer

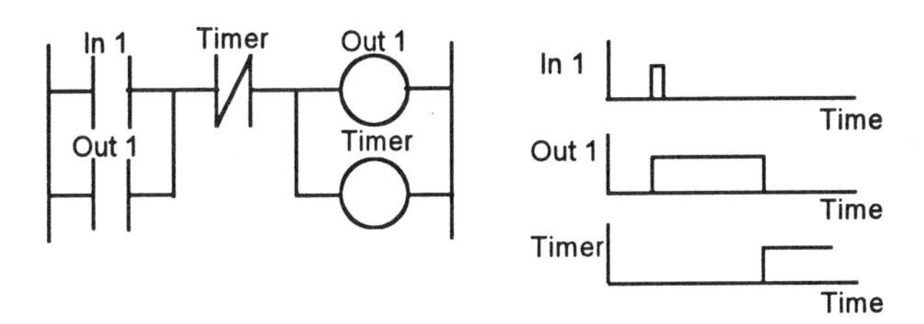

Figure 9.57 *Delay-off timer*

Figure 9.58 shows how a *one-shot timer* can be produced. Such a program produces a fixed duration output from Out 1 when there is an input to In 1. There are two outputs for the input In 1. When there is an input to In 1, there is an output from Out 1 and the timer starts. When the predetermined time has elapsed, the timer contacts open. This switches off the output. Thus the output remains on for just the time specified by the timer.

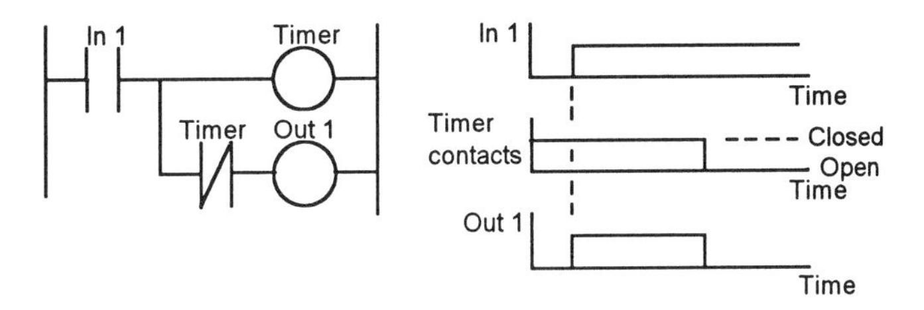

Figure 9.58 *One-shot on-timer*

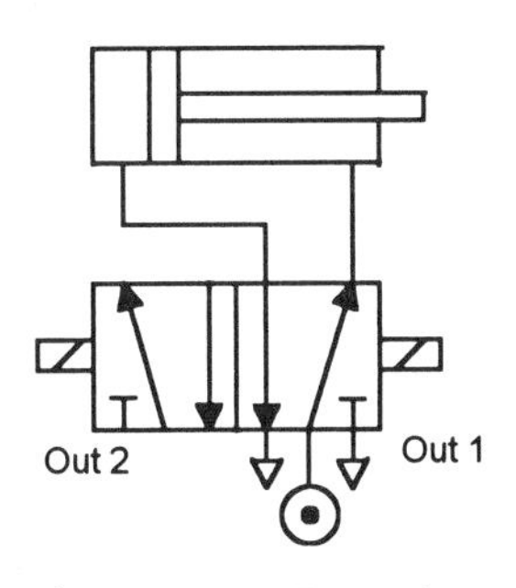

Figure 9.59 *Example*

Example

Devise a ladder program which can be used with the double solenoid valve shown in Figure 9.59 to spend 10 s extending the cylinder, then 10 s retracting it, then the sequencing repeating itself.

Figure 9.60 shows the ladder diagram and the timing sequence for the various elements, the two delay-on timers being set to 10 s. When the start contacts are closed, timer T1 starts and there is an output to Out 2, this causing the cylinder to start extending. After 10 s the normally closed T1 contacts open and stop the output Out 2. The normally open T1 contacts close, starting the timer T2 and energing the output Out 1, the cylinder then retracting. After 10 s, the T2 normally closed contacts are opened and this causes T1 to be switched off. This causes the normally closed contacts of T1 to close and so switch T2 off. The T2 normally closed contacts become closed and thus the sequence repeats itself.

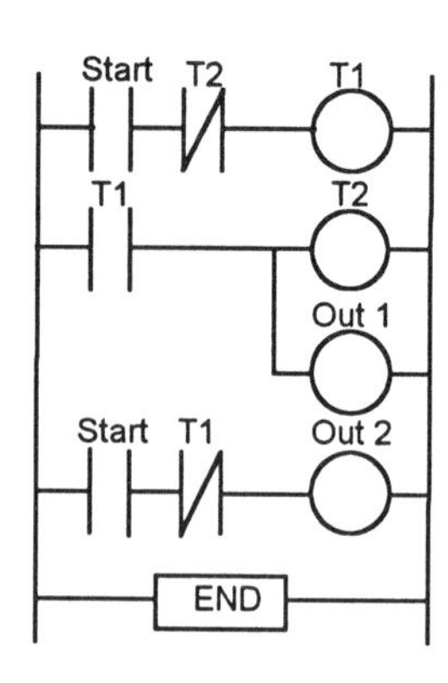

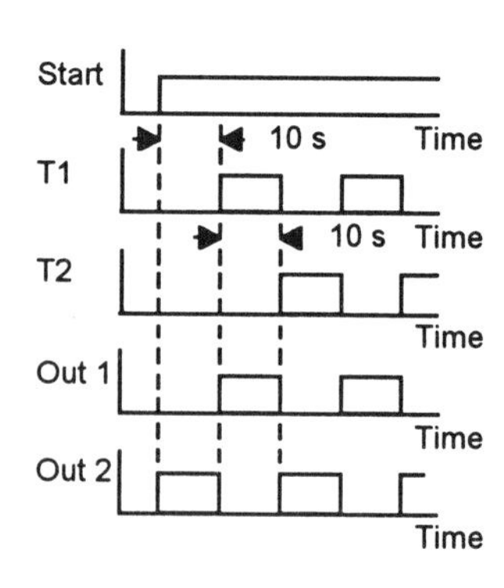

Figure 9.60 *Example*

9.6 Counters

A counter allows a number of occurrences of input signals to be counted, it being set to some preset number value and, when this value of input pulses has been received, it operates its contacts. Thus normally open contacts would be closed, normally closed contacts opened.

There are two types of counter, though PLCs may not include both types. These are down-counters and up-counters. *Down-counters* count down from the preset value to zero, i.e. events are subtracted from the set value. When the counter reaches the zero value, its contacts change state. Most PLCs offer down-counting. *Up-counters* count from zero up to the preset value, i.e. events are added until the number reaches the preset value. When the counter reaches the set value, its contacts change state.

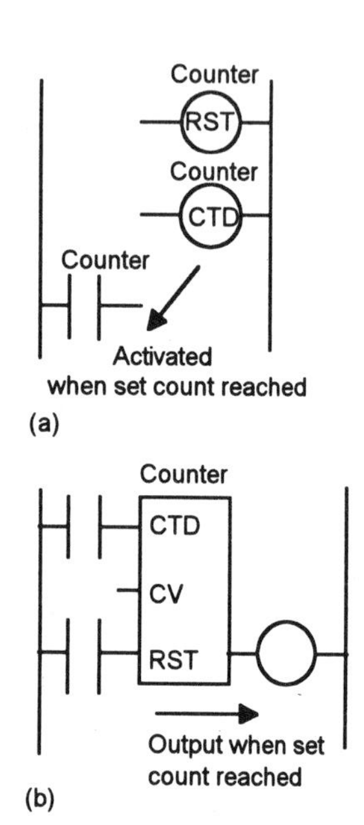

Figure 9.61 *Representing counters*

9.6.1 Programming

Different PLC manufacturers deal with counters in slightly different ways. Some count down (CTD), or up (CTU), and reset (RES), treating the counter as a rung output. In this way, counters can be considered to consist of two basic elements: one coil to count input pulses and one to reset the counter, the associated contacts of the counter being used in other rungs. Figure 9.61(a) illustrates this. Mitsubishi is an example of this type of manufacturer. Others treat the counter as an intermediate block in a rung from which signals emanate when the count is attained. Figure 9.61(b) illustrates this. Siemens is an example of this type of manufacturer. In (b), CTD indicates the count-down element, RST the reset element and CV the count value.

As an illustration, Figure 9.62 shows a basic counting circuit. When there is a pulse input to In 1, the counter is reset. When there is an input to In 2, the counter starts counting. If the counter is set for, say, 10 pulses, then when 10 pulse inputs have been received at In 2, the counter's contacts will close and there will be an output from Out 1. If at any time during the counting there is an input to In 1, the counter will be reset and start all over again and count for 10 pulses.

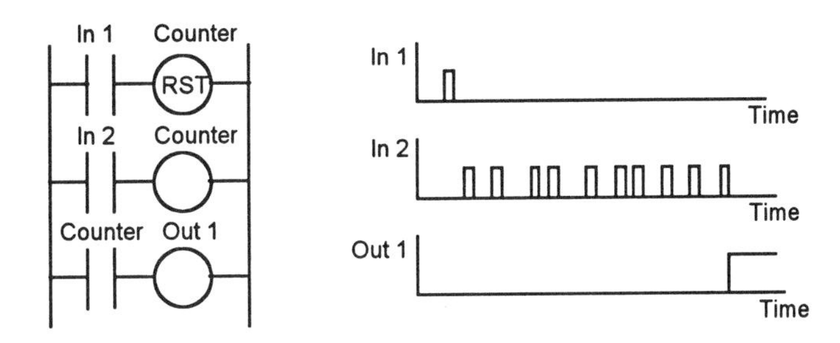

Figure 9.62 *Basic counter program*

Figure 9.63(a) shows how the above program, and its program instruction list, would appear with a Mitsubishi PLC. The reset and counting elements are combined in a single box spanning the two rungs. You can consider the rectangle to be enclosing the two counter O outputs in Figure 9.62. The count value is set by a K program instruction.

Figure 9.63(b) shows the same program, and its program instruction list, with a Siemens PLC. With this ladder program, the counter is considered to be a delay element in the output line. The counter is reset by an input to I0.2 and counts the pulses into input I0.1. The CU indicates that it is a count-up counter, a CD would indicate a count-down counter. The counter set value is indicated by the KC number.

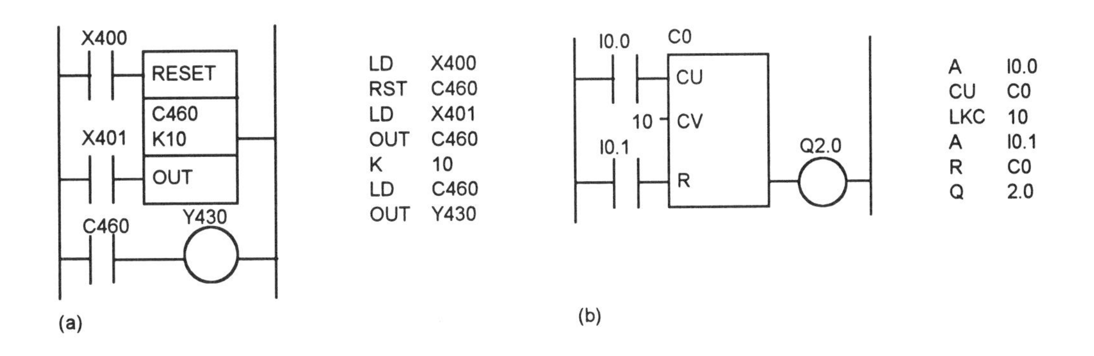

Figure 9.63 *(a) Mitsubishi, (b) Siemens program*

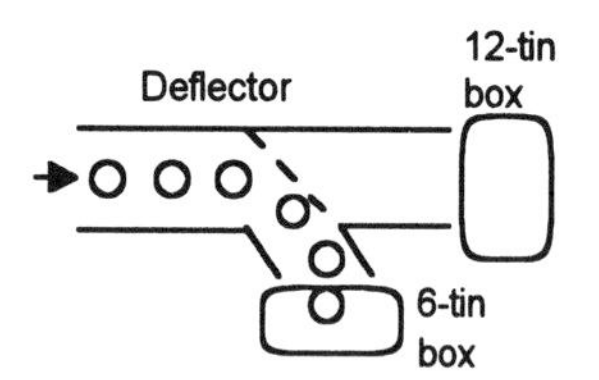

Figure 9.64 *Counting task*

As an illustration of the use that can be made of a counter, consider the problem of the control of a machine which is required to direct six tins along one path for packaging in a box and then 12 tins along another path for packaging in another box (Figure 9.64). A deflector plate might be controlled by a sensor which gives an output every time a tin passes it and activates a cylinder to move the plate. Thus the number of pulses from the sensor has to be counted and used to control the deflector. Figure 9.65 shows the ladder program that could be used. Mitsubishi notation has been used.

When there is a pulse input to X400, both the counters are reset. The input to X400 could be the push-button switch used to start the conveyor moving. The input which is counted is X401. This might be an input from a photocell sensor which detects the presence of tins passing along the conveyor. C460 starts counting after X400 is momentarily closed. When C460 has counted six items, it closes its contacts and so gives an output at Y430. This might be a solenoid which is used to activate a deflector to deflect items into one box or another. Thus the deflector might be in such a position that the first six tins passing along the conveyor are deflected into the 6-tin box, then the deflector plate is moved to allow tins to pass to the 12-tin box. When C460 stops counting it closes its contacts and so allows C461 to start counting. C461 counts for 12 pulses to X401 and then closes its contacts. This results in both counters being reset and the entire process can repeat itself.

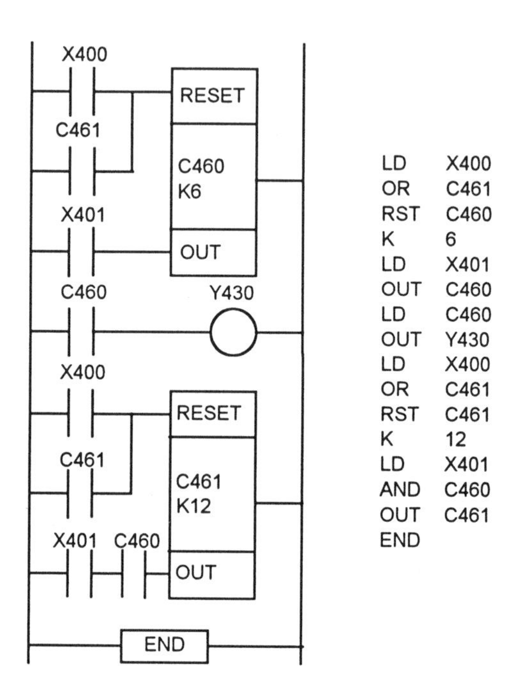

Figure 9.65 *Ladder program for Figure 9.64 task*

Example

Design a ladder program which will enable a three-cylinder, double solenoid-controlled arrangement (as in Figure 9.2) to give the sequence A+, A–, A+, A–, A+, A–, B+, C+, B–, C–.

The A+, A– sequence is repeated three times before B+, C+, B–, C– occur. We can use a counter to enable this repetition. Figure 9.66 shows a possible program. The counter only allows B+ to occur after it has received three inputs corresponding to three a– signals.

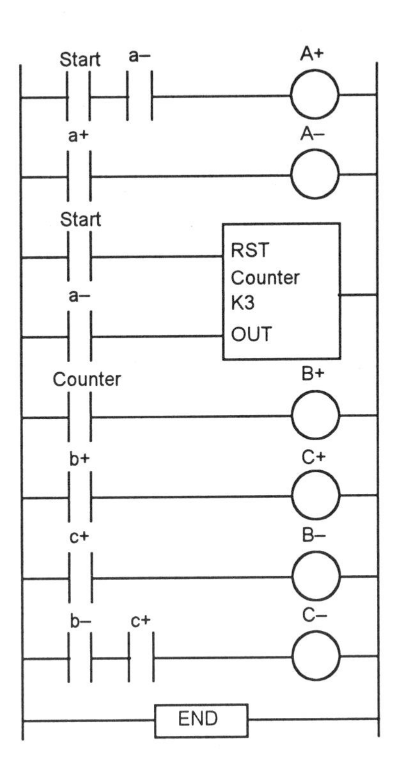

Figure 9.66 *Example*

9.7 Set and reset

Another function which is often available is the ability to set and reset an internal relay. The set instruction causes the relay to self-hold, i.e. latch. It then remains in that condition until the reset instruction is received. The term *flip-flop* is often used.

Figure 9.67 shows an example of a ladder diagram, in Mitsubishi notation, involving such a function. Activation of the first input, X400, causes the output Y430 to be turned on and set, i.e. latched. Thus if the first input is turned off, the output remains on. Activation of the second input, X401, causes the output Y430 to be reset, i.e. turned off and latched off. Thus the output Y430 is on for the time between X400 being momentarily switched on and X401 being momentarily switched on. Between the two rungs indicated for the set and reset operations, there could be other rungs for other activities to be carried out, the set rung switching on an output at the beginning of the sequence and off at the end.

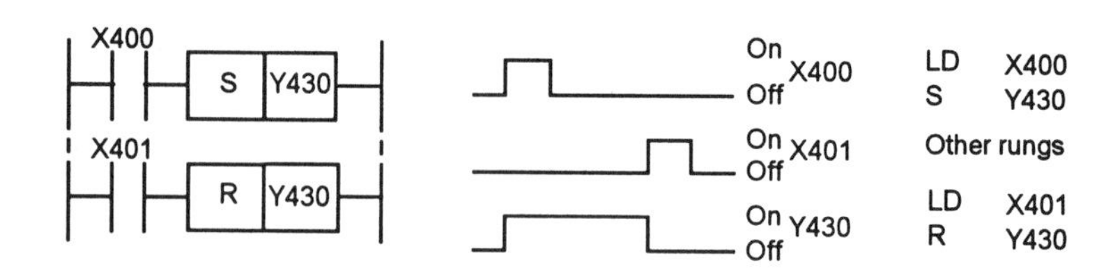

Figure 9.67 *Set and reset*

9.8 Shift registers

A register is a number of internal relays grouped together, normally 8, 16 or 32. Each internal relay is either effectively open or closed, these states being designated as 0 and 1. The term *bit* is used for each such binary digit. Therefore, if we have eight internal relays in the register we can store eight 0/1 states. Thus we might have:

Internal relays

1	2	3	4	5	6	7	8

and each relay might store an on–off signal such that the state of the register at some instant is:

1	0	1	1	0	0	1	0

i.e. relay 1 is on, relay 2 is off, relay 3 is on, relay 4 is on, relay 5 is off, etc. Such an arrangement is termed an 8-bit register. Registers can be used for storing data that originate from input sources other than just simple, single on–off devices such as switches.

With the *shift register* it is possible to shift stored bits. Shift registers require three inputs, one to load data into the first location of the register, one as the command to shift data along by one location and one to reset or clear the register of data. To illustrate this, consider the following situation where we start with an 8-bit register in the following state:

1	0	1	1	0	0	1	0

Suppose we have the input signal 0 to the register. This is an input signal to the first internal relay.

Input 0

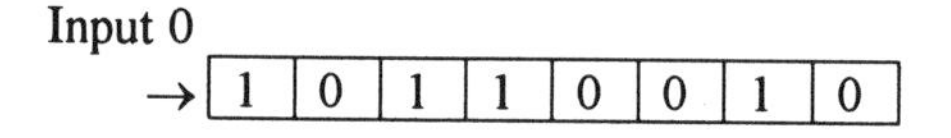

If we also receive the shift signal, then the input signal enters the first location in the register and all the bits shift along one location. The last bit overflows and is lost.

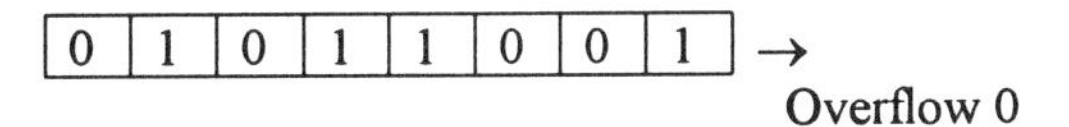

Thus a set of internal relays that were initially on, off, on, on, off, off, on, off are now off, on, off, on, on, off, off, on.

The grouping together of internal relays to form a shift register is done automatically by a PLC when the shift register function is selected. With the Mitsubishi PLC, this is done by using the programming code SFT (shift) against the internal relay number that is to be the first in the register array. This then causes a block of relays, starting from that initial number, to be reserved for the shift register.

9.8.1 Programming

Consider a 4-bit shift register in a ladder program (Figure 9.68). The input In 3 is used to reset the shift register, i.e. put all the values at 0. The input In 1 is used to input to the first internal relay in the register. The input In 2 is used to shift the states of the internal relays along by one. Each of the internal relays in the register, i.e. IR 1, IR 2, IR 3 and IR 4, is connected to an output, these being Out 1, Out 2, Out 3 and Out 4.

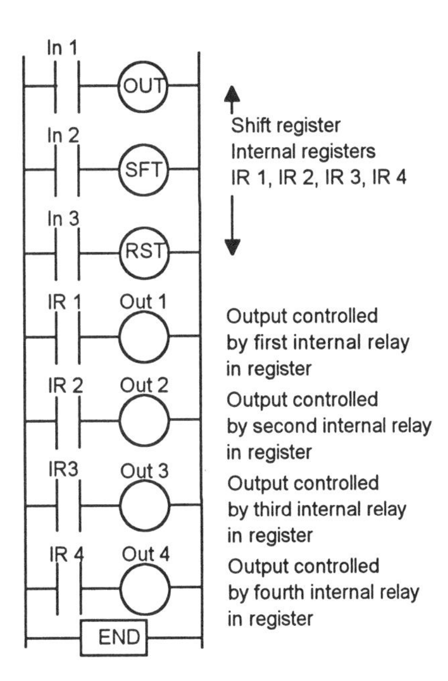

Figure 9.68 *The shift register*

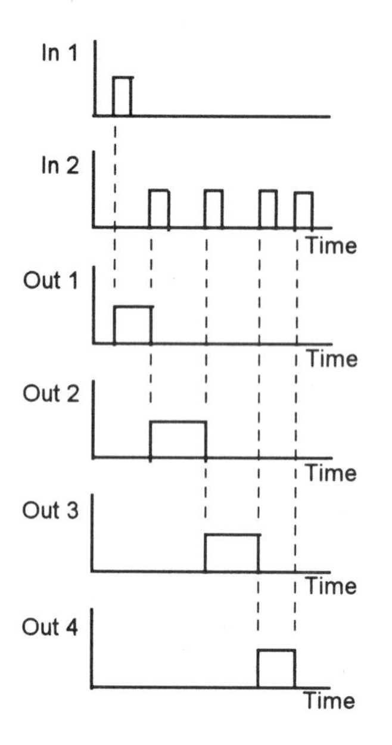

Figure 9.69 *Shift register*

Suppose we start by supplying a momentary input to In 3. All the internal relays are then set to 0 and so the states of the four internal relays IR 1, IR 2, IR 3 and IR 4 are 0, 0, 0, 0. When In 1 is momentarily closed there is a 1 input into the first relay. Thus the states of the internal relays IR 1, IR 2, IR 3 and IR 4 are now 1, 0, 0, 0. The IR 1 contacts close and we thus end up with an output from Out 1. If we now supply a momentary input to In 2, the 1 is shifted from the first relay to the second. The states of the internal relays are now 0, 1, 0, 0. We now have no input from Out 1 but an output from Out 2. If we supply another momentary input to In 2, we shift the states of the relays along by one location to give 0, 0, 1, 0. Outputs 1 and 2 are now off but Out 3 is on. If we supply another momentary input to In 2 we again shift the states of the relays along by one and have 0, 0, 0, 1. Thus now, outputs 1, 2 and 3 are off and output 4 has been switched on. When another momentary input is applied to In 2, we shift the states of the relays along by one and have 0, 0, 0, 0 with the 1 overflowing and being lost. All the outputs are then off. Thus the effect of the sequence of inputs to In 2 has been to give a sequence of outputs Out 1, followed by Out 2, followed by Out 3, followed by Out 4. Figure 9.69 shows the sequence of signals.

Figure 9.70 shows the Mitsubishi version of the above ladder program and the associated instruction list. The M140 is the address of the first relay in the register.

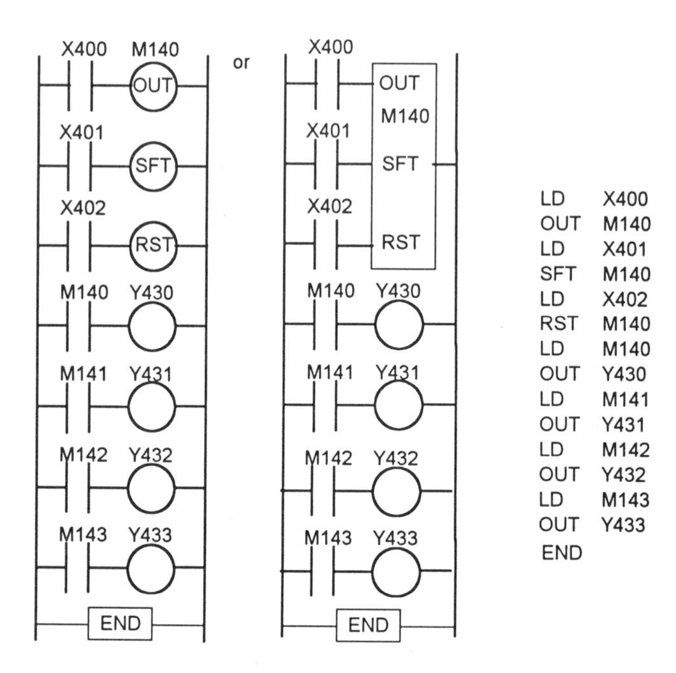

Figure 9.70 *Mitsubishi program*

Example

Write a program using a shift register for two double solenoid cylinders, the arrangement being as shown in Figure 9.28, to give the sequence A+, B+, A–, B–.

Figure 9.71 shows the program.

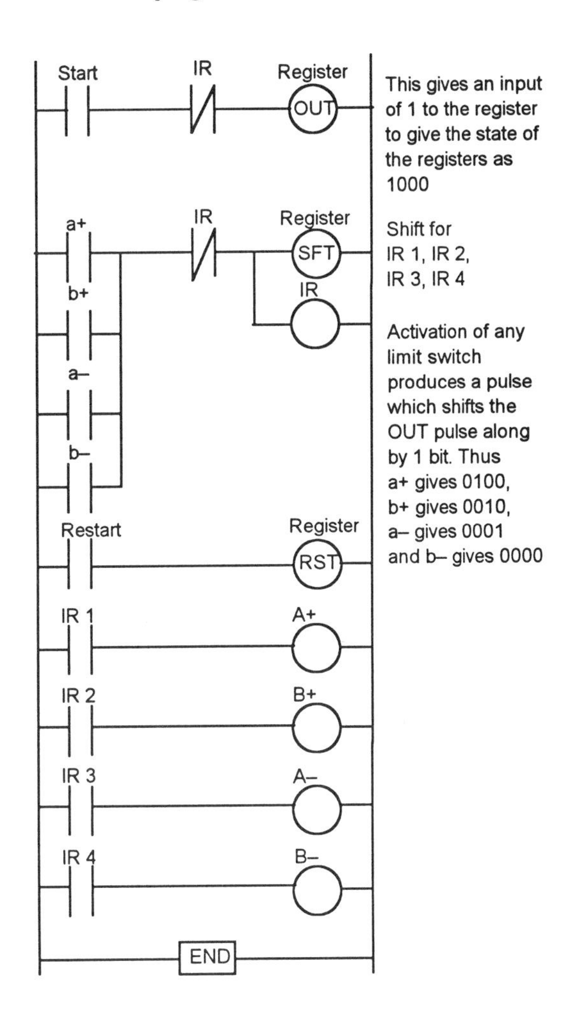

Figure 9.71 *Example*

9.9 Master control relay

When large numbers of outputs have to be controlled, it is sometimes necessary for whole sections of ladder diagrams to be turned on or off when certain criteria are realised. This could be achieved by including the contacts of the same internal relay in each of the rungs so that its operation affects all of them. An alternative is to use a *master control relay*.

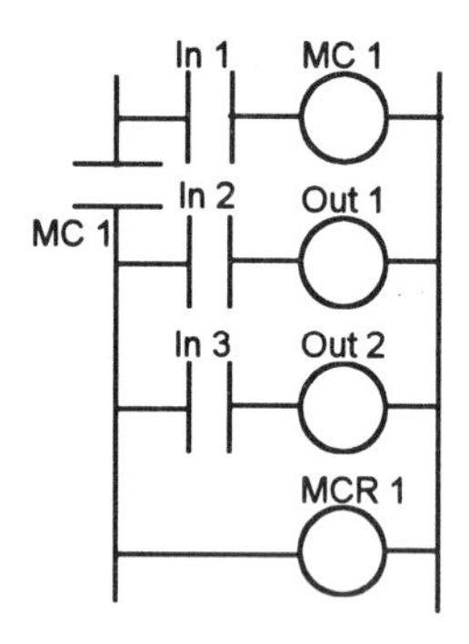

Figure 9.72 *Use of a master control relay*

Figure 9.72 illustrates the use of such a relay to control a section of a ladder program. With no input to input 1, the output internal relay MC 1 is not energised and so its contacts are open. This means that all the rungs between where it is designated to operate and the rung on which its reset MCR or another master control relay is located are switched off. Assuming it is designated to operate from its own rung, then we can imagine it to be located in the power line in the position shown and so rungs 2 and 3 are off. When input 1 contacts close, the master relay MC 1 is energised. When this happens, all the rungs between it and the rung with its reset MCR 1 are switched on. Thus outputs 1 and 2 cannot be switched on by inputs 2 and 3 until the master control relay has been switched on. The master control relay 1 acts only over the region between the rung it is designated to operate from and the rung on which MCR 1 is located.

With a Mitsubishi PLC, an internal relay can be designated as a master control relay by programming it accordingly. Thus to program an internal relay M100 to act as master control relay contacts the program instruction is:

```
MC     M100
```

To program the resetting of that relay, the program instruction is:

```
MCR    M100
```

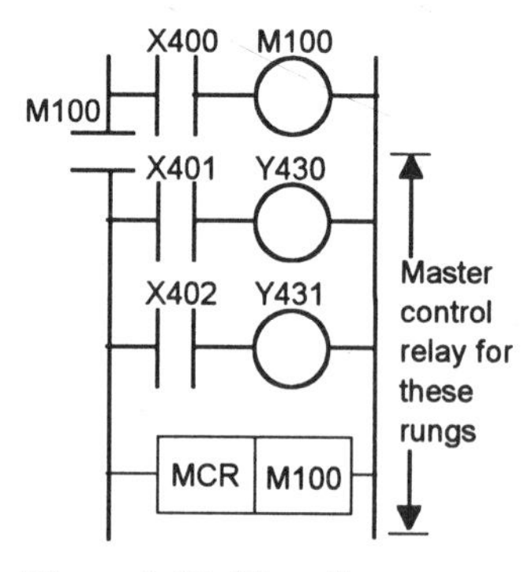

Figure 9.73 *Use of a master control relay*

Thus for the ladder diagram shown in Figure 9.73, being Figure 9.72 with Mitsubishi addresses, the program instructions are:

```
LD     X400
OUT    M100
MC     M100
LD     X401
OUT    Y430
LD     X402
OUT    Y431
MC     M100
```

A program might use a number of master control relays, enabling various sections of a ladder program to be switched in or out. Figure 9.74 shows a ladder program in Mitsubishi format involving two master control relays. With M100 switched on, but M101 off, the sequence is: rungs 1, 3, 4, 6, etc. The end of the M100-controlled section is indicated by the occurrence of the other master control relay, M101. With M101 switched on, but M100 off, the sequence is: rungs 2, 4, 5, 6, etc. The end of this section is indicated by the presence of the reset. This reset has to be used since the rung is not followed immediately by another master control relay. Such an arrangement could be used to switch on one set of ladder rungs if one type of input occurs, and another set of ladder rungs if a different input occurs.

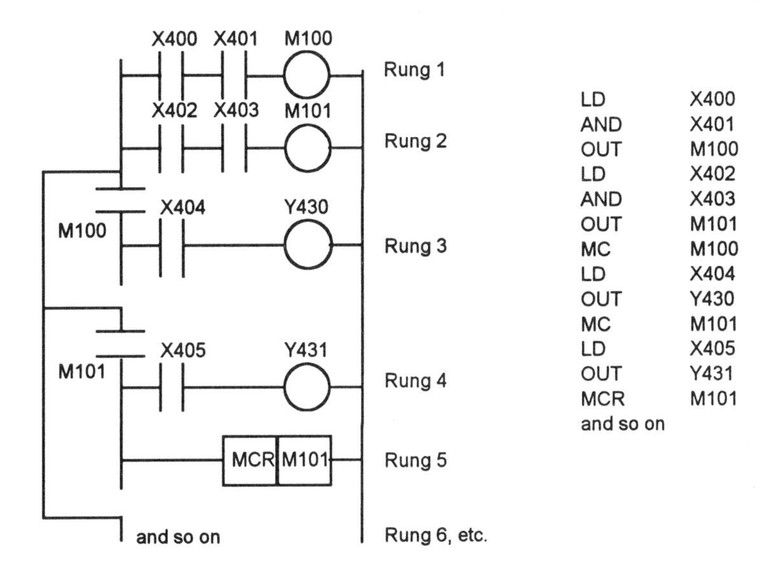

Figure 9.74 *Example showing more than one master control relay*

Example

Devise a ladder program which can be used with a pair of single-solenoid-controlled cylinders (Figure 9.75) to give, when and only when the start switch is momentarily triggered, the sequence A+, B+, A–, 10 s time delay, B– and stop at that point until the start switch is triggered again.

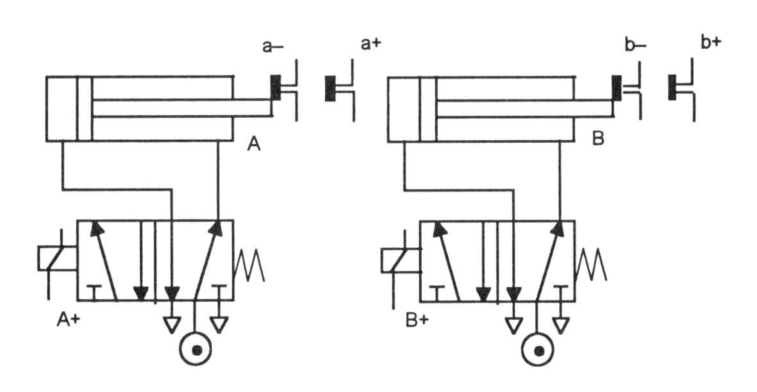

Figure 9.75 *Example*

Figure 9.76 shows how such a program can be devised using a master control relay. The master control relay is activated by the start switch and remains on until switched off by the rung containing just MCR.

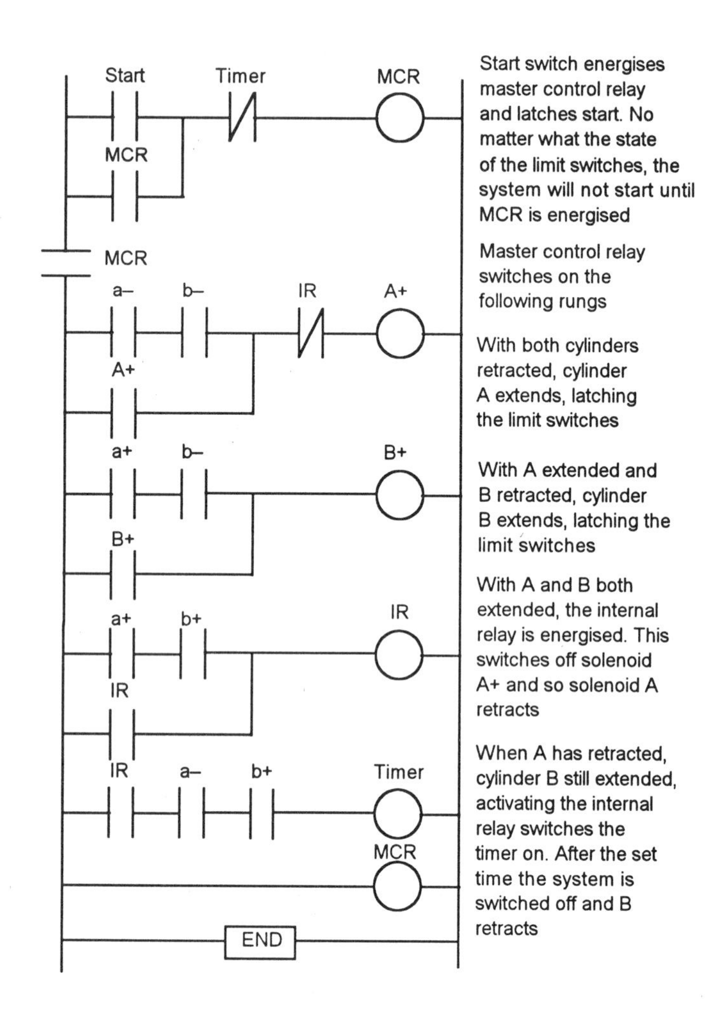

Figure 9.76 *Example*

Problems

1 For the ladder diagram rungs shown in Figure 9.77, what are the conditions that have to be realised for there to be outputs?

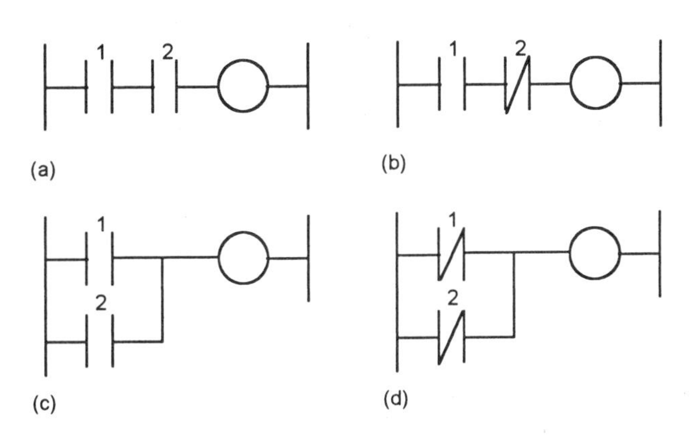

Figure 9.77 *Problem 1*

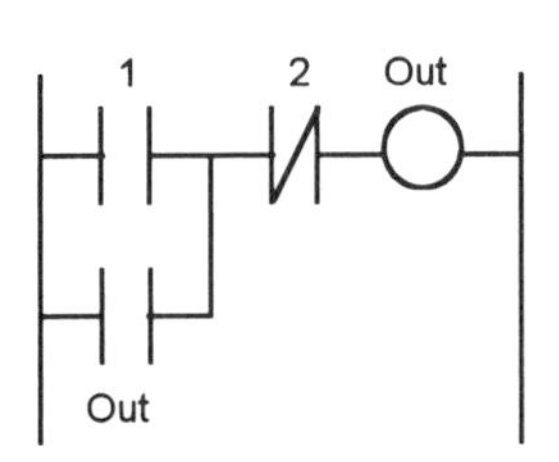

Figure 9.78 *Problem 2*

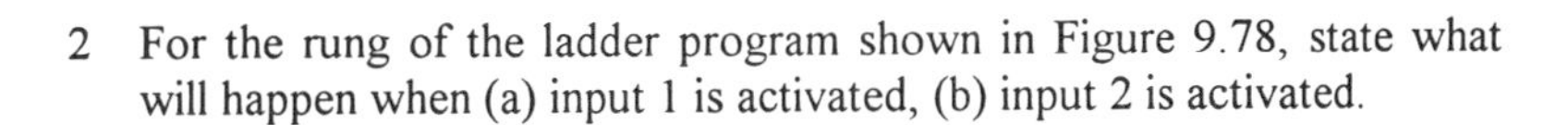

2 For the rung of the ladder program shown in Figure 9.78, state what will happen when (a) input 1 is activated, (b) input 2 is activated.

3 Draw the ladder rungs for the following programs:

(a) Two switches have to be operated before a motor will operate.

(b) Either of two switches can be operated for a solenoid to be energised.

(c) A signal lamp is to come on if a pump is off.

4 For each of the following, draw the logic circuit involved, the ladder program and the instruction list for that program:

(a) A signal lamp is to be switched on if a pump is running and the pressure is satisfactory, or if the lamp test switch is closed.

(b) A motor is to be started when the start push-button is pressed and remain on until the stop push-button is pressed.

(c) A machine has four sensors to detect when safety features are not active. When there is an input to any one of these sensors, the machine must stop.

5 The inputs from the limit switches and the start switch and the outputs to the solenoids of the valves shown in Figure 9.79 are connected to a PLC which has the ladder program shown. What is the sequence of the cylinders?

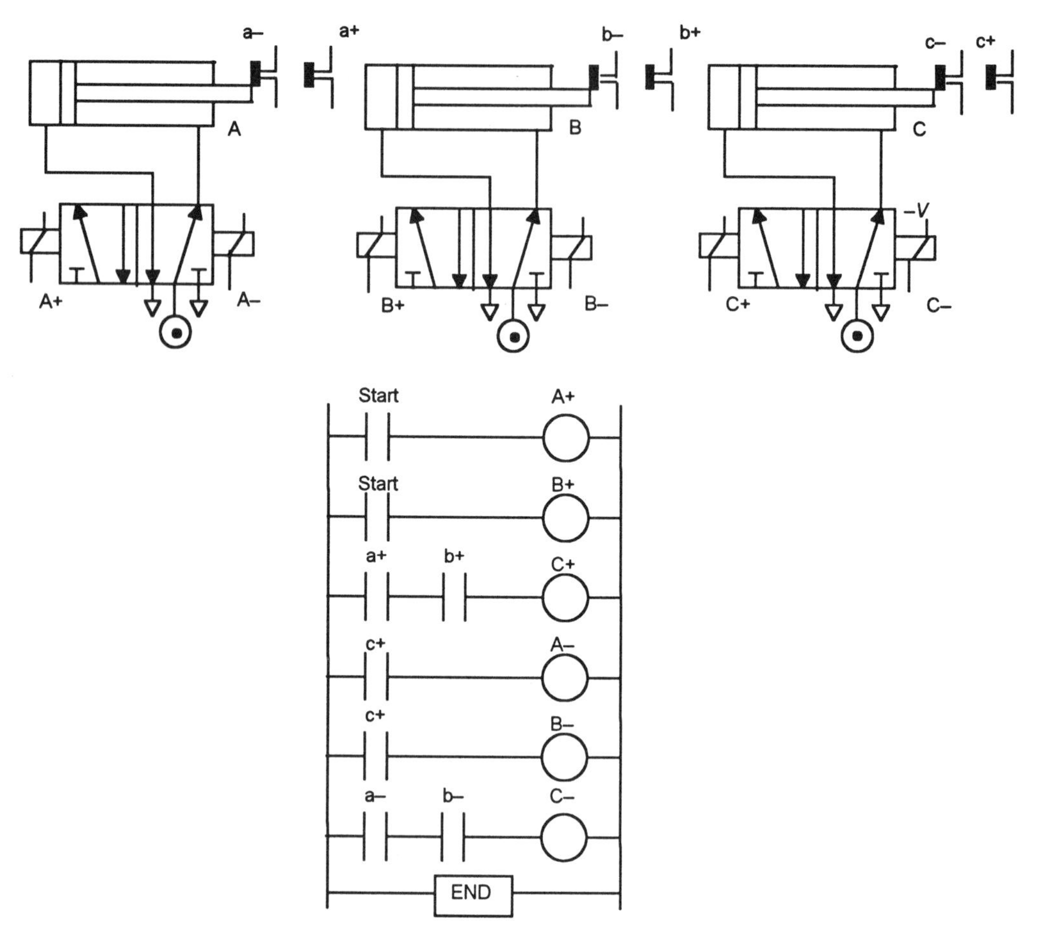

Figure 9.79 *Problem 5*

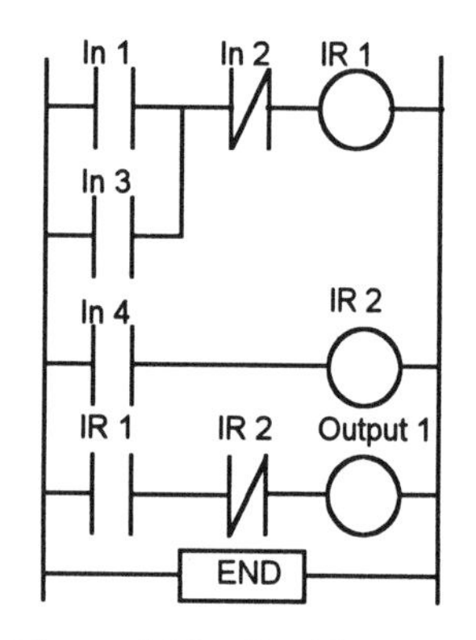

Figure 9.80 *Problem 6*

6 What switch inputs of In 1, In 2, In 3 and In 4 have to be activated for there to be an output with the ladder program shown in Figure 9.80? Internal relays are denoted by IR.

7 The inputs from the limit switches and the start switch and the outputs to the solenoids of the valves shown in Figure 9.81 are connected to a PLC which has the ladder program shown. What is the sequence of the cylinders?

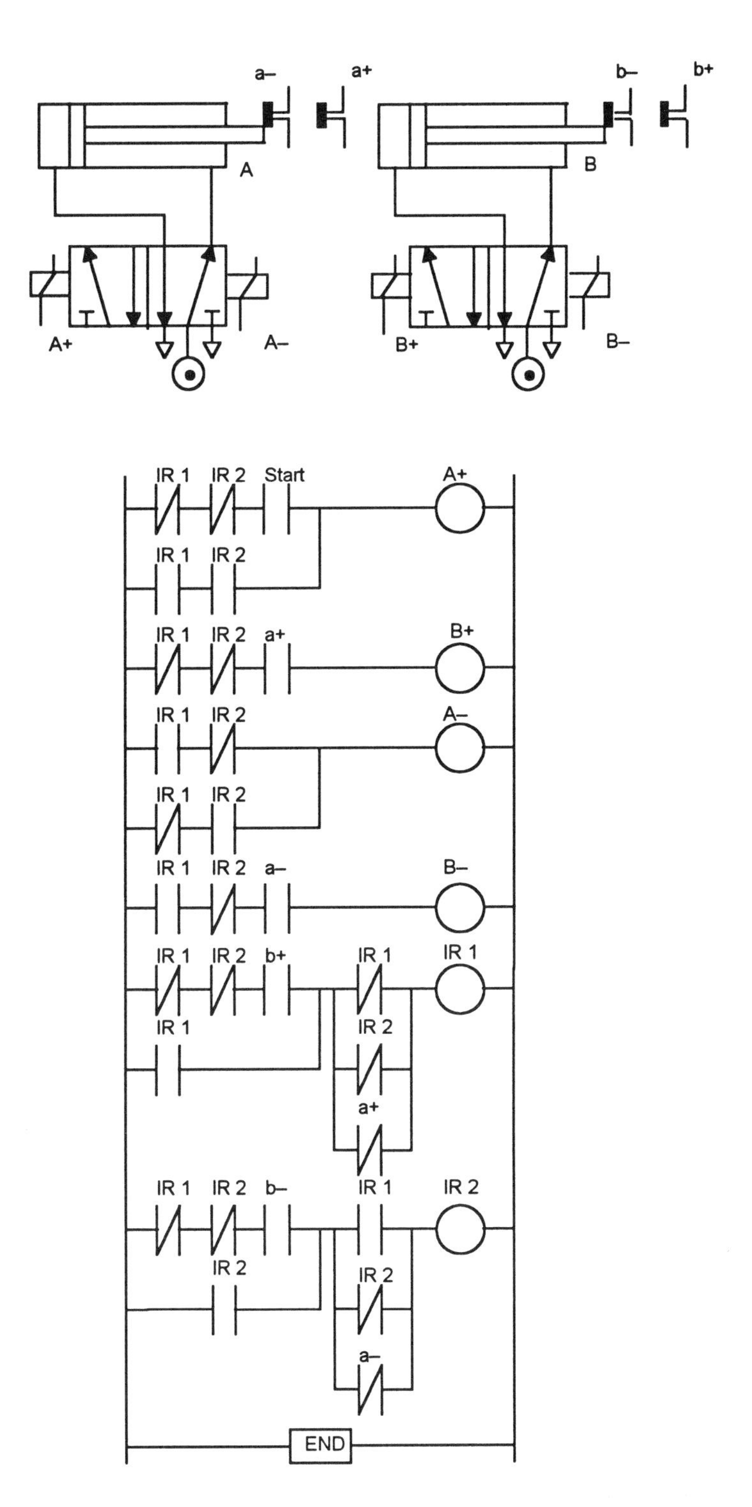

Figure 9.81 *Problem 7*

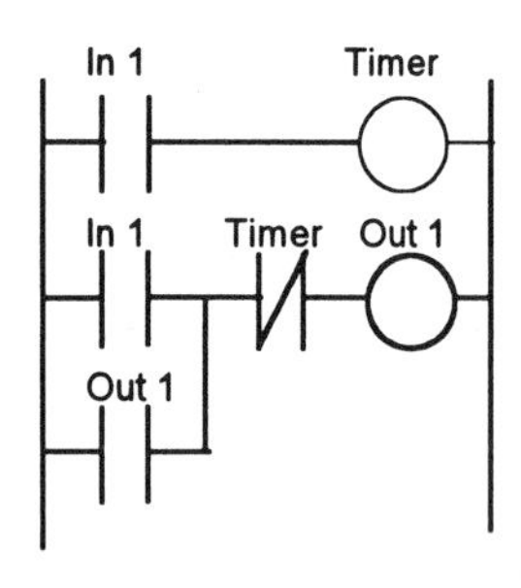

Figure 9.82 *Problem 8*

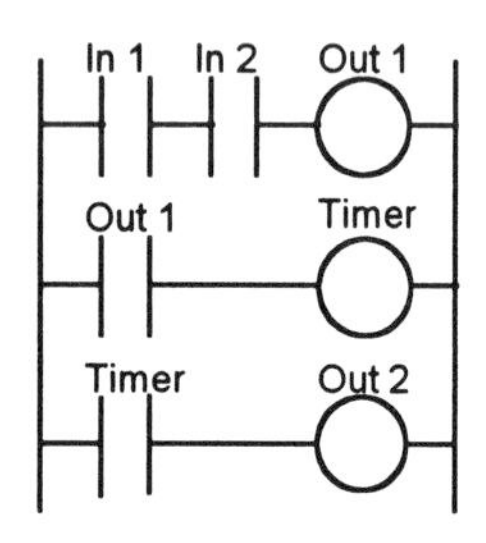

Figure 9.83 *Problem 9*

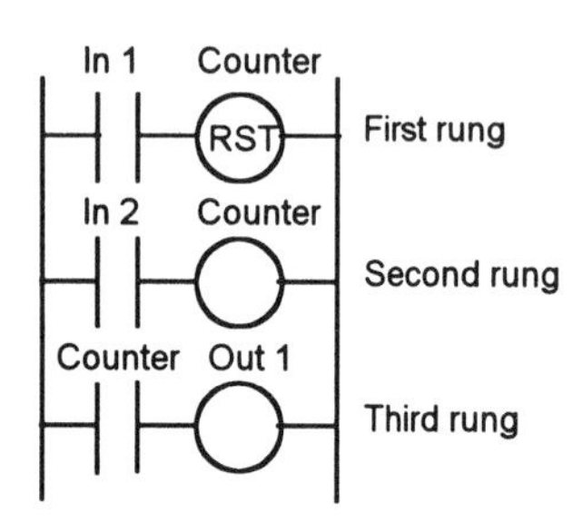

Figure 9.84 *Problem 11*

8 Figure 9.82 shows a section of a ladder diagram with a delay-on timer, an input In 1 and an output Out 1. Describe the sequence of events that occurs when there is an input to In 1.

9 For the section of ladder diagram shown in Figure 9.83 with two inputs In 1 and In 2, two outputs Out 1 and Out 2 and a delay-on timer, describe the sequence of events that occurs when there are inputs to In 1 and In 2.

10 Devise a ladder program to switch on a water pump for 100 s. It is then to be switched off and a heater switched on to heat the water for 50 s. Then the heater is switched off and another pump is used to empty the water. You can think of the sequence as being part of the domestic washing machine program.

11 For the section of ladder program shown in Figure 9.84, explain what happens when inputs In 1 and In 2 are activated, the counter having been preset to 5.

12 Figure 9.85 shows a ladder program involving a counter C460, inputs X400 and X401, internal relays M100 and M101, and an output Y430. X400 is the start switch. Explain how the output Y430 is switched on.

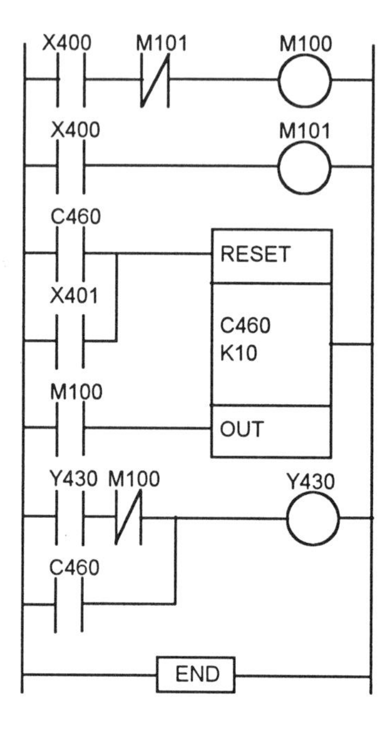

Figure 9.85 *Problem 12*

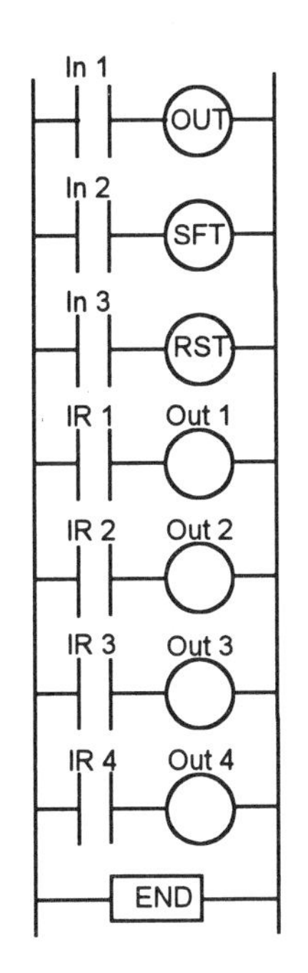

Figure 9.86 *Problem 13*

13 Figure 9.86 shows a ladder program involving a shift register. What inputs should be applied to obtain the outputs switching on in sequence and remaining on?

14 Devise a ladder program which for a sequence of four outputs such that output 1 is switched on when the first event is detected and remains on, output 2 is switched on when the second event is detected and remains on, output 3 is switched on when the third event is detected and remains on, output 4 is switched on when the fourth event is detected and remains on, and all outputs are switched off when one particular input signal occurs.

15 For a 4-bit shift register which has been reset to 0000, what will be their setting when (a) there is a pulse input to the 1 of the OUT of the shift register and then this is followed by a pulse input to SHIFT, (b) with a continuous input of 1 to the OUT, there is a pulse input to SHIFT, (c) with a continuous input of 1 to the OUT, there are two pulse inputs to SHIFT.

16 Devise a ladder program which can be used to switch on a number of rungs if certain conditions occur.

10 Maintenance and fault finding

10.1 Introduction

This chapter is a review of maintenance and fault-finding procedures that are used with pneumatic and hydraulic systems and with PLC controlled systems.

10.1 Regulations

With all such procedures, due care and attention must be given to safety and the relevant regulations followed. In Britain, all industrial processes are governed by the Health and Safety at Work Act 1974. This places duties on employers to, among other matters:

Provide and maintain safe plant and systems of work.

Make arrangements for ensuring safe use and handling of articles and substances.

Provide adequate information, instruction, training and supervision.

Maintain a safe place of work.

Provide and maintain a safe working environment.

Duties are placed on employees to, among other matters:

Safeguard himself/herself and others by using equipment in a safe manner and following safe working practices.

Cooperate with the employer in respect of health and safety matters.

Use equipment and facilities provided to ensure safety and health at work.

In addition there are regulations specific to certain types of activity. Of particular relevance in the context of this book are the Pressure Systems and Transportable Gas Container Regulations 1989. These apply to compressed air systems operating at gauge pressures above 0.5 bar. In particular the regulations:

Are intended to secure the safety of people at work.

Place a responsibility on designers and fabricators to consider at the manufacturing stage both the purpose of plant and how it will comply with the regulations, any modification or repair being such as to not give rise to danger or otherwise impair the operation of any protective devices or inspection facility.

Require designers and manufacturers to supply written information concerning design, construction, examination, operation and maintenance of equipment.

Place a responsibility on installers to ensure that equipment is fitted correctly, does not give rise to danger and does not impair the action of any protective device or inspection facility.

Require users or owners not to operate, or allow to be operated, systems until the safe operating limits have been established.

Require a suitable written scheme for the periodic examination of those parts of pressure systems in which a defect may give rise to danger, the scheme having been drawn up or certified as being suitable by a competent person.

Charge the user or owner with the responsibility for ensuring that inspections are carried out at the appropriate times and that the system is not operated outside of its inspection period.

Lay down the procedure to be followed in the event of the competent person detecting serious defects which give cause for imminent danger. The competent person has to issue a written report to the user or owner, specifying the nature of the defect and the immediate action required with the system not to be operated until the repairs have been undertaken. Additionally, the competent person has to notify the Health and Safety at Work Executive within 14 days of the nature of the defects.

Require the user or owner to provide any person operating the system with adequate and suitable instructions for safe operation of the system and the action to be taken in an emergency. This should include the manufacturer's operating manual.

Require the user or owner to ensure that the system is maintained in good repair so as to prevent danger.

Require the user and owner to keep documents relating to the pressure system.

Require users and owners to ensure that vessels having permanent outlets to the atmosphere, which if blocked could lead to over-pressurisation, to be maintained in an unblocked state.

10.2 Maintenance

The term *maintenance* is used for the combinations of actions carried out to return an item to, or restore it to, an acceptable condition. The term *preventative maintenance* is used for maintenance carried out at predetermined intervals, or to other prescribed criteria, and intended to reduce the likelihood of an item not meeting an acceptable condition. The term *corrective maintenance* is used for maintenance carried out to restore an item which has ceased to meet an acceptable condition. Thus, for example, preventative maintenance is used for a car with annual servicing or servicing at specified mileages, while corrective maintenance is used when it has to be repaired because it has broken down. Preventative maintenance is a key method of controlling the level of corrective maintenance that might be required. If a car is not regularly serviced, then it is more likely to break down. Thus preventative maintenance can reduce the amount of corrective maintenance.

In considering the best way to maintain an item, the following factors should be taken into account:

1 *Maintenance characteristics*
There are two key factors, the deterioration characteristic and the repair characteristic. With regard to deterioration, the mean life of an item before it fails is a good indicator of the need for maintenance. If the mean life of an item is less than that of the system in which it is used, then maintenance is likely to be required. Items can be classified as: replaceable items that are likely to have to be replaced during the life of the system and permanent items that are unlikely to have to be replaced. The repair characteristic is indicated by the mean time to repair or replace.

2 *Economic factors*
What is the cost of replacement prior to failure? What is the cost of unexpected failure? Failure might mean a system is out of action for some time and this could have economic implications. What is the repair cost?

3 *Safety factors*
What are the safety implications of an item failing? If an item is critical and failure would result in an unsafe situation, preventative maintenance of that item is indicated rather than waiting for it to fail and use corrective maintenance.

The *maintenance plan* consists of a schedule of preventative maintenance work and guidelines for the implementation of corrective maintenance work. The plan determines the level and nature of the maintenance workload. It can involve:

1 *Fixed-time maintenance*
The maintenance is carried out at regular intervals or after a fixed number of cycles of operation, etc. This procedure can be effective if the failure mechanism is time dependent.

2 *Condition-based maintenance*
If the approach of failure is detectable, then condition-based maintenance is effective. Thus, for example, wear of a component can be used as an indicator of the approach of failure and maintenance based on inspection for wear. Condition-based monitoring can be simple inspection involving qualitative checks based on observation, condition checking where routine measurements are made of some quantity, or trend monitoring where measurements are made and the trend of those measurements used to forecast when failure is likely to occur.

10.2.1 Maintenance examples

In hydraulic systems, oil problems are responsible for a high percentage of failures. These problems may be due to the level of oil in the reservoir or contamination by dirt, air or water. Preventative maintenance might thus involve inspection of the oil level, visual checks for oil leakage, and checking of the differential pressure across a filter element, a large drop indicating the need to change a filter before it becomes blocked. Replacement of filters might be an aspect of fixed-time preventative maintenance.

With pneumatic systems, problems can arise from contamination by dirt or water and, as with hydraulic systems, filters need to be checked and maintained. This might thus be part of preventative maintenance for such systems.

With cylinders, seals and bushing require regular inspection as part of preventative maintenance with replacement if damaged. The cylinder end caps should be removed at regular intervals and the piston assembly cleaned and inspected. Score marks can indicate dirt ingress. For long cylinder life it is essential that all dirt and moisture should be excluded from the cylinder and, in the case of pneumatic systems, the air should be adequately lubricated.

With valves, it is possible for a valve to stick due to a build-up of contamination. Cleaning might thus be required and 'O' ring seals replaced. With solenoid valves, care must be taken when replacing failed solenoids to ensure that the operating voltage of the replacement is correct.

With pumps, a noisy pump might be an indicator of air leaking into the system, a clogged or restricted intake line, a plugged air vent in the reservoir, loose or worn pump parts. Thus just listening to a pump can be a useful indicator of the need for maintenance. The oil temperature is another useful indicator. A high temperature can arise because of a clogged oil cooler, low oil, the oil used having too high a viscosity, the relief valve allowing the flow discharging with too high a pressure drop, or perhaps high internal leakages from perhaps wear. Compressors are often belt driven and the belt condition and tension requires regular checking as part of preventative maintenance.

A weekly maintenance schedule for a compressor might read:

1 Check temperature of motor. It should not be above 75°C.

2 Check noise level.
3 Clean intake filter.
4 Check V-belt tension.
5 Check compressor oil level.
6 Test safety valve.

A record of corrective maintenance for a week might include:

1 Cylinder XXX, replaced broken rod.
2 Solenoid XXX faulty and replaced.
3 Filter on airline blocked and so cleaned.
4 Faulty pressure gauge replaced.
5 Solenoid valve XXXX sticking, replaced.
6 Dirty air line cleaned out.

10.3 Installation

The following are general points which need to be taken into account in the installation of components in a system (though the discussion refers to a pneumatic system, most of the points equally apply to a hydraulic system):

1 Components waiting to be assembled should remain in their protective coverings until the last possible moment in order to reduce the chances of dust entering the components. For easier identification, all components should be labelled as on the circuit diagram.

2 Cylinders must be mounted securely. They should be checked for alignment to ensure that there are no lateral forces and that piston rods move freely through full stroke. Where piston rods actuate sensors, ensure that this actuation is correct and that no excessive loads are placed on the sensors or the movement of the rods impaired.

3 Position valves so that they cannot be damaged by coming into contact with machinery or people and check their operation. Cylinder directional control valves should be as near to the cylinder as possible in order to minimise air consumption and decrease response time.

4 Install air-line filters as close as possible to their actual point of use. Install the regulator, lubricator and main isolating valve. Bear in mind the need to be accessible for servicing.

5 Pipe up the circuit, ensuring that the air supply is not restricted by undersized pipes, kinked tubing or restrictive fittings. Where possible, for ease of identification colour-coded pipes should be used. Plastic tubing is generally used though copper is to be preferred where physical damage of heat may be encountered. 'Push-in' or 'push-on' connectors (Figure 10.1) can be used for making connections. Such fittings are reusable, interchangeable and easy to connect and disconnect. As each section is assembled, a test air line can be used to check for correct operation and air leaks.

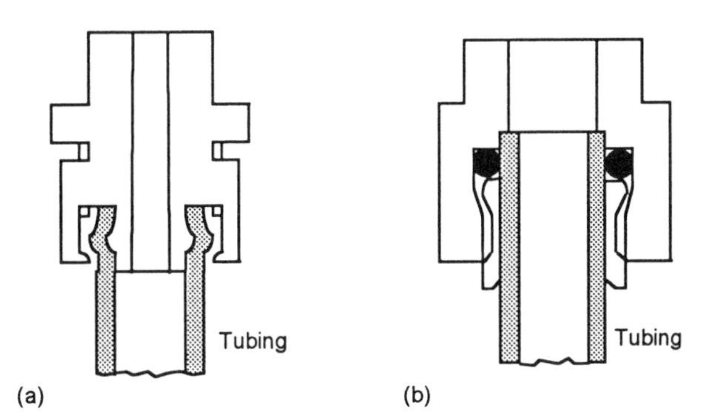

Figure 10.1 *(a) 'Push-in', (b) 'push-on' connectors*

6 Pressurised air exhausting to the atmosphere tends to be noisy. This noise can be reduced by fitting silencers.

10.3.1 Safety checks

Before a system is used, the following checks should be made to ensure safe operation of the system:

1 The effect of an air supply failure at any point in the cycle does not result in an unsafe situation.

2 The effects of variations in the air supply pressure do not result in an unsafe situation or damage. An air receiver might be necessary in the supply line immediately prior to a point where this situation can develop.

3 The effect of any cylinder jamming does not result in an unsafe situation.

4 The effect of incorrect operation of any manual control valve does not result in an unsafe situation.

5 All safety devices and guards are in place and operative.

10.3.2 Documentation

The following information needs to be available:

1 An up-to-date circuit diagram.

2 Full details of each component, each component being labelled as the component on the circuit diagram.

3 Minimum operating pressure of the system and the setting of pressure regulators.

4 Operating sequence.

5 Starting and stopping procedures.

10.3.3 Fault finding

When faults occur, determining the cause can involve:

1 Checking that the air supply is on and set to the correct pressure.

2 Checking that cylinders are in their correct positions and able to operate any sensors.

3 Manual valves are in their correct positions.

4 Pressure regulators are at their correct settings.

5 If the fault still occurs then determining the point in the cycle at which the sequence stopped by noting the position of each cylinder and comparing it with the sequence specified in the documentation for the system. Then checking through the possible ways such a fault could occur. For example, if in the sequence cylinder C (Figure 10.2) is diagnosed as not having extended, checks might be made working back from C in a logical manner:

Is there air on cylinder C extend port (1)? If air is present then the cause of the fault may be that the cylinder is jammed.

If there is no air on the cylinder extend port, check the air on the extend pilot port (2) of the control valve controlling the cylinder. If there is air then there might be failure of the control valve. Alternatively the problem might be caused by air being simultaneously on both pilot ports (3) of the valve and so locking the spool.

If there is no air, check that there is air in the outlet of the valve supplying the pilot port (4). If there is air then the air line between it and the control valve might be blocked.

If there is no air, check the air supply to the valve (5). If there is air then the valve may have failed.

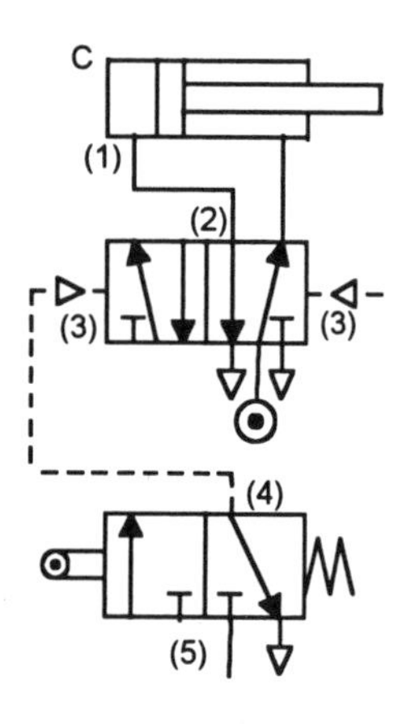

Figure 10.2 *Fault finding*

10.4 PLC systems

Commissioning of a PLC system involves:

1 Checking that all the cable connections between the PLC and the plant being controlled are complete, safe and to the required specification and meeting local standards.

2 Checking that the incoming power supply matches the voltage setting for which the PLC is set.

3 Checking that all protective devices are set to their appropriate trip settings.

4 Checking that emergency stop buttons work.

5 Checking that all input/output devices are connected to the correct input/output points and giving the correct signals.

6 Loading and testing the software.

10.4.1 Testing inputs and outputs

Most PLCs have the facility for testing inputs and outputs by what is termed *forcing*. This involves software being used to turn off or on inputs and outputs and so enable checks to be carried out on the wiring from the inputs and outputs to the PLC and whether the input and output devices are correctly working. In order to force inputs or outputs, a PLC has to be switched into the forcing or monitor mode by perhaps pressing a key marked FORCE or selecting the MONITOR mode on a screen display. Thus if an input is forced we can check that the consequential action of that input being on occurs. Care must be exercised with forcing in that an output might be forced that can result in a piece of hardware moving in an unexpected and dangerous manner.

The installed program can thus be run and inputs and outputs simulated so that they, and all preset values, can be checked. Figure 10.3 shows how inputs might appear in the ladder program display when open and closed, and outputs when not energised and energised. The display shows a selected part of the ladder program and what happens as the program proceeds. Thus at some stage in a program the screen might appear in the form shown in Figure 10.4(a). For rung 12, with inputs to X400, X401 and X402, but not M100, there is no output from Y430. For rung 13, the timer T450 contacts are closed, the display at the bottom of the screen indicating that there is no time left to run on T450. Because Y430 is not energised the Y430 contacts are open and so there is no output from Y431. If we now force an input to M100 then the screen display changes to that shown in Figure 10.4(b). Now Y430, and consequently Y431, come on.

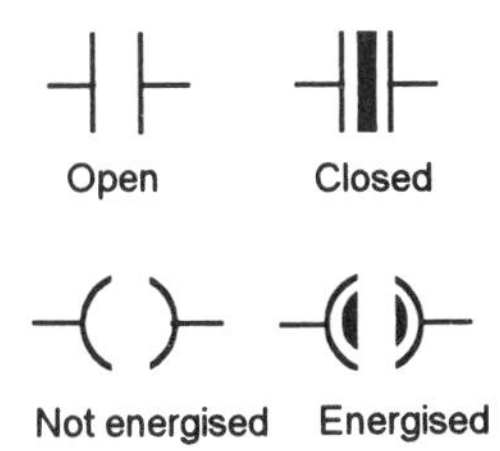

Figure 10.3 *Monitor mode symbols*

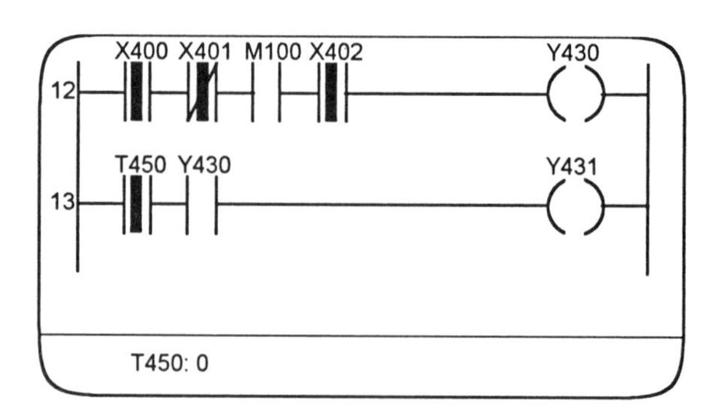

(a)

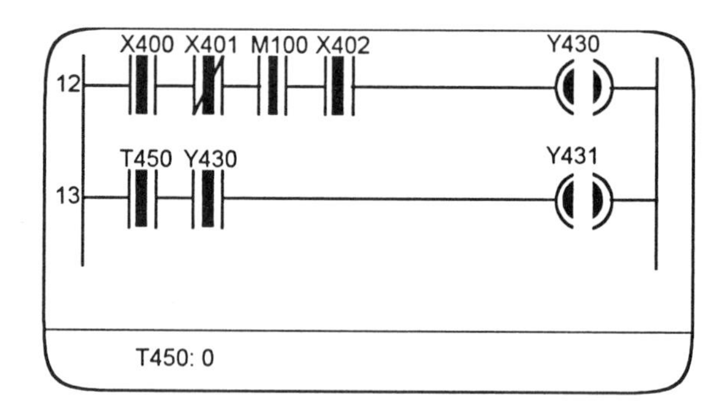

(b)

Figure 10.4 *Ladder program monitoring*

10.4.2 Program testing

Most PLCs have a software checking program. This checks through a ladder program for incorrect device addresses, and provides a list on a screen, or as a printout, of all the input/output points used, counter and timer settings, etc. with any errors detected. Thus the procedure might involve:

1 Opening and displaying the ladder program concerned.
2 Selecting from the menu on the screen Ladder Test.
3 The screen might then display the message: Start from beginning of program (Y/N)?
4 Type Y and press Enter.
5 Any error message is then displayed or No errors found.

For example, there might be a message for a particular output address that it is used as an output more than once in the program, a timer or counter is being used without a preset value, a counter is being used without a reset, there is no END instruction, etc.

As a result of such a test, there may be a need to make changes to the program. Such changes might be made by selecting from the menu displayed on screen Exchange and following through the set of displayed screen messages.

10.4.3 Fault finding

With any PLC-controlled plant, by far the greater percentage of the faults are likely to be with sensors, actuators and wiring rather than within the PLC itself. Of the faults within the PLC, most are likely to be in the input/output channels or power supply than in the CPU.

As an illustration of a fault, consider a single output device failing to turn on. If testing of the PLC output voltage indicates that it is normal then the fault might be a wiring fault or a device fault. If checking of the voltage at the device indicates the voltage there is normal then the fault is the device. As another illustration, consider all the inputs failing. This might be as a result of a short circuit or earth fault with an input and a possible procedure to isolate the fault is to disconnect the inputs one by one until the faulty input is isolated. An example of another fault is if the entire system stops. This might be a result of a power failure, or someone switching off the power supply, or a circuit breaker tripping.

PLCs provide built-in fault analysis procedures which carry out self-testing and display fault codes, with possibly a brief message, which can be translated by looking up the code in a list to give the source of the fault and possible methods of recovery. For example, the fault code may indicate that the source of the fault is in a particular module with the method of recovery given as replace that module or perhaps switch the power off and then on. The following are some of the common fault detection techniques used internally by PLCs:

1 *Timing checks*
The term *watchdog* is used for a timing check that is carried out by the PLC to check that some function has been carried out within the normal time. If the function is not carried out within the normal time then a fault is assumed to have occurred and the watchdog timer trips, setting of an alarm and perhaps closing down the PLC.

2 *Replication*
Where there is concern regarding safety in the case of a fault developing, checks may be constantly used to detect faults. One technique is *replication checks* which involves duplicating, i.e. replicating, the PLC system. This could mean that the system just repeats every operation twice and if it gets the same result it is assumed there is no fault. This procedure can detect transient faults. A more expensive alternative is to have duplicate PLC systems and compare the

results given by the two systems. In the absence of a fault the two results should be the same, a fault showing up as a difference.

3 *Expected value checks*
Software errors can be detected by checking whether an expected value is obtained when a specific input occurs. If the expected value is not obtained then a fault is assumed to be occurring.

Ladder programs can be developed to enable a PLC to supply other fault analysis data. For example, a timing watchdog program can be used to sound an alarm if some event, such as the extension of a cylinder, does not take place within the expected time. This could indicate that the control valve, the cylinder or the sensor have failed. Figure 10.5 shows such a ladder program.

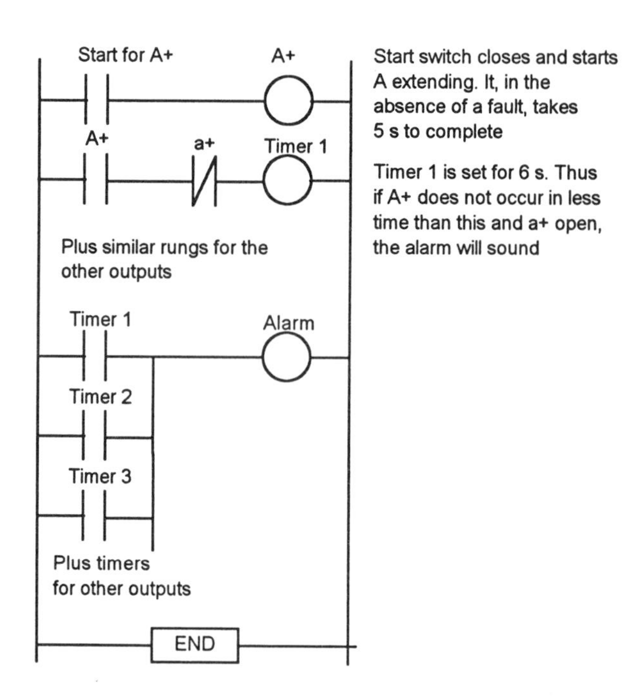

Figure 10.5 *Timing watchdog program*

As another example, in a pneumatic system involving cylinders operated in a sequential pattern a program can be developed to indicate at which point in the cycle a fault occurred. Figure 10.6 shows such a program, this being added to the program used to execute the cylinder sequence. Each of the cylinder movements has a light-emitting diode associated with it. The last cylinder movement to occur is indicated by its LED being illuminated.

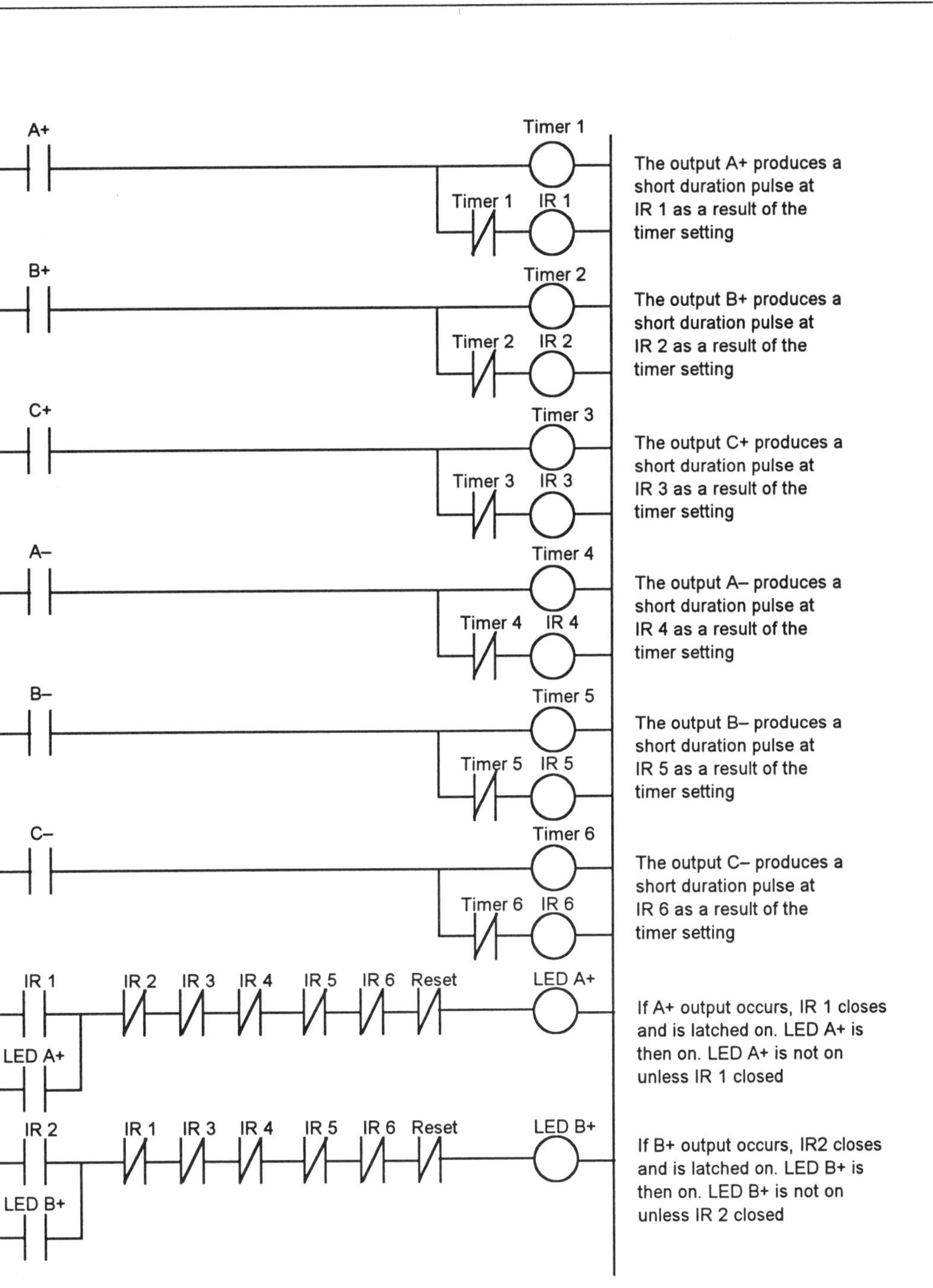

Ladder diagram continued on next page

Figure 10.6 *Diagnostic program for last cylinder action (continued on next page)*

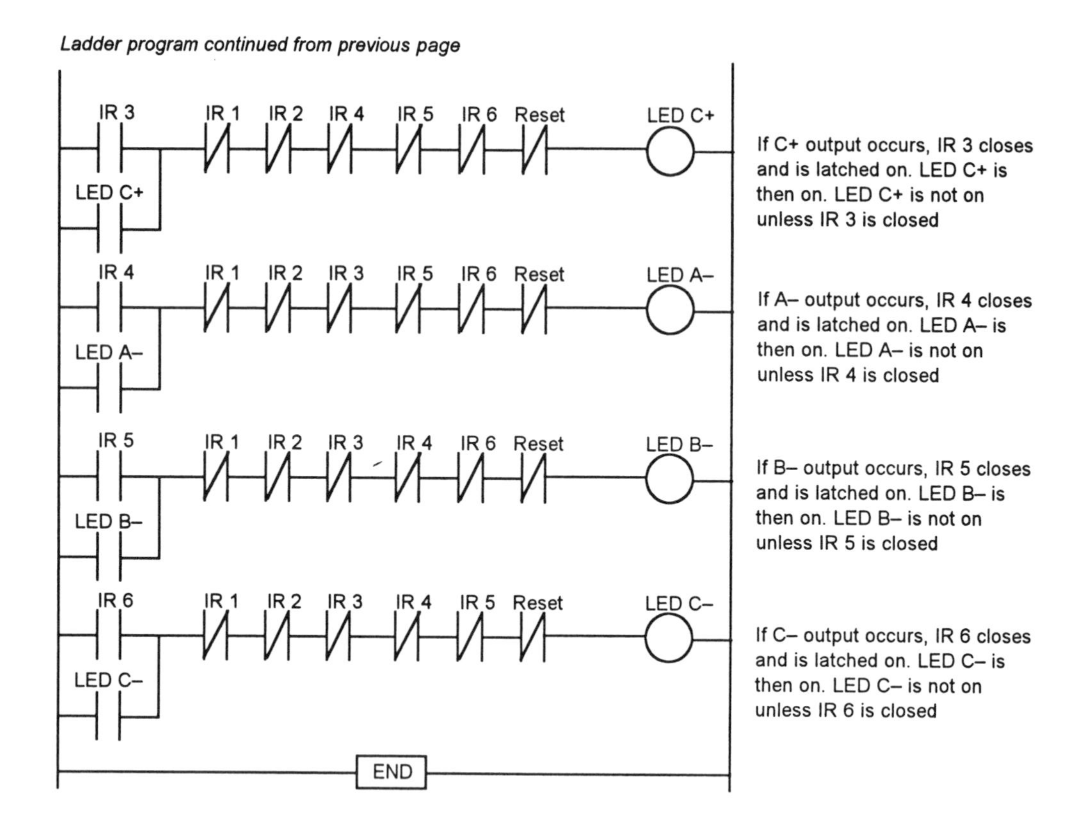

Figure 10.6 *Diagnostic program for last cylinder action (continued from previous page)*

10.4.4 Documentation

The documentation that should be developed for a PLC installation should include:

1 A description of the plant.
2 Specification of the control requirements.
3 Details of the programmable logic controller.
4 Electrical installation diagrams.
5 Lists of all input and output connections.
6 Application program with full commentary on what it is achieving.
7 Software backups.
8 Operating manual, including details of all startup and shutdown procedures and alarms.

Problems

1. State the different between preventative maintenance and corrective maintenance, illustrating your answer with examples.

2. State the difference between fixed term-based maintenance and condition-based maintenance.

3. The pneumatic circuit shown in Figure 10.7 suffers a fault and stops with B fully retracted and cylinder A fully extended. Determine the possible fault if (a) checks show that no air is present on cylinder B extend port, no air is present at the extend pilot port of valve X but air is present at the outlet port of Y,(b) checks show that air is present on cylinder B extend port.

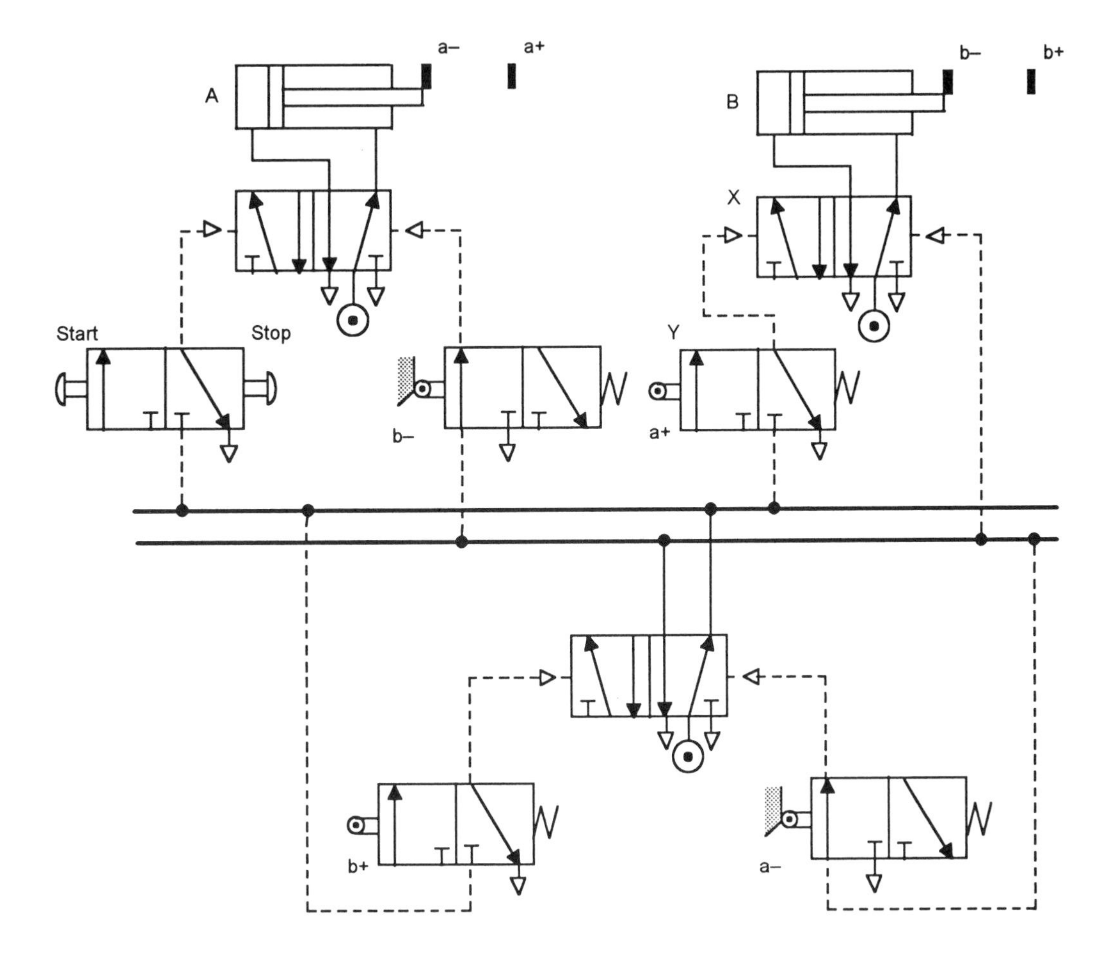

Figure 10.7 *Problem 3*

4 Explain how, using forcing, the failure of an input sensor or its wiring can be detected.

5 State the type of information typically supplied by the software-checking program of a PLC.

6 Devise a timing watchdog program to be used to switch off a machine if faults occur in any of the systems controlling its actions.

7 Design a pneumatic system for control by a PLC to give the cylinder sequence A+, B+, B–, A– and which will give a LED display indicating, in the presence of a fault such as a sticking cylinder, at which point in the cycle the fault occurred. Explain the action of all elements in the system.

Appendix: Symbols

The following are the standard, commonly used symbols for pneumatic, hydraulic and logic components.

Basic symbols

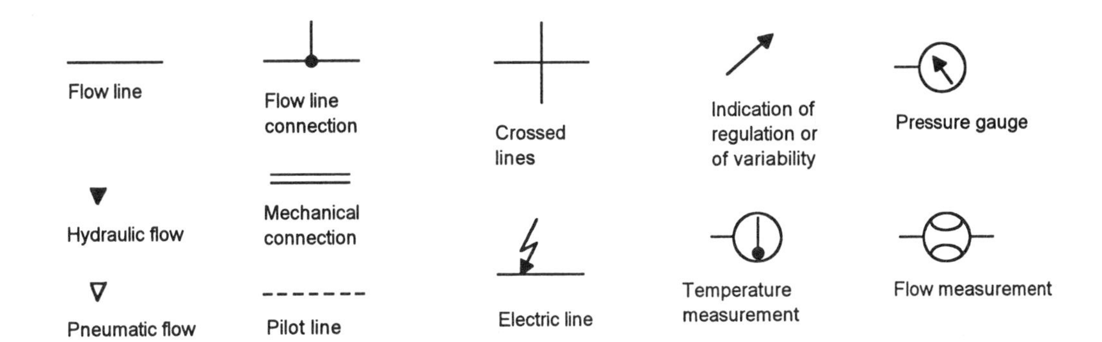

Pumps, compressors, motors, reservoirs and accumulators

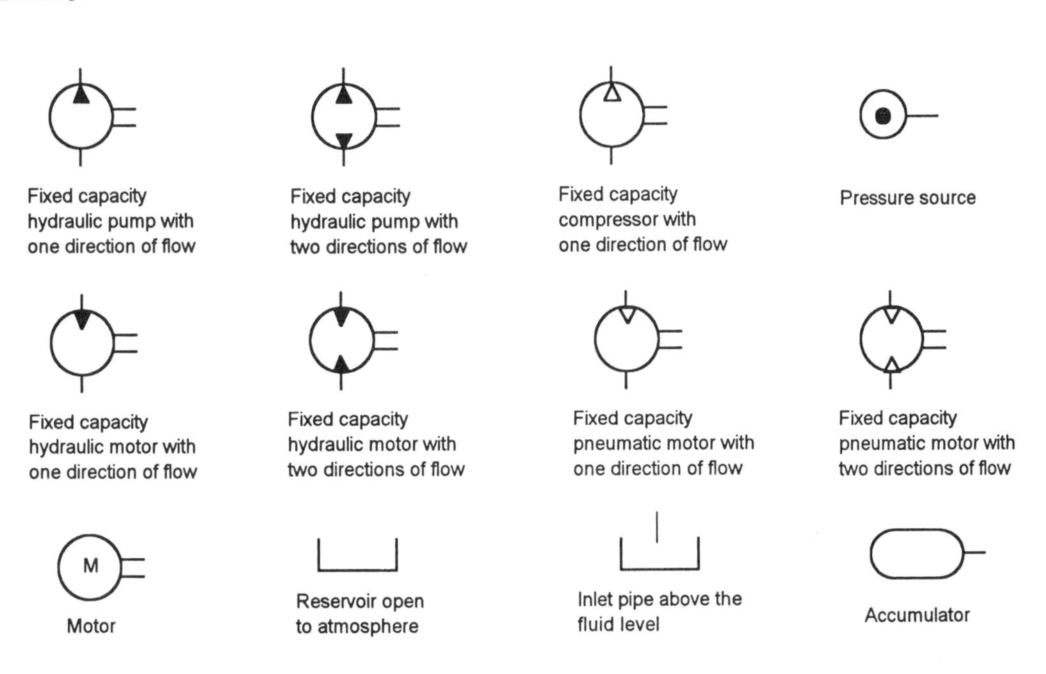

Conditioning

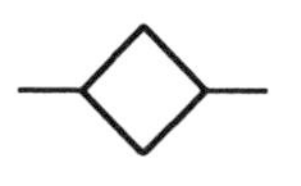
Conditioning apparatus

Cooler

Heater

Filter

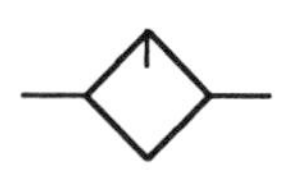
Lubricator

Water trap with manual control

Water trap with automatic control

Filter with water trap with manual control

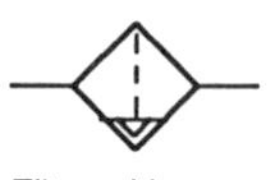
Filter with water trap with automatic control

Dryer

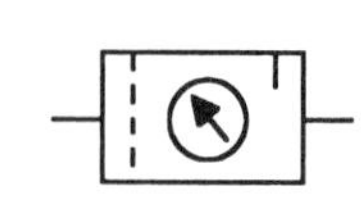
Conditioning unit

Directional control valves

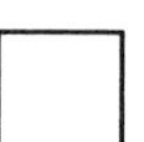
Unit for controlling flow or pressure

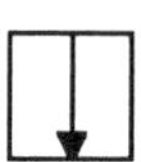
One flow path

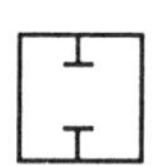
Two closed ports

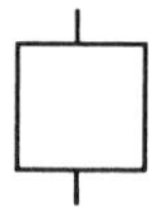
Input connections

Two or more squares indicate a directional control valve having as many position as there are squares

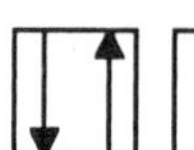
Two flow paths

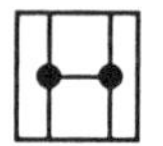
Two flow paths with cross connection

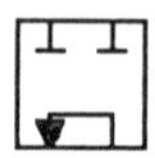
One flow path in a by-pass position, two closed ports

Non-return, shuttle and rapid exhaust valves

Non-return valve

Opens if inlet pressure more than outlet pressure

Opens if inlet pressure more than outlet and spring pressures

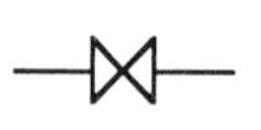
Shut-off valve

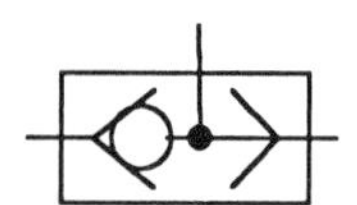
Shuttle valve

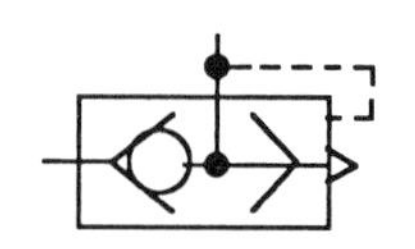
Rapid exhaust valve

Pressure control valves

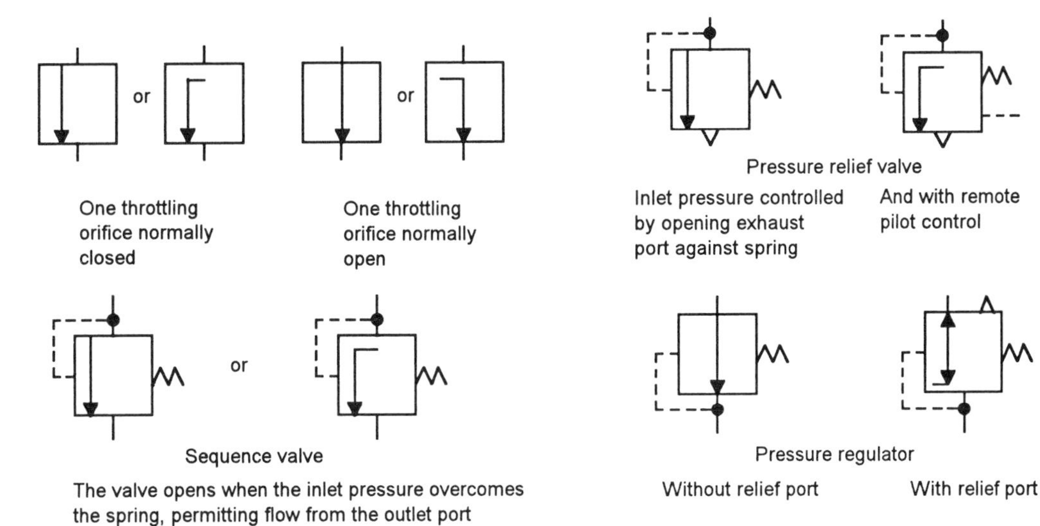

One throttling orifice normally closed

One throttling orifice normally open

Inlet pressure controlled by opening exhaust port against spring

And with remote pilot control

Sequence valve

The valve opens when the inlet pressure overcomes the spring, permitting flow from the outlet port

Without relief port

With relief port

Control methods

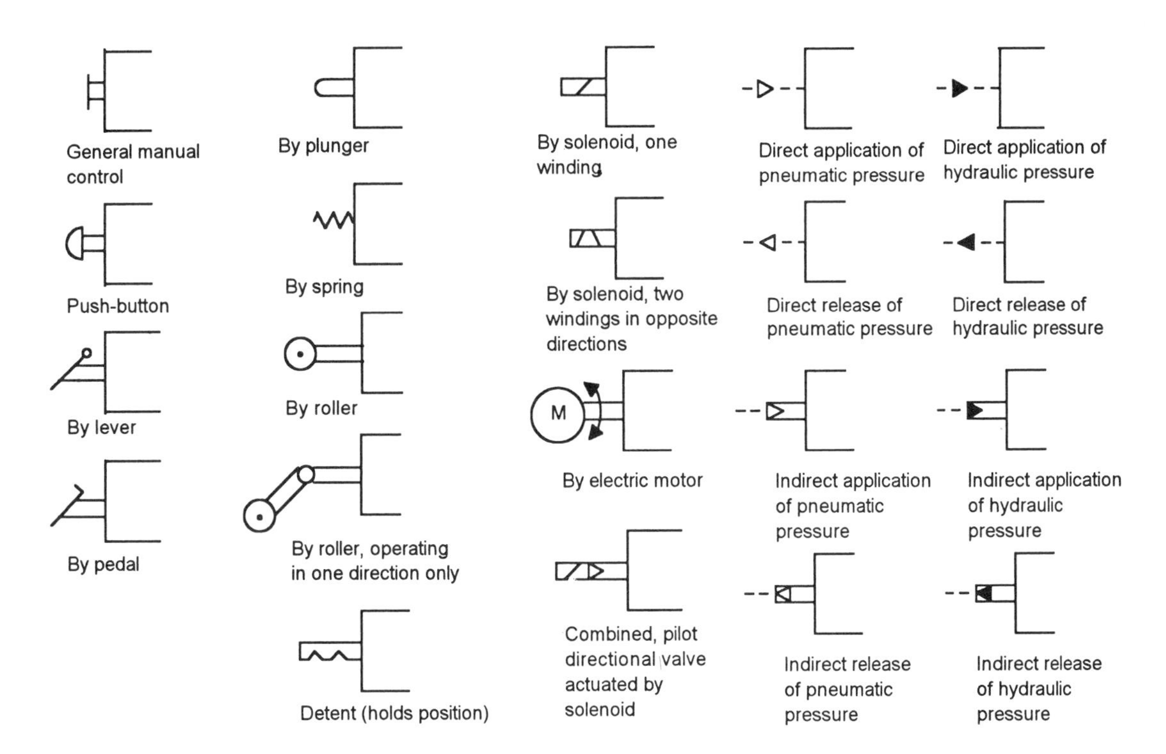

General manual control

By plunger

By solenoid, one winding

Direct application of pneumatic pressure

Direct application of hydraulic pressure

Push-button

By spring

By solenoid, two windings in opposite directions

Direct release of pneumatic pressure

Direct release of hydraulic pressure

By lever

By roller

By electric motor

Indirect application of pneumatic pressure

Indirect application of hydraulic pressure

By pedal

By roller, operating in one direction only

Combined, pilot directional valve actuated by solenoid

Indirect release of pneumatic pressure

Indirect release of hydraulic pressure

Detent (holds position)

Cylinders

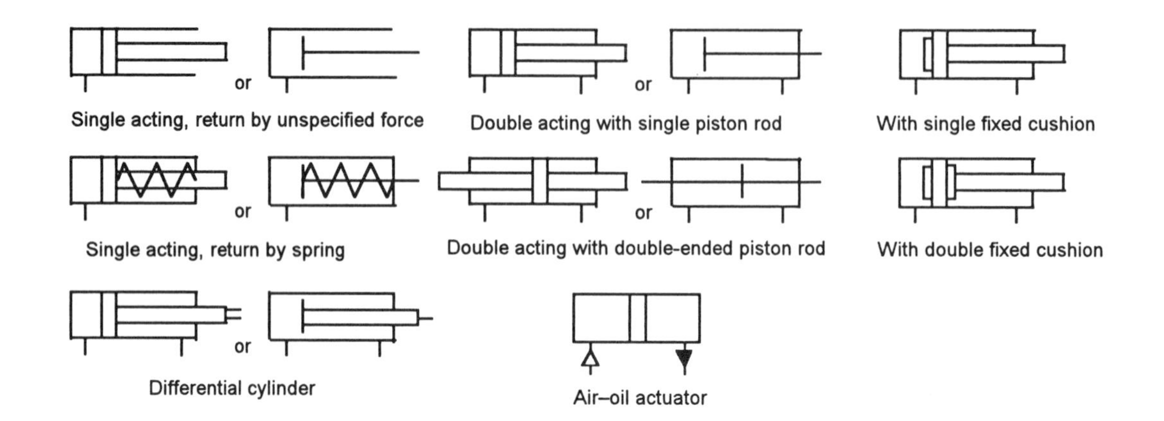

Logic gates

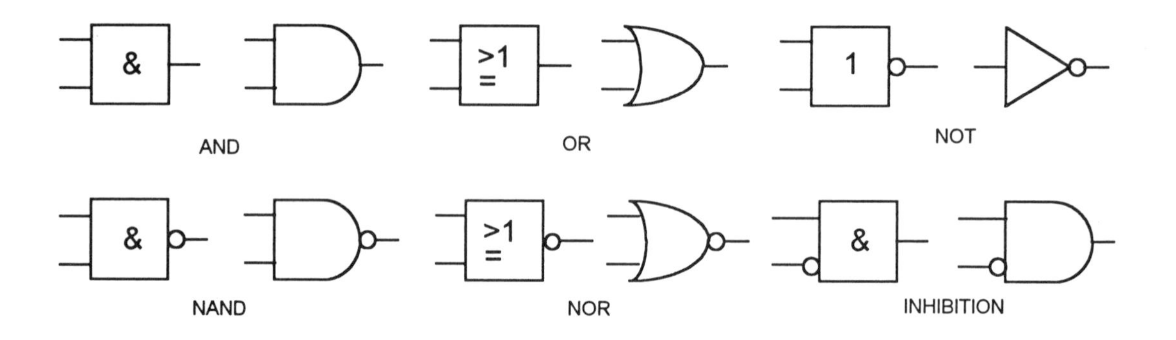

Ladder programs

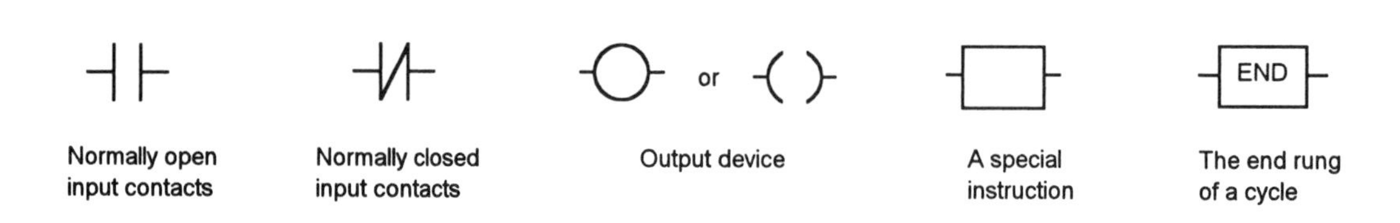

Answers

Chapter 1

1 0.4 MPa
2 10.4 kN
3 729 kPa
4 11.1 kg
5 1.3 MPa
6 0.047 m^3
7 0.23 kg/s
8 50.4 mm
9 53.2 mm
10 25.7 m/s

Chapter 2

1 0.18 m^3/min
2 3.55 m^3/min
3 (a) Less power required, (b) less pulsing, more efficient
4 Sealing clearance gaps between vanes and casing and vanes and rotor, also heat dissipation
5 9.6 g
6 6.8 g/m^3
7 About 5 m^3
8 12.9 m^3
9 16 m^3
10 21
11 See Section 2.4
12 See Figure 2.26
13 Drainage
14 About 80 mm
15 About 60 mm
16 23.6 m
17 0.12 m^3/min
18 0.72
19 0.81
20 See Section 2.7.4
21 144 dm^3
22 0.32 m^3
23 11.25 dm^3
24 9.3 dm^3
25 1.5 dm^3

Chapter 3

1 (a) Pressure to 2, exhaust closed; push-button pressed, 2 exhausts, pressure closed; spring return, (b) pressure closed, 2 exhausts; solenoid activated, pressure to 2, exhaust closed; spring return, (c) pressure

exhausts, outputs closed; pressure activated, pressure to 4, 2 exhausts; pressure activated, pressure to 2, 4 exhausts; spring return, (d) all closed; plunger pressed, pressure to 2, 4 exhausts; solenoid activated, pressure to 4, 2 exhausts.

2 (a) See Figure A.1(a), (b) see Figure A.1(b)
3 (a) See Figure A.2(a), (b) see Figure A.2(b)
4 (a) and (b) Piston moves to right, then left
5 See Figure A.3
6 See Figure 3.18 and associated text. Piston moves when X exceeds set pressure.
7 In the rest position, the cylinder in (a) is left at pressure and in (b) it is exhausted. With (a), as a result of pressure leakage in the rest condition, the cylinder piston will have some drift in either direction, with (b) this will not be the case. (b) also allows the piston in the rest position to be manually repositioned.
8 Retract rapidly, extend with adjustable speed.

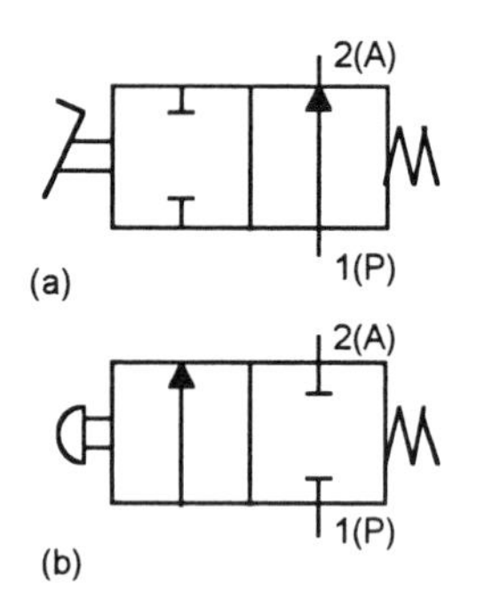

Figure A.1

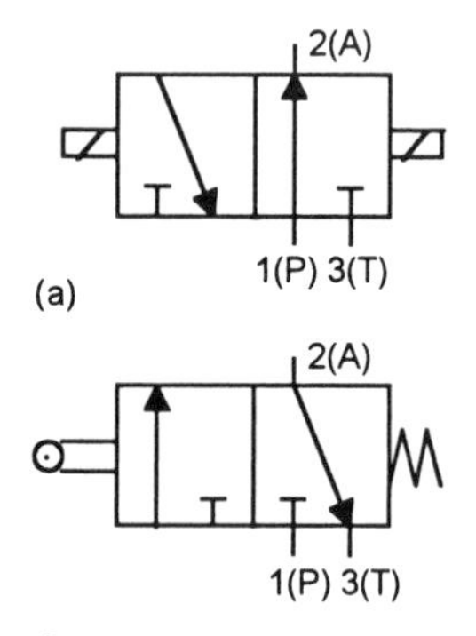

Figure A.2

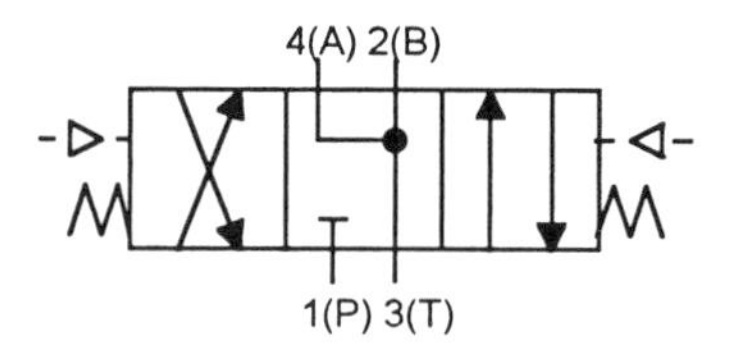

Figure A.3

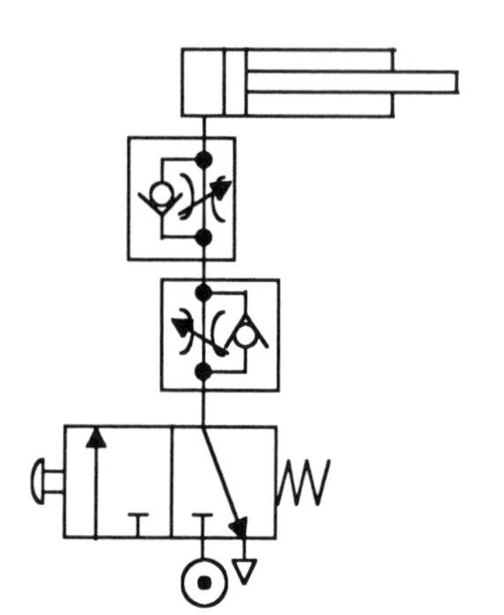

Figure A.4

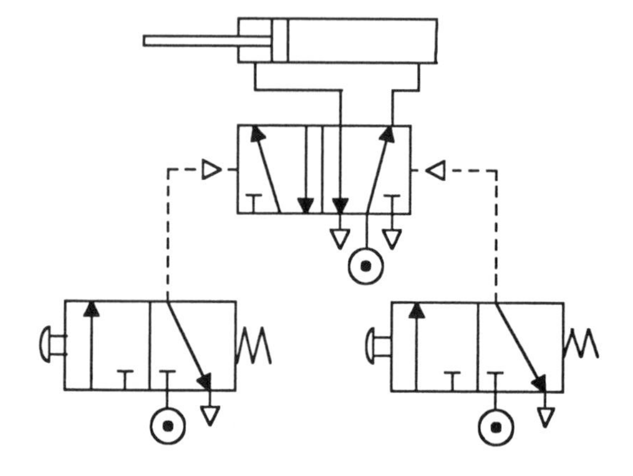

Figure A.5

Chapter 4

1 884.2 kPa
2 0.11 m^3/min
3 For example, see Figure A.4

4 94.5 mm diameter
5 See Figure A.5
6 880 N, 740 N
7 1.56
8 112 mm
9 84 mm
10 65 mm
11 47.7 rev/min, 2.35 N m, 11.7 W

Chapter 5

1 See Section 5.3
2 For example, as in Figure 5.2
3 For example, as in Figure 5.9
4 Cam-operated switch, with cam rotated at a constant rate
5 For example, (a) plunger/roller-operated switch, (b) reed switch

Chapter 6

1 See Section 6.2
2 (a) Solenoid directly actuates valve, (b) solenoid controls a pilot which actuates valve. This reduces the force required from the solenoid and hence the solenoid current.
3 Right-hand end of solenoid activated
4 Cylinder reciprocates back and forth
5 To start S1, to stop S2
6 See Figure 6.11 or 6.17

Chapter 7

1 (a) Piston floats, (b) piston can be held at intermediate position but will creep.
2 Meter-out speed control, (a) cylinder extends with controlled speed, (b) cylinder retracts at full speed.
3 For example, as in Figure 7.11 with the flow control for the extension only allowing a slow rate of oil flow, while the other flow control for the retraction gives a faster rate flow.
4 50 bar
5 36 mm
6 Time delay circuit, so time taken for extension to start.
7 For example, see Figure 7.21 or 7.23
8 (a) A, (b) B or C
9 A+, B+, A–, B–, see Figure 7.25
10 A+, B+, B–, A–
11 See Figure A.6
12 A+, B+, A–, B–
13 See Figure A.7
14 Two: Group I A–, Group II A+, B–, C–, Group I B+, A–. See Figure A.8

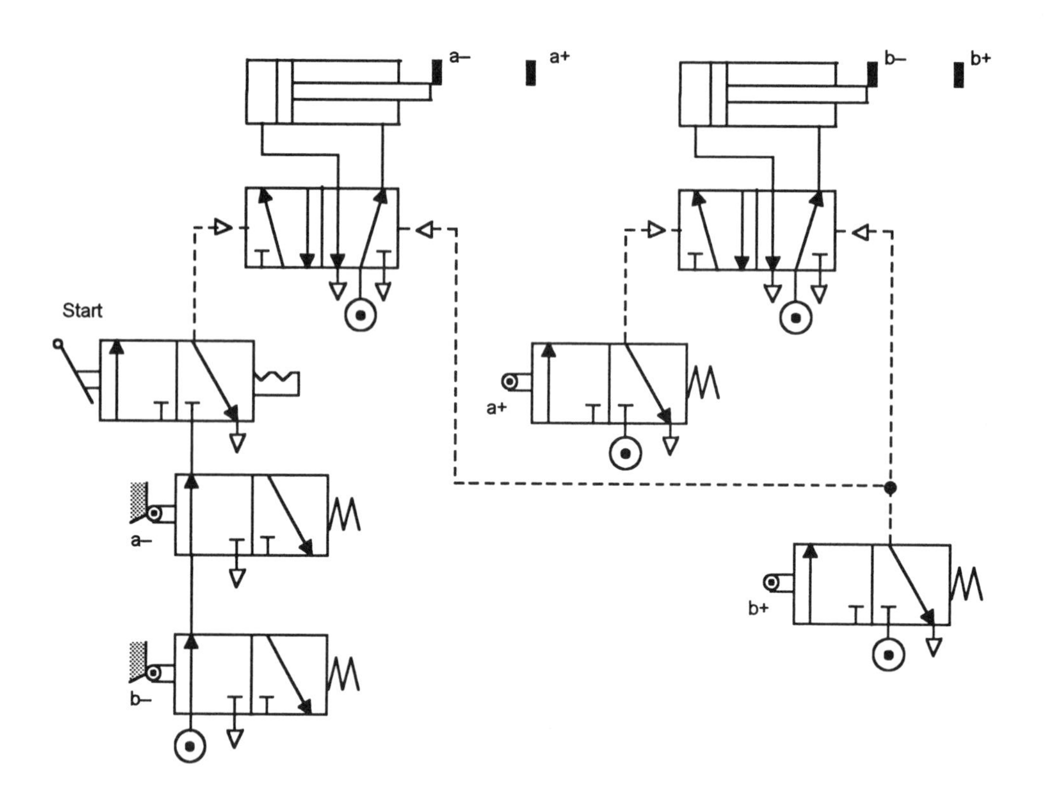

Figure A.6

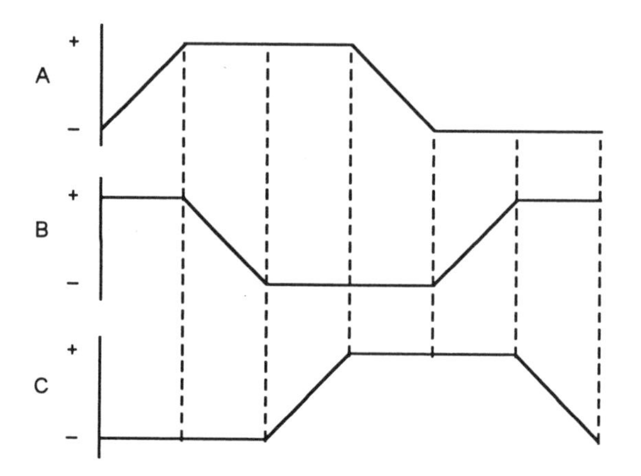

Figure A.7

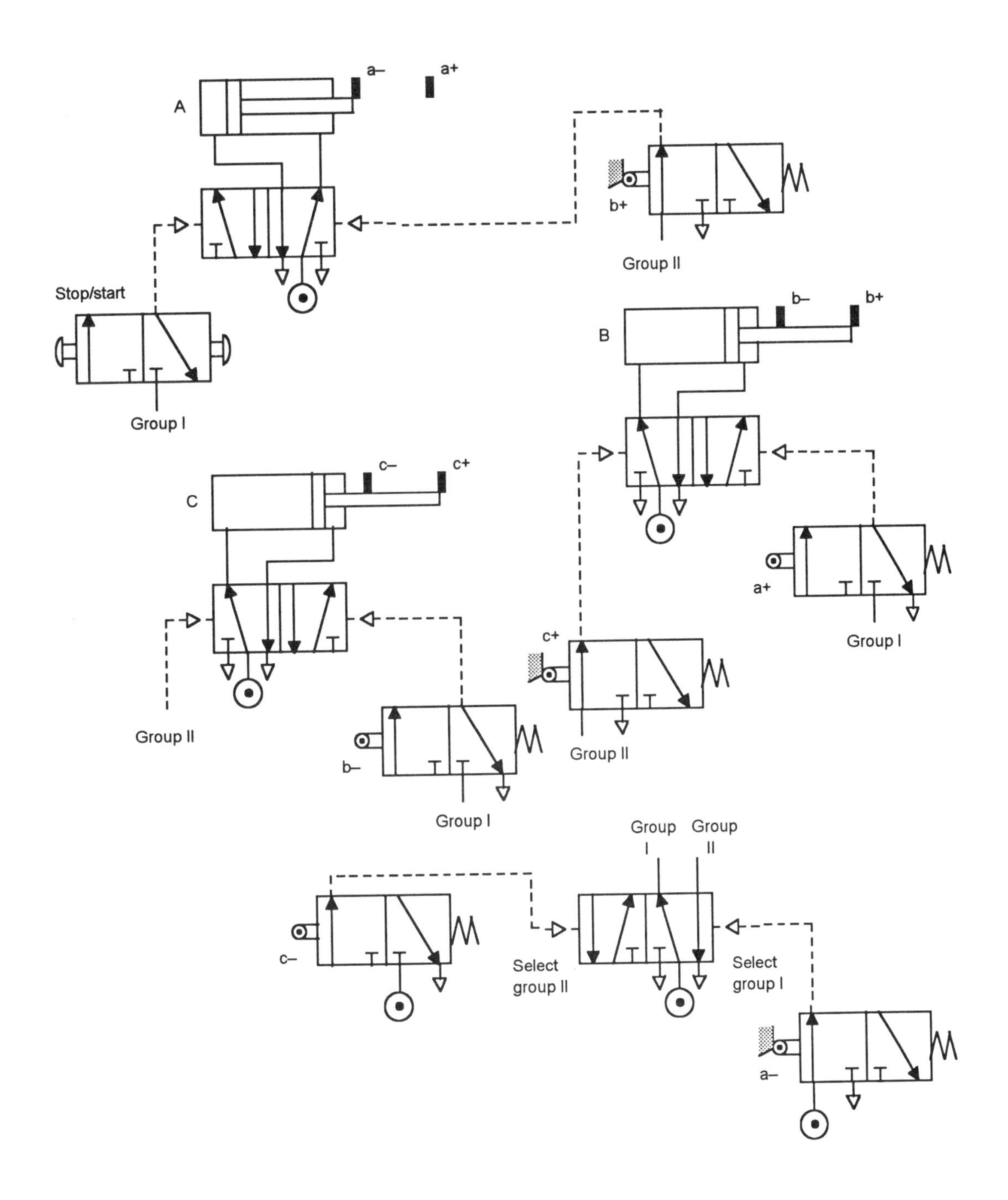

a–
a+
A
b+
Group II
Stop/start
Group I
b–
b+
B
c–
c+
C
a+
Group I
c+
Group II
Group II
b–
Group I
Group I
Group II
c–
Select group II
Select group I
a–

Figure A.8

Chapter 8

1. a_0, start 1 and start 2 must all be activated.
2. When 1, 2 or 3 are pressed, the cylinder retracts.
3. OR system, e.g. Figure 8.12.
4. (a) $A \cdot C \cdot D + B \cdot C \cdot D$, (b) $A + \bar{B} + C$, (c) $B \cdot \bar{C}$, (d) $A + \bar{B} \cdot C$, (e) $\bar{A} + \bar{B} + \bar{C}$, (f) $a \cdot (b + c)$
5. (a) $A \cdot (B + C)$, (b) $A + B + \bar{A} + \bar{B}$, (c) $\bar{A} \cdot B + A \cdot \bar{B} + A \cdot B$, (d) $C \cdot (A \cdot (B + \bar{C}) + \bar{A} \cdot B \cdot C)$
6. (a)

A	B	C	$(A + \bar{B}) + (A + \bar{C})$
0	0	0	1
0	0	1	0
0	1	0	0
0	1	1	0
1	0	0	1
1	0	1	1
1	1	0	1
1	1	1	1

(b)

A	B	$B \cdot A$	$\bar{B}$	$A \cdot B + \bar{B}$	$\bar{A}$	$\bar{A} \cdot (A \cdot B + \bar{B}) \cdot \bar{B}$
0	0	0	1	1	1	1
0	1	0	0	0	1	0
1	0	0	1	1	0	0
1	1	1	0	1	0	0

7. (a) See Figure A.9(a), (b) see Figure A.9(b), (c) see Figure A.9(c), (d) see Figure A.9(d)
8. (a) $\bar{A} \cdot \bar{B} \cdot C + A \cdot B \cdot \bar{C}$, (b) $C \cdot (A \cdot B + \bar{A} \cdot B)$
9. (a) $(A \cdot B + C \cdot D) \cdot E$, (b) $\bar{C} \cdot (A + \bar{B})$
10. (a) See Figure A.10(a), (b) see Figure A.10(b), (c) see Figure A.10(c), (d) see Figure A.10(d)

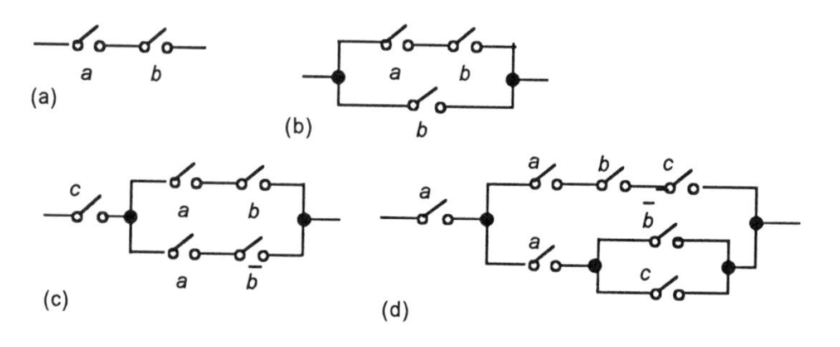

Figure A.9

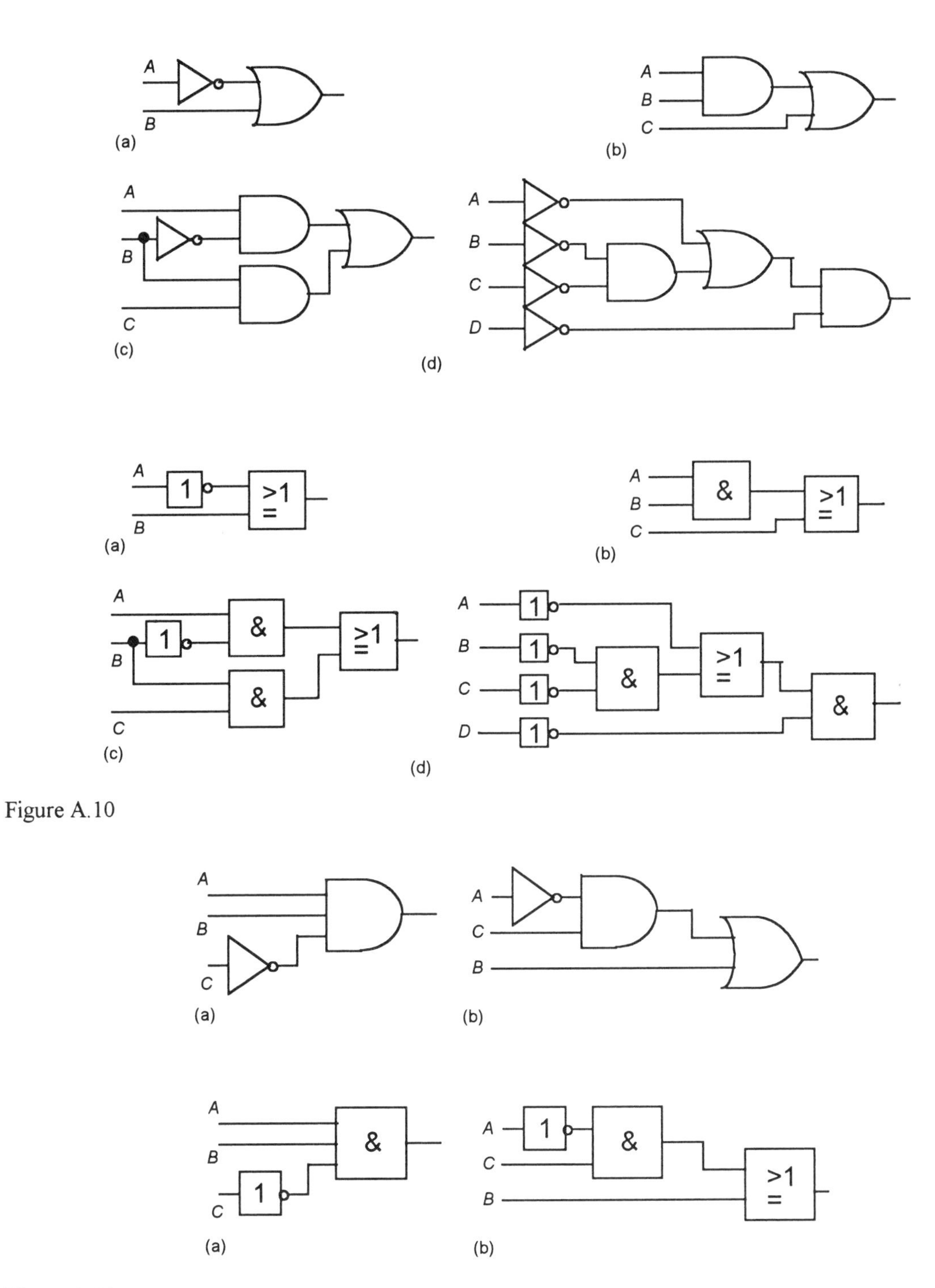

Figure A.10

Figure A.11

13 (a) $A \cdot B \cdot \overline{(\bar{A}+B \cdot C)}$, $A \cdot B \cdot \bar{C}$, Figure A.11(a), (b) $(A \cdot B+\bar{A} \cdot C)+(\bar{A} \cdot B+B \cdot C)$, $B+\bar{A} \cdot C$, Figure A.11(b)

14 See Figure A.12, (a) $A + B$, (b) $C+\bar{A} \cdot B$, (c) $A \cdot C+\bar{A} \cdot \bar{C} \cdot \bar{D}$

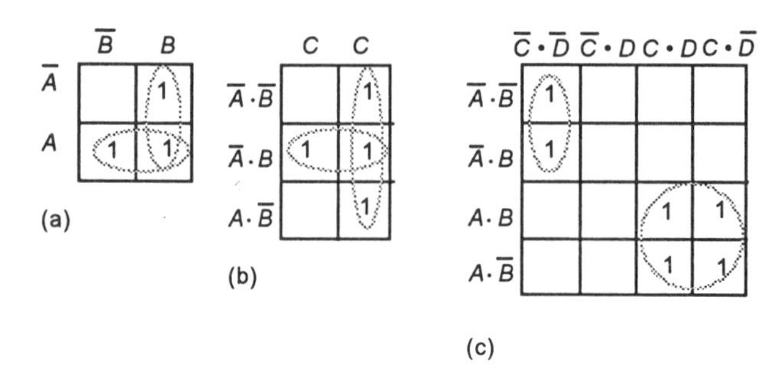

Figure A.12

15 (a) $\bar{A} \cdot \bar{B}$, (b) $\bar{B}$, (c) $\bar{A} \cdot \bar{B}+\bar{C}$, (d) $\bar{A} \cdot \bar{B}+\bar{A} \cdot \bar{C}$, (e) $\bar{A} \cdot \bar{D}$, (e) $\bar{A} \cdot \bar{B} \cdot C+A \cdot \bar{B} \cdot \bar{D}$

16 See Figure A.13

17 See Figure A.14, $\bar{A}+B \cdot C$

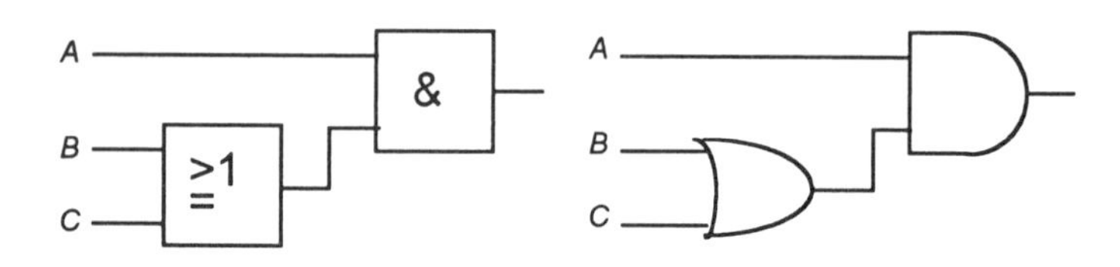

Figure A.13

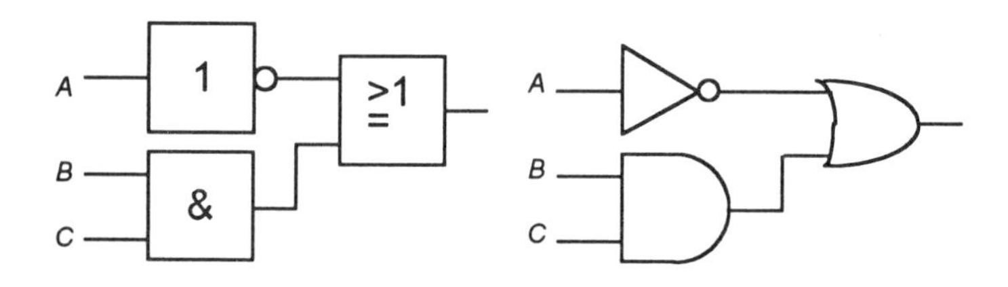

Figure A.14

Chapter 9

1 (a) 1 and 2 activated, (b) 1 but not 2 activated, (c) 1 or 2 activated, (d) on until 1 or 2 activated

2 (a) When momentarily activated, output latched on, (b) output switched off

3 (a) AND, see Figure 9.18, (b) OR, see Figure 9.20, (c) NOT, see Figure 9.22

4 (a) Figure A.15, (b) Figure A.16, (c) Figure A.17

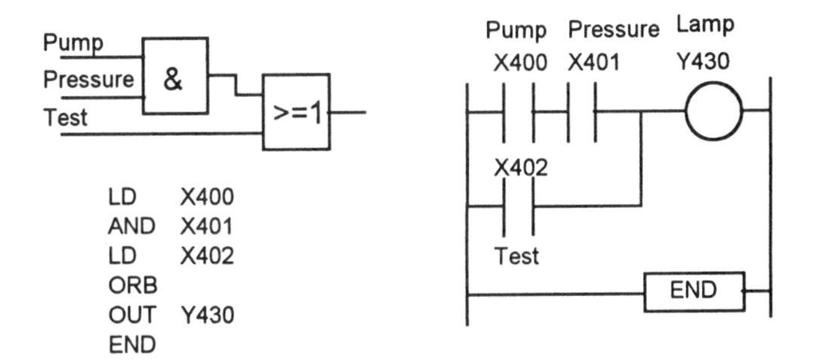

Figure A.15

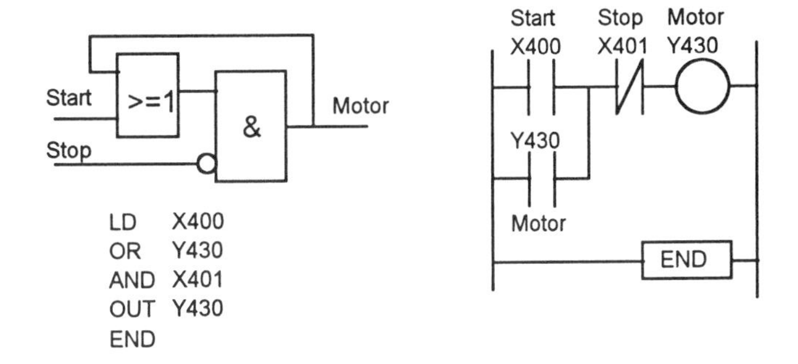

Figure A.16

Figure A.17

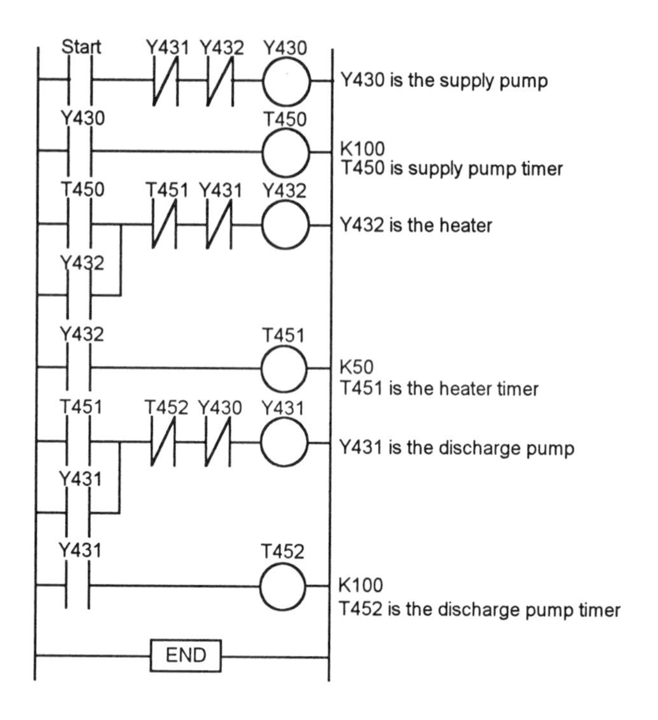

Figure A.18

5 A+ and B+, C+, A– and B–, C–
6 In 1 or In 3, not In 2, not In 4
7 A+, B+, A–, B–, A+, A–
8 Out 1 on and switched off after time delay
9 Out 1 on, followed after time delay by Out 2
10 Figure A.18
11 In 1 resets the counter and starts the count of In 2. After 5 In 2 inputs, the counter's contacts close and the output occurs.
12 M100 and M101 activated. This switches off M100. Ten pulses on X401 counted. Then output.
13 A continuous input to In 1 followed by three pulse inputs to In 2.
14 As in Figure 9.68 with a continuous input to In 1, so entering a 1 at each shift.
15 (a) 0100, (b) 1100, (c) 1110
16 As in Figure 9.72

Chapter 10

1 See Section 10.2
2 See Section 10.2
3 (a) Air line between Y and X blocked, (b) cylinder jammed or valve X failed.
4 See Section 10.4.1
5 See Section 10.4.2
6 Figure A.19
7 Figure A.20

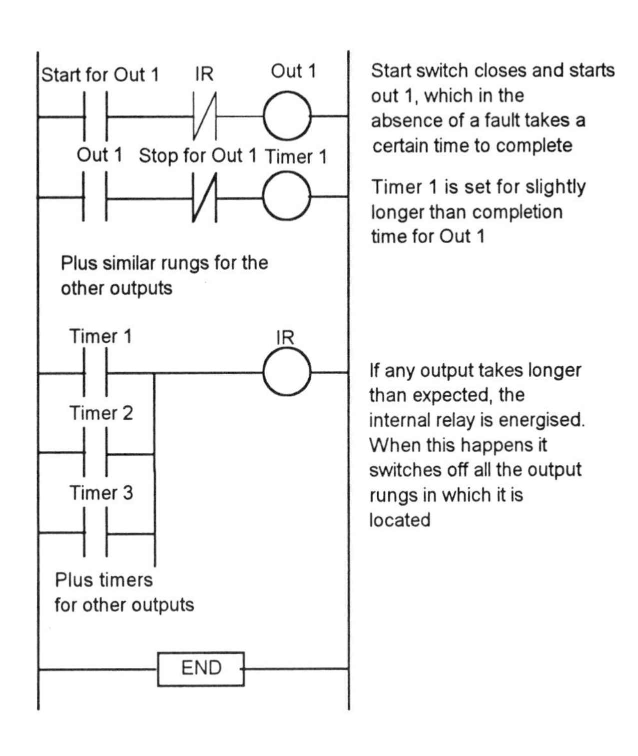

Figure A.19

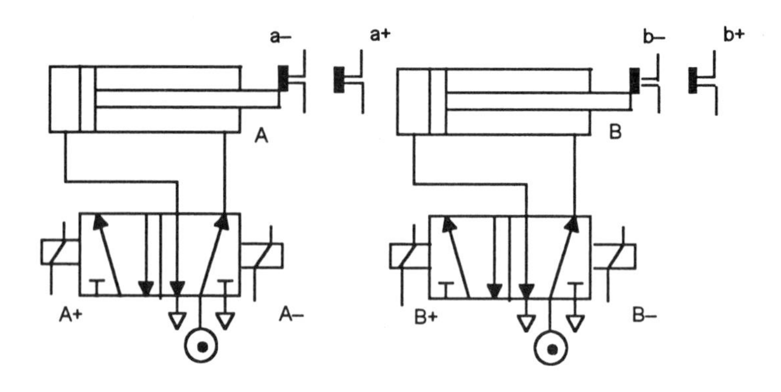

(a) The pneumatics

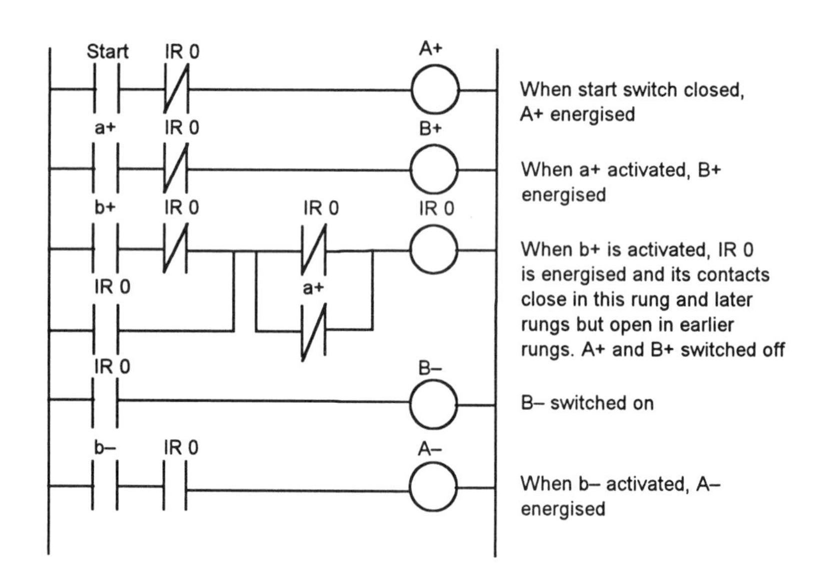

(b) The part of the program to give the sequence A+, B+, B–, A–

Figure A.20 *Continued on next page*

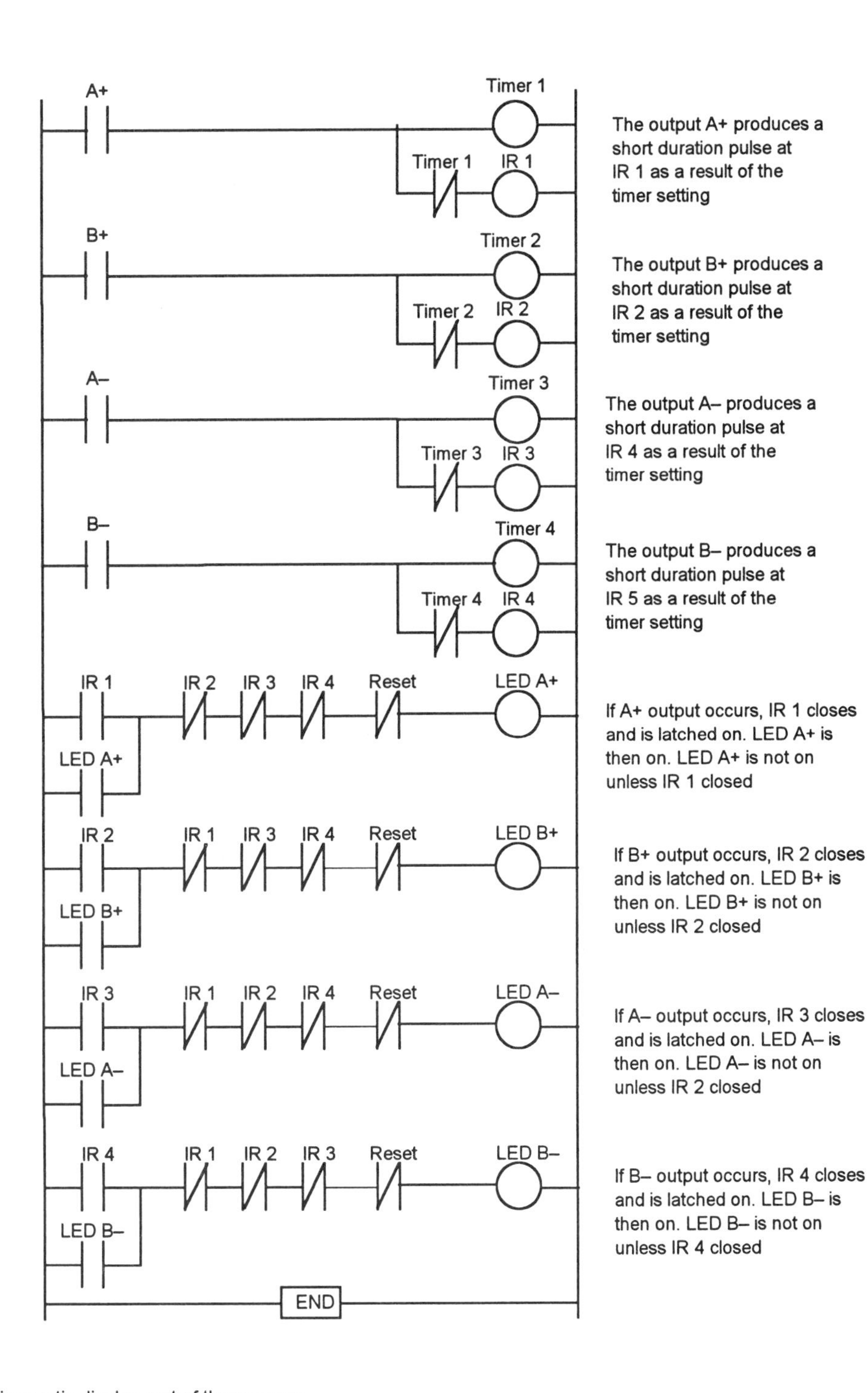

(c) The diagnostic display part of the program

Figure A.20 *Continued from previous page*

Index